DATE DUE

~~OC 15'99~~			
~~DE 2 0'00~~			

DEMCO 38-296

High-Definition Television

High-Definition Television

A Global Perspective

Michel Dupagne and Peter B. Seel

Iowa State University Press ■ *Ames*

School of Communication at
n mass communications from
iunication technologies, inter-
................ics.

Peter B. Seel is assistant professor in the Department of Journalism and
Technical Communication at Colorado State University. He holds a Ph.D.
in mass communications from Indiana University. Dr. Seel has produced
and directed television programs in California and currently is interested
in the use of the Internet as a medium for audio and video broadcasting.

Cover design by Robert Mickey Hager, Hager Design

Authorization to photocopy items for internal or personal use, or the internal or personal use of specific clients is granted by Iowa State University
Press, provided that the base fee of $.10 per copy is paid directly to the
Copyright Clearance Center, 27 Congress Street, Salem, MA 01970. For
those organizations that have been granted a photocopy license by CCC, a
separate system of payments has been arranged. The fee code for users of
the Transactional Reporting Service is 0-8138-2925-9/97 $.10.

♾ Printed on acid-free paper in the United States of America

First edition, 1998

International Standard Book Number: 0-8138-2925-9

Library of Congress Cataloging-in-Publication Data

Dupagne, Michel
     High-definition television: a global perspective/Michel Dupagne and
Peter B. Seel—1st ed.
          p. cm.
     Includes bibliographical references and index.
     ISBN 0-8138-2925-9
     1. High-definition television—Government policy—Case studies.
  I. Seel, Peter Benjamin. II. Title.
HE8700.73.D87 1998
     384.55—dc21   97-28423

Last digit is the print number: 9 8 7 6 5 4 3 2 1

To our mothers,
Lucienne Chatelain Dupagne
and
Cheryl Hayes Wilson

■ Contents

■ Abbreviations

ACATS	Advisory Committee on Advanced Television Service
ACT	Association of Commercial Television
ACTV	Advanced Compatible Television
ADTT 1187	EUREKA 1187 Advanced Digital Television Technologies
AEA	American Electronics Association
ANSI	American National Standards Institute
AP	Action Plan
ARIB	Association of Radio Industries and Businesses
ASC	American Society of Cinematographers
ATEL	Advanced Television Evaluation Laboratory
ATP	Advanced Television Publishing
ATRC	Advanced Television Research Consortium
ATSC	Advanced Television Systems Committee
ATTC	Advanced Television Test Center
ATV	Advanced Television
BS	Broadcasting Satellite
BSB	British Satellite Broadcasting
BSkyB	British Sky Broadcasting
BSR	ANSI Board of Standards Review
BSS	Broadcasting Satellite Service
BT	British Telecom
BTA	Broadcasting Technology Association
CA	Conditional Access
CCIR	International Radio Consultative Committee
CDR	Common Data Rate
CEC	Commission of the European Communities
CEMA	Consumer Electronics Manufacturers Association
CFA	Consumer Federation of America
CICATS	Computer Industry Coalition on Advanced Television Service
CIF	Common Image Format
COFDM	Coded Orthogonal Frequency Division Multiplexing
CP	Common Position
CRT	Cathode Ray Tube
CS	Communication Satellite
CSPP	Computer Systems Policy Project
DARPA	Defense Advanced Research Projects Agency
DBS	Direct Broadcast Satellite
DG	Directorate-General
DGA	Directors Guild of America
DIVINE	Digital Video Narrow-band Emission
DoD	Department of Defense

DRAM	Dynamic Random Access Memory
DSS	Digital Satellite System
DTH	Direct-to-Home
DTI	Department of Trade and Industry
DTTV	Digital Terrestrial Television
DTV	Digital Television
DVB	Digital Video Broadcasting
DVD	Digital Versatile Disk
EBU	European Broadcasting Union
EC	European Community
ECU	European Currency Units
EDTV	Enhanced-Definition or Extended-Definition Television
EEIG	European Economic Interest Grouping
EIA	Electronic Industries Association
ELG	European Launching Group
EP	European Parliament
ESC	Economic and Social Committee
ETSI	European Telecommunications Standards Institute
EU	European Union
EU-95	EUREKA EU95
FCC	Federal Communications Commission
FPD	Flat Panel Display
FPDTF	Flat Panel Display Task Force
FSS	Fixed Satellite Service
GA	Grand Alliance
GATT	General Agreement on Tariffs and Trade
GI	General Instrument
HBO	Home Box Office
HDS	High Definition Systems
HDTC	High Definition Television Committee (Broadcasting Technology Association)
HDTV	High-Definition Television
HPA	Hi-Vision Promotion Association
HVPC	Hi-Vision Promotion Council
IBA	Independent Broadcasting Authority
IDTV	Improved-Definition Television
IEEE	Institute of Electrical and Electronics Engineers
INTV	Association of Independent Television Stations
ISO	International Standards Organization
IT	Information Technology
ITU	International Telecommunication Union
ITU-R	Radiocommunication Sector
ITU-T	Telecommunication Standardization Sector
IWP	Interim Working Party
JDB	Japan Development Bank
JSB	Japan Satellite Broadcasting
KLT	Krasnow, Longley, and Terry

LPTV	Low-Power Television
LR	Living Room
MAC	Multiplexed Analog Components
MAP	Media Access Project
MEP	Member of the European Parliament
MIT	Massachusetts Institute of Technology
MITI	Ministry of International Trade and Industry
MNC	Multinational Corporation
MoU	Memorandum of Understanding
MPEG	Moving Picture Experts Group
MPT	Ministry of Posts and Telecommunications
MSO	Multiple System Operator
MST	Association of Maximum Service Telecasters
MSTV	Association for Maximum Service Television
MUSE	Multiple Sub-Nyquist Sampling Encoding
NAB	National Association of Broadcasters
NAFTA	North American Free Trade Agreement
NCRA	National Cooperative Research Act
NCTA	National Cable Television Association
NERA	National Economic Research Associates
NHK	Nippon Hoso Kyokai
NII	National Information Infrastructure
NIRI	Nikkei Industry Research Institute
NIST	National Institute of Standards and Technology
NOI	Notice of Inquiry
NPRM	Notice of Proposed Rule Making
NTIA	National Telecommunications and Information Administration
NTSC	National Television System Committee
OECD	Organization of Economic Co-operation and Development
OFDM	Orthogonal Frequency Division Multiplexing
O.J.	Official Journal of the European Communities
OMA	Orderly Marketing Agreement
ORC	Opinion Research Corporation
PAL	Phase Alternating Line
PBS	Public Broadcasting Service
PC	Personal Computer
PPV	Pay-Per-View
PTT	Ministry of Posts, Telephones, and Telegraphs
RA	Radiocommunication Assembly
R&D	Research and Development
R&O	Report and Order
RBOC	Regional Bell Operating Company
RCA	Radio Corporation of America
RRNA	Robert R. Nathan Associates
RSO	Regional Standards Organization
SC	Shopping Center
SDTV	Standard-Definition Television

SEA	Single European Act
SECAM	Séquentiel Couleur à Mémoire
SEM	Single European Market
SES	Société Européenne de Satellites
SMPTE	Society of Motion Picture and Television Engineers
S/N	Signal/Noise
TG	Task Group
TTC	Telecommunications Technology Council
TWF	Television Without Frontiers
UDTV	Ultra-Definition Television
UHF	Ultra High Frequency
USDC	U.S. Display Consortium
VCR	Videocassette Recorder
VSB	Vestigial Sideband
VHF	Very High Frequency
VTR	Video Tape Recorder
WARC	World Administrative Radio Conference
WGHDEP	Working Group on High-Definition Electronic Production
WP	Working Party
WRC	World Radiocommunication Conference
WT	Walk-Through

■ Preface

We're going to have a bigger, better picture. Whether it's called advanced television or high definition is irrelevant.
—*Randall Dark, President of HD Vision, 1995*

Television, that ubiquitous means of electronic communication, is second only to radio in its global reach. Arguably it exerts a greater impact than any other mass medium on world cultures, political systems, and the international exchange of ideas. The estimated number of television sets exceeds 1 billion worldwide, with over 200 million sets in the United States alone (Cookson, 1995; Stern, 1995b). For better or for worse, there are very few places on earth without television service of some kind. However, for over 30 years the world has been split between three discrete television systems—NTSC, PAL and SECAM—that are incompatible with each other and inhibit international program exchange.

In 1986 it seemed that the new high-definition television (HDTV) system might transcend these barriers and all nations would adopt a common medium for electronic media production and distribution. Not only did this not take place, but we are again faced with the prospect of multiple television systems vying for acceptance amongst the nations of the world. The story of how this situation occurred is a fascinating blend of television technology, global industrial politics, and trade protectionism. At the same time, the world has been inundated by new digital communication technologies that may subsume television broadcasting and possibly make these national incompatibilities irrelevant. It is an interesting and tumultuous period in the history of international telecommunications.

The purpose of this book is to make sense of these technological and political changes by using the development of high-definition television as a case study. The international battles over HDTV standardization make it ideal for examining the larger economic issues with which high-technology societies are wrestling. We provide a comprehensive, non-technical survey of high-definition television and related advanced television (ATV) systems in North America, Europe, and Asia. Given the wealth of information on HDTV technology and the related development of international industrial policies (Beltz, 1991; Sudalnik and Kuhl, 1994; Tyson, 1993), there is a need to synthesize this literature and draw preliminary conclusions concerning the development of HDTV and its influence on global high-technology policymaking. This book is organized around nine key themes, each corresponding to one chapter.

Chapter 1 looks at the global evolution of HDTV from its inception to mid-1997. Main topics include Japanese development of HDTV technologies, the debate over a single worldwide HDTV production standard, the HDTV transmission standardization process in the United States, European multiplexed analog components (MAC) policies, Japan's reconsideration of multiple sub-nyquist sampling encoding (MUSE) broadcasting, and enhanced-definition television (EDTV) systems. Michel Dupagne argues that the rejection of the Japanese/U.S.-backed 1125/60 production system at the CCIR Plenary Assembly in Dubrovnik (1986) was inevitable because so many political and economic considerations conspired against its adoption as a single worldwide production standard. Political pragmatism dominated the debate and ultimately prevailed over a technical compromise.

Chapter 2 provides an overview of worldwide HDTV production formats (Japanese/U.S. 1125/60 and European 1250/50) and explores the relationship between HDTV production and global ATV standardization conflicts. Peter Seel concludes that the production efforts to date have created fundamental aesthetic differences between widescreen HDTV formats and conventional television systems. HDTV production trials, especially Olympic Game telecasts, have boosted the credibility of Japanese and European HD standards. In Europe and Japan, the production of HDTV software is seen as an important part of the diffusion process of the technology. Seel also examines the role of motion picture film production in the United States for conversion to widescreen high-definition television programming.

Chapter 3 reviews the history of Hi-Vision, MUSE, and Clear-Vision technologies in Japan from the mid-1960s to the early 1990s and analyzes how Nippon Hoso Kyokai (NHK), broadcast electronics manufacturers, the Ministry of Posts and Telecommunications (MPT), the Ministry of International Trade and Industry (MITI), and private broadcasters have influenced these developments. Special attention is paid to the characteristics of the 1125/60 studio standard, standard-setting activities, and promotional policies. Scott Elliott and Dupagne point out that NHK developed Hi-Vision and MUSE not just for improving television technology but also for social and economic reasons, and that private broadcasters spearheaded extended-definition television efforts to counter NHK's competitive advantage with Hi-Vision and MUSE.

Chapter 4 retraces the development of the European Union (EU)'s HDTV standards policy from 1986 to 1995 and analyzes the interrelationships between the EU institutions and market forces in the audiovisual field. In 1993, the EU officially abandoned its controversial standards strategy and its exclusive support of the MAC system, leaving only the widescreen 16:9 format as the only remaining common factor. Concurrently, it gave up its competitive rhetoric and entered into an explicit policy alliance with its former rivals, the United States and Japan. Sophia Kaitatzi-Whitlock argues that the fundamental flaw of the EU MAC/HDTV policy was to create a loophole for low- and medium-power satellite services that defeated efforts to enforce a common HDTV standard.

Chapter 5 analyzes the rise and fall of HDTV industrial policy initiatives in the United States between 1988 and 1990. Dupagne concludes that many of the protagonists involved in this controversy (White House, Congress, Secretary of Commerce Robert Mosbacher, American Electronics Association, DARPA Director Craig Fields) bear some responsibility in this debacle for failing to pull in the same direction and set common, realistic goals. In light of the history of the U.S. consumer electronics industry, he also questions whether HDTV industrial policy instruments would have been successful in developing a substantial U.S. manufacturing base.

Chapter 6 evaluates the involvement of five key actors (Congress, the Federal Communications Commission, broadcast industries, the White House, and citizen groups) in the U.S. ATV policymaking process from 1983 to mid-1997. Seel uses game theory as an analytical tool to identify winners and losers in the policymaking process. He concludes that while Congress talked a great deal about HDTV policy, the FCC has assumed the lead role in managing an industry-dominated Advisory Committee that developed an innovative digital ATV transmission system. The Commissioners of the FCC, especially the Chairman, played an important role in deciding key ATV policy issues in the U.S. standard-setting process.

Chapter 7 examines how the FCC's Advisory Committee on Advanced Television Service (ACATS) was selected, how the Committee utilized hundreds of volunteer participants to create a competitive testing process, and how the Committee's decisions influenced the FCC's ATV decisionmaking. Seel argues that the 25 members of the blue-ribbon executive committee of ACATS were very dependent on the technical expertise of the larger committee and the Advanced Television Test Center. Richard Wiley, ACATS Chairman, played a pivotal role in managing the work of the Committee, designing a decisionmaking structure, and negotiating a merger of the proponent companies into a Grand Alliance consortium.

Chapter 8 explores three HDTV economic issues: (1) station conversion costs; (2) market potential; and (3) consumer acceptance. The third and main section reviews 29 HDTV-related consumer studies that have been conducted in Japan, North America, and Europe between the 1980s and the mid-1990s according to five diffusion-related variables. These include: HDTV awareness, preference for HDTV over conventional TV, interest in HDTV, desirability of HDTV attributes, and purchase intent and willingness to pay for HDTV. The chapter concludes with a discussion of the HDTV versus standard-definition television (SDTV) implementation issue and the formulation of six empirical propositions about the potential adoption of HDTV. Despite lukewarm consumer reactions to HDTV, Dupagne contends that there is little reason to believe that HDTV would not diffuse as rapidly as color TV in the United States, which by all accounts had a long gestation period before gaining wide consumer acceptance.

Finally, Chapter 9 examines HDTV policymaking on a global basis by integrating national or regional models into a worldwide system that reflects international influences in the process. Seel proposes an integrative model that links the regional political entities involved in HDTV standardization. Existing national/regional models exclude important policymaking actors and ignore the important role of regional standards organizations and multinational corporations.

We do not confine our analysis to high-definition television technology. We also chronicle the evolution of other forms of advanced television, such as non-HD digital television (DTV) and enhanced-definition television (EDTV) systems. Therefore, the scope of the book is more inclusive than its title might suggest. The book is intended for all readers with an interest in advanced television and does not require any special technical or policymaking knowledge.

Yet, we anticipate that not everyone will agree with the focus of the book in light of the growing convergence of modes, a term coined by late MIT political scientist Ithiel de Sola Pool to describe the fusion of traditionally separate media industries (see Pool, 1983). For instance, MIT Media Lab Director Nicholas Negroponte has argued that HDTV is now "clearly irrelevant" (Jacobi, 1995, p. 21). "There is no TV-set industry in the future. It is nothing more or less than a computer industry: displays filled with tons of memory and lots of processing power" (Negroponte, 1995, p. 47). Futurist George Gilder echoes this sentiment in his writings, predicting that computer technology will ultimately replace TV technology (e.g., Gilder, 1989). For these and other critics this book may seem anachronistic and passé. We take the view that it is premature to trumpet the demise of television—not with an installed base of over 1 billion sets. Perhaps it is too early to predict that all television sets throughout the world will soon be replaced with PC-TVs (Brinkley, 1997b). But more importantly, whether the traditional television receiver as a piece of hardware disappears from the home in the next century, as some have forecast, or sees itself reconfigured as a PC-TV hybrid, we argue that the television experience and the viewer interest in television programming is unique and will not vanish any time soon. Rather it will necessitate more technologically advanced features, such as digital interactivity, to compete with other media. As such, we view HDTV more as an evolutionary concept with a series of unique attributes (digital image/sound processing; sharper and wider pictures; multichannel audio) than a time-bound technology.

Due to the central role of television as the primary means of information distribution in modern society, any proposed changes in the production, transmission, and reception (or related broadcast policies) of this medium are worthy of our attention. The industrialized nations of the world are now planning for the transition from the analog technologies of first-generation television systems to the digital display formats that will be part of the global information networks of the 21st century. Digital HDTV systems will be a fundamental part of those networks and their development is a central theme of this book.

Numerous individuals from organizations in the United States, Europe, and Japan have contributed their insights to this work, and we are indebted to them. Without their support, the completion of this book would not have been possible. In particular, the authors and contributors would like to acknowledge the assistance of Peter Fannon (former President of the Advanced Television Test Center), Joseph Flaherty (CBS Senior Vice President for Technology), Robert Hopkins (former Executive Director of the Advanced Television Systems Committee), Paul Misener (Manager of Computer Technology Policy at Intel Corporation), John Reiser (U.S. Chairman, ITU-R Study Groups 10 and 11, Federal Communications Commission), Jean-Luc Renaud (Editor-in-Chief of *Advanced Television Markets*, London), Richard Smith (Chief, FCC Office of Engineering and Technology), Michael Wagner (Legal Advisor to the European Broadcasting Union), and Richard Wiley (Partner at Wiley, Rein & Fielding and former Chairman of the Advisory Committee on Advanced Television Service). On repeated occasions, they have provided us with valuable background information and comments on the chapters. To those we omitted mentioning, we apologize and express our heartfelt thanks.

Peter Seel would like to recognize his dissertation committee at Indiana University for their guidance and feedback in the project that formed the basis for his chapters in this book. Michael McGregor chaired the committee that consisted of Christine Ogan, Martin Siegel, and Kathy Krendl. Many faculty members in the Department of Telecommunications at Indiana were helpful with research advice: Don Agostino, Louise Benjamin, Michael Curtin, Walter Gantz, and Herb Terry to name just a few. His doctoral thesis was supported by a 1994 Jagdish Sheth Dissertation Award from the Center for Telecommunication Management at the University of Southern California and a travel grant from the Indiana University Department of Telecommunications donated by former professor Kenji Kitatani. He would also like to thank Frank Moakley of San Francisco State University for access to his extensive high-definition television archives, and graphic designer Greg Nelson of the Office of Instructional Services at Colorado State University for his assistance with charts and diagrams for this book.

We are grateful to the staff of Iowa State University Press for their assistance and patience throughout the publication process, especially Laura Moran, former acquisitions editor, Judi Brown, acquisitions project manager, and Jane Zaring, editor. Our appreciation also goes to Corey Carbonara of Baylor University and Steven Wildman of Northwestern University who reviewed this manuscript prior to publication and provided many helpful comments. However, any errors or omissions are our responsibility.

We thank the School of Communication at the University of Miami, the Department of Journalism and Technical Communication at Colorado State University, and the Broadcast and Electronic Communication Arts Department at San Francisco State University for their moral and administrative support during the writing of this book.

We would like to thank our parents for giving us the necessary emotional and financial support to allow us to enter and complete graduate school. Without them, this book, as well as our academic careers, would not have materialized. Finally, our loving appreciation goes to our spouses, Alyx Lin-Dupagne, Nanci Seel, and Richard Whitlock-Blundell, who have had the courage to put up with our erratic (and yes, we admit, sometimes irrational) behavior during the 12-month writing process. Their constant encouragement helped make this book a reality.

High-Definition Television

1 Thirty Years of HDTV Technology and Policy

The problem is that both sides think they will lose more than they gain if they depart from their current field rates. They both want a single worldwide standard, as long as it doesn't cost more (as they see it) than it gains.
—*David Wood, European Broadcasting Union, and Jose Tejerina, Retevision, 1990*

This chapter retraces the evolution of high-definition television (HDTV) and advanced television variants in the United States, Europe, and Japan (the so-called triad powers) from the mid-1960s to mid-1997, with a special emphasis on international HDTV standardization matters. It reviews the technological and policy aspects of HDTV in both a chronological *and* a thematic order. Main topics include: NHK's development of HDTV technologies, the story of the "Dubrovnik rejection," the establishment of the U.S. DTV transmission standard, the abandonment of European MAC policies, Japan's reconsideration of MUSE broadcasting, and the search for terrestrial EDTV systems. This overview concludes with some thoughts about the international standardization process of high-definition television.

The need for international technical standards in the telecommunications sector is self-evident. "In engineering terms, standards are design specifications shared by the industry [telecommunications] to determine the degree and means of interoperability between both networks and the component on which they are based" (Drake, 1994, p. 71). Nations come together to propose, discuss, and ultimately agree—often by compromise—on a set of technical specifications to be used worldwide in order to avoid the emergence of multiple de facto and/or incompatible standards and facilitate the introduction of new telecommunication products and services in world markets. Therefore, adoption of international standards has direct economic implications for equipment manufacturers because it will ultimately define their production and marketing strategies (e.g., economies of scale). Not surprisingly, this process is delicate and politically-laden. "Set a standard too soon or too rigidly and important technical advances may be precluded. Set it too late, and a proliferation of incompatible systems will enter the market-place" (Nickelson, 1990, p. 302).

International television standardization matters are further complicated by the nature of the technology, which involves several interrelated areas each requiring standardization. In the case of television, at least two television standards are necessary: a studio standard to produce programs and a transmission standard to broadcast them to the viewer. It is therefore possible that the studio standard be distinct from the transmission standard, and this is in fact the case for HDTV. In addition, controversy has arisen to determine which component of the HDTV process, production or transmission, ought to be first standardized. Should the parameters of the production standard be first developed, followed by those of the transmission standard, or should it be the reverse (Nickelson, 1990)? As noted below and in Chapter 2, these two issues are intertwined and have affected the outcome of international HDTV standardization negotiations.

The 1960s and 1970s: Japanese Genesis

Although it is unanimously recognized that the Japanese, under the aegis of Nippon Hoso Kyokai (NHK), the Japan Broadcasting Corporation, have pioneered HDTV technology, it is difficult to pinpoint the exact year when they formally began HDTV research (see Chapter 3). Dates vary from 1964 to 1970, although many Japanese authors cite 1970 as the official year for the beginning of full-scale HDTV development. When NHK engineers initiated HDTV (labeled "Hi-Vision"[1] by NHK since February 1985) development, their primary motivation was to improve the quality of the color 525-line NTSC (National Television System Committee) standard that Japan adopted in December

1956. They first conducted studies to test fundamental physical and psychovisual requirements (e.g., optimal viewing distance) in 1964 and then undertook research on display and camera equipment in 1970 (Nippon Hoso Kyokai [NHK], 1996). In May 1970, NHK scheduled its first HDTV exhibition in Japan. So shortly after beginning color television broadcasting (in September 1960), NHK had already set its sights on exploring a new set of technical parameters for the next generation of television technology. Since the early 1960s, NHK has actively invested in and experimented with new communication technologies, such as direct broadcast satellite (DBS) in 1978, multichannel television sound in 1978, and teletext in 1983. No more was this evident than when NHK Science and Technical Research Laboratories showcased close to 30 different technologies at the 1991 convention of the National Association of Broadcasters (NAB) in Las Vegas, even though much of the technology presented there was prototypical. NHK's capability of innovation was and still is breathtaking. NHK's long-standing interest in new media falls in line with official Japanese broadcast policy as the Corporation is mandated by law "to conduct researches and investigations necessary for the improvement and development of broadcasting and the reception thereof" (*Japanese Legislation*, 1991, p. 10). It is also a strategy to avert technological obsolescence and protect its revenue base (see Chapter 3).

Although NHK's investigations marked the beginning of post-NTSC advanced television research, they did not constitute the first attempt at designing a high-resolution television system. In 1939-1940, the American Allen DuMont proposed a monochrome television system capable of displaying almost 800 lines (Freeman, 1984). Although this system was ultimately rejected by the National Television System Committee due to smear problems, DuMont's invention can be viewed as one of the first efforts to produce high-definition pictures. From 1970 to 1980, NHK quietly crafted HDTV studio standards, developed HDTV hardware, and experimented with satellite transmission of HDTV signals (see Fujio et al., 1982).

By the early 1980s, NHK was poised to unveil its HDTV system to the rest of the world. This production system comprised seven major features. First, with 1125 lines, it more than doubled the number of scanning lines used in conventional NTSC (525 lines), thereby offering sharper images. Second, Hi-Vision's scanning structure was 2:1 interlaced, in accordance with the NTSC television standard in which the electron beam of the camera first scans the odd-numbered lines (1, 3,

5, … up to 525) and then the even-numbered lines (2, 4, 6, … up to 524). The interlaced scanning structure was designed to minimize the flicker effect (i.e., sudden changes in brightness) and save bandwidth. Therefore, an HDTV picture or frame consisted of two fields of 562.5 lines. Third, again consistent with NTSC,[2] each HDTV frame was scanned at 1/30th second and each HDTV field was scanned at 1/60th second (hence the 60 Hz field rate). Fourth, the aspect ratio (i.e., the ratio of the screen width to the screen height) was widened from NTSC's 4:3 to 16:9[3] to give viewers a greater sense of immediacy and depth (see Figure 2.1 in Chapter 2). Fifth, given its larger and wider screen, Hi-Vision also provided a greater viewing angle—30 degrees compared to 10 degrees for conventional NTSC television—creating a greater sense of personal involvement similar to a movie theater experience (see Figure 2.2 in Chapter 2). Sixth, the optimal viewing distance for HDTV was estimated to be three times the screen height (3H), as opposed to seven times for NTSC (7H).[4] For instance, if the screen height of an NTSC receiver were 20 inches, the optimal viewing distance would be about 10.5 feet (20 x 7) or 3.5 meters. On the other hand, an HDTV set with the same screen height would yield an optimal viewing distance of about 4.5 feet (20 x 3) or 1.5 meters. Consequently, viewers could sit closer to their receivers without noticing the scanning lines or experiencing acute eye fatigue. Finally, Hi-Vision offered digital multichannel sound using pulse code modulation, analogous to the sound quality of a compact disc (NHK, 1996). The NHK HDTV system was not without drawbacks. First, it was designed for HDTV *production*, not transmission. It required a video bandwidth of 20 MHz and consequently was impractical for terrestrial transmission purposes—Hi-Vision's video bandwidth was about five times larger than that of NTSC's (4.2 MHz) (NHK Science and Technical Research Laboratories, 1993). Satellite communications was probably the only electronic media outlet capable of transmitting the full HDTV bandwidth without significant compression.[5] Second, it was incompatible with existing NTSC receivers.

Meanwhile in the United States, the Society of Motion Picture and Television Engineers (SMPTE), a professional organization that submits film and broadcast standard proposals to the American National Standards Institute (ANSI), began expressing interest in HDTV by establishing a committee to study the technology in May 1977. The creation of the Study Group on HDTV represented the first major involvement of the United States in high-definition television. Its role was to investigate the application of technical requirements of HDTV systems in the home, theaters, and motion picture studios. It held seven meetings between May 1977 and June 1979 and recommended four key technical specifications for HDTV: (1) 2:1 interlaced scanning at 60 fields per second (30 frames per second) for all applications except motion picture production; (2) an aspect ratio of at least 5:3 and preferably 2:1; (3) 1100 lines per frame and possibly 1500 lines in the future; and (4) separate transmission of chrominance (image color) from luminance (image brightness) to enhance color rendition, instead of NTSC's interweaving of chrominance and luminance on the same signal (Fink, 1980a, 1980b). Not surprisingly, these recommendations coincided with Hi-Vision's specifications given that both the United States and Japan had adopted the NTSC color television system. Interestingly, the study group also proposed that "digital video signal techniques be used for HDTV to the fullest possible extent" (Fink, 1980a, p. 89), a recommendation that may seem futuristic in the primarily analog television world of the late 1970s but that will take its full meaning 10 years later with General Instrument's announcement of the first all-digital HDTV transmission system (DigiCipher).

The Early 1980s: Enter the United States

Like the 1970s, the 1980s were fertile in HDTV-related innovations, but unlike the 1970s, this decade was controversy-prone. In February 1981, NHK demonstrated its 1125-line HDTV system, described above, at the annual SMPTE meeting in San Francisco. This was the first major HDTV exhibition in the United States. Some attendees raved over the picture quality. "It was just outstanding," proclaimed Lew Wetzel, NAB Vice President of Engineering. "You could stand three feet in front of the monitor and not see any lines at all" ("Clear Advantages," 1981, p. 30).

In September 1983, the CBS Technology Center announced the development of a two-channel compatible system to transmit HDTV signals using DBS, cable, or multipoint distribution service ("CBS Breakthrough," 1983). The main technical parameters included 1050 lines, 30 frames per second, 2:1 interlaced scanning, 5:3 aspect ratio, and stereophonic sound. The CBS compatible HDTV system was not suitable for terrestrial transmission, because it necessitated a video bandwidth of 16 MHz. Furthermore, it had limited compatibility with NTSC requiring viewers to purchase special set-top units to convert the 1050-line picture into an "improved" 525-line picture. This proposed HDTV transmission system was never implemented but would have worked as follows. Channel 1, the compatible channel, would have displayed a 525-line signal with a 4:3 aspect ratio on conventional NTSC receivers with the help of a converter, while Channel 2, the augmentation channel, would have carried an additional 525 lines with the 5:3 aspect ratio. HDTV receivers would then have combined the information from both channels and reproduced the 1050-line signal with a 5:3 aspect ratio ("CBS Breakthrough," 1983; Taylor, 1983). The CBS proposal must be remembered as one of the first attempts, and the first in the United States, to produce a "true" HDTV transmission system.

In 1984, NHK developed MUSE (multiple sub-nyquist sampling encoding), an HDTV transmission standard that was compatible with the 1125/60 studio standard. Simply stated, this system compressed the video bandwidth of Hi-Vision from the original 20 MHz to 8.1 MHz through complex video sampling and subsampling procedures, so that the HDTV signal could be more easily used for DBS transmission. Like Hi-Vision, MUSE was incompatible with NTSC. It was first demonstrated in the United States in January 1987. Audience reactions to this first working HDTV transmission over standard TV channels were highly positive. Such attendees as NAB President Eddie Fritts called MUSE "the next generation of TV" and Knight-Ridder Broadcasting President Daniel Gold referred to the difference between NTSC and HDTV as the "difference between an etching and a watercolor." Gold predicted: "I'm absolutely convinced the American public will not only perceive the difference, but once the cost is made reasonable, will embrace it" ("HDTV: Efforts," 1987, p. 138).

1986: The "Dubrovnik Rejection"

"The final failure of the C.C.I.R.'s work in this field is now well known; it was not possible to recommend one unique system for 625-line colour television," lamented one observer after the conference's unsuccessful attempt to secure an international agreement on color television standardization (Hansen, 1966, p. 138).[6] By June 1966, the world was irreversibly divided into three major incompatible television systems: NTSC, SECAM (Séquentiel Couleur à Mémoire), and PAL (Phase Alternating Line).[7] Twenty years later, HDTV would face the same destiny for similar reasons. Before examining how the international standardization process of color television paralleled that of

HDTV and chronicling the chain of events that led to the "Dubrovnik rejection," a brief description of the CCIR and its international standard-setting procedures is necessary.

The primary function of the International Radio Consultative Committee (CCIR), a permanent organ of the Geneva-based International Telecommunication Union (ITU) until it was replaced with the Radiocommunication Sector (ITU-R) in 1993, was to recommend global standards for radio communications, ranging from mobile radio service to high-definition television. The CCIR held a Plenary Assembly every four years (originally every three years). At each Assembly participants submitted a series of technical and operational Questions, which were then assigned to the appropriate Study Groups. CCIR Study Groups were composed of technical experts from different countries, standards organizations, and operating agencies. If necessary, they established Working Parties (WPs) and Task Groups (TGs; for urgent matters) to study one or several Questions. WPs and TGs drafted Recommendations and submitted them to the Study Groups who in turn considered and adopted them before forwarding them to the Plenary Assembly for approval (see Macpherson, 1990). In sum, most of the CCIR work was conducted through the Study Groups. Although CCIR Recommendations were nonbinding and therefore did not carry force of treaty, countries generally endorsed them (Savage, 1989). CCIR delegates often sought to overcome objections and reach consensual decisions, even though unanimity was not necessarily required for approving a Recommendation. CBS Senior Vice President for Technology Joseph Flaherty, one of the early advocates of HDTV development in the United States, stressed that global acceptance of a CCIR Recommendation depends upon the type of Recommendation and upon the country or countries taking a reservation during the approval procedure. He explained:

> The United States took a reservation during the WARC-83 [World Administrative Radio Conference] on the satellite power. There have been recommendations put through with 30 or 40 people taking reservations. So it's really very unclear to say in advance what constitutes a success. It's not a black and white situation, and it changes during the dynamics of a meeting. ("High Definition TV," 1986, p. 137)

The case of color television standardization is strikingly similar to that of HDTV standardization because both illuminate the challenge of hammering out an international accord when so many internal and external factors thwart that goal. For the record, the die was cast at the last session of the CCIR XIth Plenary Assembly held in Oslo, Norway, from June 22 to July 22, 1966, when the delegates failed to recommend unanimously a single standard for color television. Instead the CCIR was only able to issue a Report that described the characteristics of the different systems proposed for a worldwide color TV standard (Herbstreit and Pouliquen, 1967).

How is it conceivable that the CCIR Plenary Assembly at Oslo, which was all too well aware of the impact of multiple monochrome television systems on program exchange, was unable to reach a consensus for the standardization of color television? Crane (1979) blamed protectionism, and in particular France's manipulation of the SECAM system to support its television manufacturing industry, for the CCIR's inability to recommend a worldwide color television standard. Certainly, national agendas and industrial policies played a major role in derailing the CCIR negotiations from their original purpose. From a French point of view, there were few pros to accepting the NTSC (or PAL) standard. The French government with the full support of General Charles de Gaulle viewed SECAM as a "national champion" capable of promoting the French technological savoir-faire at home and abroad. Had the French decided to embrace NTSC instead of SECAM, they would have been required to pay royalty fees to American companies holding the NTSC patents and would not have been able to recoup their research and development (R&D) investment in SECAM (Crane, 1979).

But these politico-economic factors were not the only ones. At the organizational level, lack of authority, poor coordination between international entities, formation of regional blocs, and language differences all contributed to impede the negotiations and agreement on standardization of color television (Crane, 1979). In addition, Herbstreit and Pouliquen (1967) suggested that none of the three proposed systems was eminently superior to the other two and therefore there was no clear winner around which CCIR delegations could have rallied. Had one of the systems showed a distinct *technical* advantage, they argued, the CCIR could have recommended that system.

While being concerned by the adverse effects of Oslo 1966 and agreeable to remedy them, the CCIR delegates at the XVIth Plenary Assembly in Dubrovnik, Yugoslavia, were unable, once again, to surmount the political and economic obstacles that had hampered the worldwide standardization process of color television and recommend a single production standard for high-definition television. Like its Oslo predecessor, the Dubrovnik rejection exemplifies the complexity of establishing a common standard when different electrical systems are in operation and when national interests are at stake.

In 1972, the Japanese Administration submitted a proposal to the CCIR for a new study program "to determine what standards should be recommended for HDTV systems intended for broadcasting to the general public" (Krivocheev, 1993, p. 914). Soon after, the CCIR began work on HDTV with the adoption of Question No. 27/11 (standards for high-definition studio and for international program exchange), which spawned a vast array of questions, study programs, resolutions, and decisions regarding HDTV in subsequent years. In September 1983, Study Group 11 (television broadcasting) adopted Decision 58 to establish Interim Working Party (IWP) 11/6 and charged it with the mission of preparing "within the present study period [1982-1986] a draft recommendation for a single worldwide high-definition standard for the studio and for international program exchange" (U.S. Department of State, 1985, p. 2).

Meanwhile, the United States began taking a more active role—and positioning itself—in international HDTV standardization proceedings. In May 1983, SMPTE, the NAB, the National Cable Television Association (NCTA), the Electronic Industries Association (EIA), and the Institute of Electrical and Electronics Engineers (IEEE) formed the Advanced Television Systems Committee (ATSC), an industry-led ATV standards organization.[8] Not only was the ATSC's mission "to explore the need for and, where appropriate, to coordinate development of voluntary national technical standards for Advanced Television Systems," but its charter also instructed it "to develop a proposed national position for presentation to the Department of State for purposes of developing a United States position with the International Radio Consultative Committee (CCIR) and with other international organizations as appropriate."

In April 1984, the ATSC formed three groups to study several ways to improve video technology: the improved NTSC group, the enhanced television group, and the HDTV group.[9] The improved NTSC group adopted an evolutionary approach by investigating methods to perfect the NTSC signal by means of refinements in studio and transmission equipment. The enhanced television group explored new production and transmission techniques that would produce superior pictures while maintaining the number of lines and aspect ratio of NTSC. Finally, the HDTV group examined technical specifications, similar to those of the NHK HDTV system, which would produce pictures with twice as many scanning lines and an aspect ratio of at least 5 to 3 ("Improving," 1984).

In March 1985, the HDTV Technology Group of the ATSC adopted by a three-to-two margin a document specifying NHK-based technical parameters for an international HDTV *production* standard. Specifications included 1125 lines, 60 fields per second, 2:1 interlaced scanning, and a 5.33:3 aspect ratio ("U.S. Industry," 1985). ATSC Chairman E. William Henry urged "the rest of the world to recognize the many benefits of a worldwide HDTV standard" and join the United States in this endeavor ("ATSC Recommends," 1985, p. 1). The ATSC Executive Committee approved the document in April 1985 and forwarded it the same month to the U.S. Department of State (CCIR Study Groups, 1985). The Department of State then submitted the proposal to the national CCIR committee, an advisory panel consisting of technical experts from the public and private sector, which adopted the ATSC recommendation as the official U.S. position for the next meeting of Study Group 11 (R. Hopkins, Advanced Television Systems Committee, personal communication, November 15, 1995). On May 30, 1985, the United States submitted its proposal for a worldwide HDTV production standard to the CCIR's Study Group 11. Japan did the same independently. In September 1985, IWP 11/6 examined the U.S./Japanese-backed recommendations and then sent them to Study Group 11 (Streeter, 1987). In the submission, a number of countries manifested some reticence to the adoption of a 60 Hz field rate. In October 1985, Study Group 11, with delegates from over 50 countries, approved a new draft Recommendation to be considered by the forthcoming CCIR Plenary Assembly in May 1986. The Recommendation contained the same technical parameters: 1125 lines, 60 Hz field rate, 2:1 interlaced scanning, and 16:9 aspect ratio. Again, some countries expressed reservations (Streeter, 1987).

So the dispute surrounding the acceptance of the NHK-based HDTV production standard did not erupt suddenly at the Dubrovnik Plenary Assembly; rather it was "finalized" there. As early as 1983, the Europeans through the European Broadcasting Union (EBU)'s Specialist Group V1/HDTV—set up in 1981 to study current HDTV technology and prospects—voiced concerns on a number of issues, such as the selection of field rate and the relationship of an HDTV standard to the

4:2:2 digital standard adopted by the CCIR in 1982 (Recommendation 601)[10] (Habermann and Wood, 1986). In 1984, CCIR IWP 11/6 met in the United Kingdom to draft a first Recommendation for HDTV parameters. At the meeting, some delegates from North America reported that "they believed that it was unlikely that the United States could agree to a field-rate other than 60 Hz" (Habermann and Wood, 1986, p. 270). They also predicted that the 80 Hz field rate, seriously considered by the ATSC's HDTV group in 1984,[11] would not be an acceptable alternative, leaving IWP 11/6 little choice but to work on 50-60 Hz standards conversion systems. One week before Dubrovnik, the CCIR delegates from the 12 EC member states met in Brussels and unanimously agreed to postpone a decision on the HDTV production standard for at least two years to "conduct further studies" ("On Eve," 1986, p. 3). Yet the Europeans were themselves divided. While the EC Commission, the member states, and the European consumer electronics companies advocated that the NHK-based HDTV Recommendation needed more study, the EBU announced on May 1, 1996 that it would support a worldwide HDTV standard and that the NHK proposal appeared to be the most likely candidate (Hughes, 1986).

From May 12 to 24, 1986, delegations from over 70 countries converged on Dubrovnik, Yugoslavia, to participate in the XVIth CCIR Plenary Assembly. As anticipated, the U.S./Japanese-backed standard proposal did not attract unanimity among the member nations, and the HDTV Recommendation was referred back to Study Group 11 for further study. Annex I of *Report 801-2* briefly explained that after extensive technical objective and subjective testing the 1125/60/2:1 proposal was found to satisfy the requirements for a single worldwide HDTV production standard as outlined by IWP 11/6 (CCIR, 1986). It also stated that "many other administrations, using the 625/50 standard, have reservations about, and are not able to accept, the 1125/60/2:1 parameter values at this time" (p. 53). These administrations urged to reconsider some of the parameters presented in Annex II of the Report and devote more time to unresolved issues, such as field rate and the system's relationship to the digital standard 4:2:2. As mentioned above, these problems were identified as early as 1983. Given strong opposition from some countries spearheaded by European television equipment manufacturers such as Philips (Netherlands) and Thomson (France), the Dubrovnik Plenary Assembly decided to defer the decision on an HDTV production standard until the next Plenary Assembly in 1990. However, it adopted a resolution scheduling a meeting of Study Group 11 before the next Plenary Assembly to reach a consensus on a draft HDTV Recommendation.

The reasons for this postponement are hardly detailed in official CCIR documents. Fortunately, trade press and journal articles shed some light on the issues involved in the Dubrovnik controversy. There were numerous objections to the HDTV Recommendation, but they can be broadly classified into three categories: technical, economic, and philosophical (Roizen, 1986).

Technical Issues

Undoubtedly, field rate (50 Hz versus 60 Hz) was the principal technical tug of war between the Americans/Japanese and other CCIR countries, especially the Europeans. The United States argued that 60 Hz was a preferred field frequency because it would minimize flicker and provide better motion portrayal. Moreover, U.S. officials claimed that standard conversion from 1125/60 HDTV to 625/50 was no longer an obstacle thanks to continuously improved converters (U.S. Department of State, 1985). Despite Japan's extensive research on 60 Hz-to-50 Hz conversion systems, European broadcasters and manufacturers were still concerned about the consequences of a 60 Hz-based HDTV production standard, such as possible picture degradation, flickering reflections, and obsolescence of European TV receivers.

American engineers retorted that a lower field rate would create severe problems in lighting operations, slow motion on videotape, and intolerable flicker levels. The flicker effect, which consists of visible alternations of light and dark, depends heavily on the field rate. The lower the field frequency, the more conspicuous the flicker. At the 50 Hz rate, the flicker worsens when picture brightness is turned up.[12] For this reason, PAL and SECAM receivers operate at a lower level of brightness than their NTSC counterparts (Fink, 1980b).

Although the Japanese demonstrations of HDTV-to-PAL converters revealed little image degradation, some Europeans still contended that the high cost of converters constituted a severe drawback of the 60 Hz field rate, thereby placing their manufacturing industries at a competitive disadvantage. Furthermore, they felt that the NHK system would unfairly favor 60 Hz over 50 Hz countries—considering that 75 percent of the world's population receive television images in 50 Hz (Roche, 1987). The Europeans campaigned for an evolutionary, not revolutionary, approach. They asserted that given the incompatibility of the proposed NHK system with existing receivers, the adoption of this standard would render obsolete the estimated 700 million TV sets in the world. This exchange of arguments clearly highlights the difficulty of separating technical and economic issues in international standardization debates. At Dubrovnik, they were often one and the same.

In addition, the triad powers differed in their implementation plans for HDTV. While the United States and Japan had strived to develop an HDTV production standard (although Japan was also working on the MUSE transmission system), European countries had focused their attention on the MAC (multiplexed analog components) systems since the early 1980s. This family of transmission standards (C-MAC, D-MAC, and D2-MAC) were designed for DBS and cable television delivery, and for ensuring compatibility with existing PAL/SECAM receivers and future HDTV service (the so-called backward and forward compatibility concept). MAC systems were engineered to improve video quality through new signal processing techniques (e.g., time multiplexing transmission of chrominance, luminance, and sound signals), offer programmers scrambling protection, and provide multiple digital audio channels for stereophony and multilingual sound. They are best described as intermediary enhanced-definition television (EDTV) systems paving the way to true HDTV (HD-MAC). Indeed, the bedrock of the MAC approach was its *evolutive compatibility*, that is, its ability to be compatible every step of the way, from PAL/SECAM to HD-MAC. SECAM (or PAL) receivers would be able to display D2-MAC, widescreen D2-MAC, and HDTV images via a set-top decoder, and HDTV pictures would be directly receivable on D2-MAC and widescreen D2-MAC sets (Niblock, 1991). In 1986, France envisioned a MAC blueprint leading to HDTV in three main stages. First, D2-MAC in 4:3 format would begin as the transmission standard for French DBS in 1987, followed by the introduction of widescreen (16:9) D2-MAC receivers in 1990 and HDTV in 1995 (*An Approach*, 1986; Roche, 1987).

Given Europe's commitment to the MAC program, it is not surprising that European and other Administrations objected to the lack of serious consideration given to transmission when drafting the HDTV production Recommendation (Habermann and Wood, 1986). Adam Watson Brown (1987), at the time a consultant with KMG Thomson McLintock, explains:

By accepted engineering criteria in Europe, MUSE is not HDTV; its quality is equivalent to European MAC, which has not been promoted

as HDTV. ... The tactic hinges on blurring the distinction between production and transmission; the two components are being collectively
promoted as Hi-Vision. However, the lobbying campaign was designed to have the production standard accepted without mentioning
MUSE, or transmission at all. ... The upshot is that the evolution of European MAC towards HDTV transmission would have been stymied
completely. *Acceptance of the NHK production system as world standard
would dictate a dependent transmission system* [emphasis added]. (p. 6)

Therefore, some Europeans viewed the 1125/60 proposal not just as a
production standard, but as a concerted effort to impose its associated
transmission technology on the rest of the world. They reasoned that
if the NHK system were adopted, MUSE would derail the MAC initiative.

Economic Issues

More than anything else the United States claimed that the Europeans rejected the HDTV proposal on economic grounds—to protect
their national consumer electronics industries (e.g., Philips, Thomson)
("CCIR Puts," 1986; "U.S. Faces," 1986). U.S. delegates and observers
maintained that European opposition was not guided so much by
technical considerations as by political and economic agendas. It is
clear from trade press accounts that European nations, under the leadership of the Commission of the European Communities and European consumer electronics companies, acted against the HDTV Recommendation to gain time for developing a competing HDTV system,
thereby thwarting, in their opinion, a sure domination of the television
manufacturing industry by the Japanese in Europe. At the time of
Dubrovnik, about 80 percent of the receivers in Europe were sold by
either Philips or Thomson (Roche, 1987),[13] and some European nations,
especially France, were not bashful to admit publicly or privately that
they would combat a proposal that would jeopardize the development
of their consumer electronics sector, for example, TV sets and video-
cassette recorders ("European Counterattack," 1988). American film-
maker Francis Ford Coppola (1992) recounted an encounter with "an
extremely highly placed person in France" who told him: "Listen, we
have 24,000 people making television sets. We are not going to risk
that—even if it means the world standard—because we just cannot"
(p. 95). While the Americans viewed this approach as plainly protec-
tionistic, the Europeans perceived it as being vitally strategic.

Philosophical Issues

Last, there were some philosophical differences between the United States/Japan and many other CCIR countries. The United States and Japan urged the CCIR delegates to approve the HDTV standard at *this* Plenary Assembly to avoid the development of de facto multiple standards. Responding to the need for further investigations of the proposed NHK production standard, Renville McMann, Head of the ATSC's HDTV group, stated that the "ATSC can't wait. We've already waited and studied and waited for 10 years" (Hughes, 1986, p. 9). Yozo Ono, an NHK engineering manager, echoed this view: "We could not wait 4 years. ... We will not be able to be patient" ("Digitally Assisted TV," 1986, p. 4). This prodding for immediate action bordering on impatience antagonized European and other delegations (Roizen, 1986).

Dubrovnik is widely recognized as a European victory. But how did the European delegates manage to dominate CCIR meetings and rally other delegates to their cause? Certainly, the objections articulated above explain in great part how these CCIR participants could have been swayed. In addition, it has been argued that the Europeans already represented a formidable force to reckon with, and were far more comfortable and acquainted with the structure and style of the CCIR than their American and Japanese counterparts.

Late 1980s: The Aftermath of Dubrovnik

Following the impasse at Dubrovnik, Europe, the United States, and Japan went their separate ways. The Europeans developed a competing HDTV production system through EUREKA 95; the United States after much controversy decided *not* to adopt the 1125/60 production system as a voluntary standard for domestic use; and Japan unflinchingly and staunchly continued to promote its system at home and abroad.

Soon after Dubrovnik (on June 30, 1986), under the initiative of four European consumer electronics manufacturers (Bosch, Philips, Thomson, and Thorn EMI), the ministers of 19 European nations[14] approved the creation of an industry-led consortium, called EUREKA EU95, whose role was to define the parameters of a compatible European HDTV system for the 1990 CCIR Plenary Assembly (see EUREKA EU95, 1988).[15] The four founders were soon joined by about 25 European manufacturers, broadcasters, and research organizations. This number grew to more than 80 participants by the second half of 1993. By the end of 1993, estimated investment in the EUREKA 95 project amounted to ECU 730 million ($855 million; $1 = ECU 0.854) (HDTV Directorate, 1994). National governments contributed between 25 percent and 50 percent to the project. Contrary to popular wisdom, EUREKA projects are not funded by the European Union. Instead EUREKA participants themselves assume most of the financial burden, with each national government providing only partial reimbursement for these costs. In June 1987, the Europeans formally notified the CCIR at the IWP 11/6 meeting of the parameters for their worldwide HDTV production standard (EUREKA EU95, 1988). Main features included 1250 scanning lines, 50 Hz field rate, 1:1 progressive scanning,[16] and 16:9 aspect ratio (CCIR Study Groups, 1987).

Meanwhile in the United States, efforts went underway to adopt the 1125/60 system as a national voluntary HDTV production standard. In July 1986, the ATSC asked SMPTE to document the technical parameters of the 1125/60 production system (CCIR Study Group, 1989). In June 1987, it reiterated its support for the 1125/60 HDTV system in a letter addressed to the Department of State and urged the U.S. delegation to uphold this position in CCIR negotiations ("Word From ATSC," 1987).

After seeking comments from all its members according to its standard-setting procedures, SMPTE's Working Group on High Definition Electronic Production approved on August 11, 1987 a technical document ("Signal Parameters of the 1125/60 High Definition Television Production System") describing the parameters of what would become known as the SMPTE 240M standard. In November 1987, SMPTE published the document in the *SMPTE Journal* for public comment. Finally in February 1988, before submission to the American National Standards Institute (ANSI), SMPTE's Executive Committee for Standards Approval ratified the proposed standard as a SMPTE standard without dissenting votes (Society of Motion Picture and Television Engineers, 1989).

On October 6, 1988, the ANSI Board of Standards Review approved SMPTE 240M as an American national standard. But then a week later, NBC submitted its 1050/59.94 system (ACTV-E) to SMPTE for consideration as a second HDTV production standard. Michael Sherlock, NBC President for Operations and Technical Services, asserted that the NBC system "departs from other proposals and will be uniquely suited for any transmission system that intends to be compatible with the existing United States system, NTSC" ("NBC Unveils," 1988, p. 31). But other broadcast engineers, such as CBS's Joseph Flaherty, were deeply saddened by the introduction of a second HDTV production standard. Flaherty stated: "There is only one tiny window to get a world production standard. Introducing a new standard at this stage of the game can only make that prospect more difficult. It can only lead to further division worldwide" (p. 31). The unveiling of a second production standard in the United States marked another turning point in the history of the international HDTV production standardization process, because since that time broad U.S. support for the 1125/60 system has waned domestically and internationally—and deteriorated further with the development of fully digital advanced television systems in the early 1990s.

In December 1988, Capital Cities/ABC filed an appeal with the ANSI Board of Standards Review (BSR) to reverse the approval of the standard on grounds that (1) SMPTE had not achieved industry consensus; (2) SMPTE 240M was unsuitable for domestic use; and (3) SMPTE had not followed due process procedures in promulgating the standard (ABC Broadcast Operations and Engineering, 1988). This appeal was denied on February 2, 1989. But then on February 22, Capital Cities/ABC filed a second appeal with the ANSI Appeals Board to reverse the BSR decision, and this time the appeal was upheld. On April 18, 1989, after hearing oral statements, the Appeals Board reversed the BSR action and nullified the approval of SMPTE 240M on grounds of inadequate industry consensus (American National Standards Institute, 1989). Then on August 15, 1989, at the request of SMPTE, the Appeals Board reheard a second appeal but reaffirmed, on the same grounds, its previous decision to uphold Capital Cities/ABC's appeal and disapprove the SMPTE 240M standard. Therefore, *SMPTE 240M was never approved as a voluntary U.S. standard* until recently when SMPTE resubmitted its proposal to ANSI in October 1995, which eventually approved it in December 1995. The ANSI/SMPTE seal of approval is important to manufacturers and users of HDTV production equipment to ensure compatibility among the different pieces of hardware in the production process. "Cameras need to be able to match with high-definition videotape recorders, switchers and other signal processors" (Carbonara, 1992, p. 21).

At the CCIR level, positions on the HDTV studio standard also shifted considerably. During a meeting of a CCIR working party in February 1989, participants discussed three courses of action that would change the direction of HDTV standardization. The three options were: (1) adoption of a single unique standard; (2) adoption of a dual standard including two picture rates; and (3) the adoption of standards based on a "common image" format (CIF) leading to the future adoption of a single standard (CCIR Study Groups, 1989; "'Common Image,'" 1989). The first two alternatives were discounted—the establishment of a worldwide standard, either 1125/60 or 1250/50, was considered unlikely and the adoption of *both* standards was considered undesirable. The third alternative attracted some support, though. The CIF approach was a "compromise" production standard in which CCIR members would seek to maximize commonalities between the Japanese/U.S. 1125/60 and the European 1250/50 proposed production standards without reaching complete agreement. Specifically, the Administrations would agree on setting as many common parameters for HDTV (e.g., colorimetry, resolution, aspect ratio) as possible, except field rate, and facilitating conversion from 50 Hz to 60 Hz and vice versa for the international exchange of HDTV programs. While *time* (50 Hz vs. 60 Hz) could vary by country, *space* (image aspect ratio and the number of active lines per frame) would remain constant. "It will also help the eventual achievement of a completely unique standard because there will be fewer scanning parameters remaining which are different" (Wassiczek, Waters, and Wood, 1990, p. 316).

In May 1989, during an extraordinary meeting of CCIR Study Group 11, participants approved the U.S. delegation's suggestion to delay the adoption of a production standard until 1994.[17] The United States also defeated a European motion to vote on the adoption of the 1250/50 system at the meeting ("U.S. Gets," 1989). This time, it was the United States that was seeking to gain extra time. By early 1989, the once-unflinching U.S. consensus on the adoption of the NHK production standard was in a state of disarray. An inquiry by the National Telecommunications and Information Administration (NTIA), the federal agency in charge of advising the White House on domestic and international telecommunication policy matters, had revealed a serious rift between broadcasters, program producers, and other concerned parties, with a plurality of filings criticizing the establishment of the 1125/60 high-definition production system as a worldwide standard ("HDTV Production," 1989). Following this outcome, in April 1989, the ATSC sent a letter to the Department of State urging it to postpone the decision on a worldwide HDTV standard, which in turn echoed this view at the extraordinary meeting.

In May 1990, the XVIIth Plenary Assembly of the CCIR, the last major CCIR conference before its dissolution in 1993, met in Dusseldorf, Germany, and approved 23 HDTV parameters, including colorimetry and 16:9 aspect ratio (CCIR, 1990; "CCIR Sets," 1990). Specifically, it adopted five Recommendations: Recommendation 709 dealt with the basic parameters for the HDTV production standard; Recommendation 710 focused on subjective assessment, which determines the methodological approach to rate picture quality; and Recommendations 713, 714, and 716 proposed solutions to HDTV program exchange problems (Krivocheev, 1991). CCIR participants, however, failed to agree on field rate and scanning lines but pledged to pursue their discussions during the 1990-1994 study period to recommend a unified HDTV production standard using common image format (CIF) and common data rate (CDR) approaches. CDR implied a dual standard approach in which HDTV standards based on 50 Hz and 60 Hz fields would be adopted while maximizing commonalities among other parameters such as line frequency and sampling frequency, based on the principles of Recommendation 601 (CCIR Study Groups, 1989).

After the reorganization of Study Group 11 activities in 1990, the formulation of an HDTV studio standard was reassigned from IWP 11/6 to Task Group (TG) 11/1 (Krivocheev, 1993). In March 1992, Study Group 11 established TG 11/3 to study digital terrestrial television broadcasting. By the end of 1992, neither the 1125/60 nor the 1250/50 system had mustered the necessary consensus to become a worldwide production standard, although there was enough support for both to encourage manufacturers to produce equipment.

1987: HDTV Transmission Standardization in the United States: Going and Going and ...

When the Federal Communications Commission (FCC) issued its first *Notice of Inquiry* (*NOI*) on advanced television (ATV) systems in July 1987, HDTV appeared to be just around the corner. Unlike the Japanese and Europeans who have sought to design an HDTV transmission system primarily for satellite broadcasting, the Americans prioritized the development of HDTV for terrestrial delivery. But it took nearly 10 years for the FCC to approve a standard for domestic terrestrial transmission. It finally did so on December 24, 1996. For all intents and purposes, the technical part of the HDTV standard-setting process ended on November 28, 1995, but not without controversy (e.g., inter-laced versus progressive scanning; see Chapter 7), when the member-ship of the Advisory Committee on Advanced Television Service rec-ommended that the FCC adopt the "ATSC Digital Television Standard" as the U.S. ATV terrestrial broadcasting standard.

It is important to note that the Commission has taken a broad ap-proach to advanced television by defining it as "any system that re-sults in improved television audio and video quality, whether the methods employed improve the existing NTSC transmission system or constitute an entirely new system" (Federal Communications Com-mission [FCC], 1988, p. 6544). ATV encompasses not only true high-de-finition television but also improved-definition television (IDTV), which refers to systems that correct some of the NTSC defects through such techniques as progressive scanning and pre-combing (i.e., use of special filters to separate the luminance signal from the chrominance signals prior to encoding) without an increase in bandwidth, as well as enhanced-definition television (EDTV), which refers to systems that involve major transformations of the NTSC standard (FCC, 1987a). Therefore, the FCC uses ATV as a generic term to refer to television technology surpassing NTSC quality.

The formal policymaking[18] process of HDTV transmission stan-dard-setting, and FCC involvement, began on February 17, 1987, when the Association of Maximum Service Telecasters (MST) and 57 other broadcasting organizations petitioned the Commission to initiate an NOI that would explore issues pertaining to ATV (see Chapter 6). In July 1987, the FCC issued an NOI (First Inquiry) soliciting comments on a vast array of ATV-related matters, including what criteria (e.g., band-width, compatibility) should be used to evaluate ATV systems and whether ATV service should be solely implemented in the UHF band (FCC, 1987a). In connection to this inquiry, the FCC decided to freeze all new (UHF) television station applications and allotment requests in 30 metropolitan markets in order to guarantee sufficient spectrum space for the use of advanced television systems (FCC, 1987b). In No-vember 1987, the Commission formed the Advisory Committee on Ad-vanced Television Service (ACATS), a 25-member advisory body com-prised of broadcasters, cable operators, manufacturers, and government officials, to recommend policies for the introduction of advanced television service in the United States (FCC, 1987c). Richard Wiley, a Washington attorney and former FCC Chairman, was ap-pointed to head the Committee. Most of the ACATS work was carried out by three subcommittees (Planning, Systems, and Implementation), which in turn delegated responsibility to working parties for exploring specific technical and economic ATV issues (see Chapter 7). In essence, ACATS supervised the HDTV technical transmission standardization process and advised the FCC about it on a regular basis. Naturally, it did not have the authority to *approve* a broadcasting standard, a function reserved by law to the FCC. Between November 1987 and November 1995, when it was officially disbanded, ACATS submitted five Interim Reports, an ATV System Recommendation, and a Final Report and Recommendation to the Commission that suggested recommendations and detailed progress on testing and selection of an HDTV transmission standard. The FCC then considered the findings of these reports in formulating policies for the implementation of HDTV in the United States.

In March 1988, seven[19] broadcast networks and organizations (CBS, Capital Cities/ABC, NBC, Public Broadcasting Service [PBS], the Association of Maximum Service Telecasters [MST], the National Association of Broadcasters [NAB], and the Association of Independent Television Stations [INTV]) established the Advanced Television Test Center (ATTC)[20] to "conduct broad, thorough, and impartial tests on ATV systems." Between 1988 and 1995, ATTC expenditures totaled $26.1 million.[21] Testing of proposed systems was due to finish in late 1991, with a recommendation made to the FCC by September 1992. The FCC would then set a transmission standard in 1993 ("Testing," 1989).

In its *Tentative Decision and Further Notice of Inquiry (Second Inquiry)* of September 1988, the Commission determined that (1) terrestrial ATV use would benefit the public; (2) spectrum for ATV would be allocated within the existing television frequency bands; (3) the existing NTSC service should continue to be provided to viewers, at least during a transitional phase, through either NTSC/ATV compatible or simulcasting techniques; and (4) independent development of nonterrestrial ATV services (e.g., cable) would be in the public interest (FCC, 1988). These provisions effectively ruled out the Japanese MUSE technology as a potential ATV transmission standard contender, because this 8.1 MHz HDTV system would have exceeded the allocated 6 MHz channel bandwidth. A compatible system implies that an ATV signal can be received on an existing NTSC receiver without extra equipment or significant picture deterioration. On the other hand, simulcasting refers to the transmission of the same program in dual formats. Broadcasters would send two over-the-air feeds, one in NTSC and another in HDTV.

Originally some 23 systems, all analog, were submitted to ACATS for ATV broadcasting in the United States (Advisory Committee on Advanced Television Service [ACATS], 1995). By October 1988, 12 companies had proposed 18 systems to ACATS ("Twelve," 1988). Most of these proposals were receiver compatible, and featured 1,000 or more scanning lines and a 16:9 aspect ratio. Half relied on progressive scanning, and about a third used channel augmentation.[22] The number of proponents and systems were winnowed to seven and nine by October 1989 ("Testing," 1989) and then to six and six by November 1990 ("HDTV Transmission," 1990), respectively.

HDTV technology changed radically on June 1, 1990, when General Instrument (GI) announced that it would design an all-digital HDTV system, DigiCipher. This declaration must be remembered as one of the most important actions in HDTV history, because it not only altered the course of the transmission standardization process in the United States but it also affected HDTV development programs in Europe and, to a lesser extent, in Japan. Indeed, some engineers felt that digital transmission would not be feasible until the 21st century (Brinkley, 1997a). Soon after the first digital entry, other companies redirected their efforts for devising a digital rather than analog system and made similar announcements. In September 1990, the FCC adopted its *First Report and Order*, by selecting a simulcast high-definition television approach for ATV use. Furthermore, the Commission stated that it would consider an EDTV system only after a decision was reached on a "true" HDTV standard, effectively "deprioritizing" research on EDTV systems (FCC, 1990a).

In July 1991, after numerous delays, ATTC laboratory and field tests of six systems finally went under way. The six systems were: Advanced Compatible TV (Advanced Television Research Consortium), Narrow MUSE (NHK), DigiCipher (GI/MIT), Digital Spectrum Compatible HDTV (Zenith/AT&T), Advanced Digital HDTV (Advanced Television Research Consortium), and Channel Compatible DigiCipher (GI/MIT). The number of contenders became five when the Advanced Television Research Consortium withdrew ACTV, an NTSC-compatible, analog EDTV system, in March 1992. By that time, all certified systems but Narrow MUSE (see Chapter 3) were fully digital.

In April 1992, the Commission adopted its *Second Report and Order/Further Notice of Proposed Rule Making* in which it proposed a timetable for ATV implementation. Existing full-power broadcasters would have two years to apply for HDTV spectrum once the Commission approved the standard, three years for constructing the new HDTV facility once that spectrum was allocated, seven years to simulcast 50 percent of their programming, nine years to simulcast 100 percent of their programming, and 15 years to fully convert their stations from NTSC to HDTV. After this 15-year conversion period, broadcasters would be required to surrender one of the two broadcast channels and cease broadcasting in NTSC (FCC, 1992a). Broadcasters' reactions to the FCC timetable ranged from skepticism to displeasure to outright anger. Whereas some broadcasters called the rules premature, others complained that even if they reequipped their facilities, thereby incurring significant costs, HDTV would not yield a "satisfactory return on investment"[23] (Flint, 1992, p. 14). In the early 1990s, estimated costs of converting television facilities ranged from a high of $12 million to a low of $750,000, depending on station size and type of conversion (see Chapter 8).

In July 1992, the Commission decided to assign a separate HDTV channel to all existing (full-power) broadcasters and place all but 17 HDTV channels in the UHF band. Therefore, 99 percent of all HDTV allotments would be made on UHF frequencies (FCC, 1992b). In September 1992, on reconsideration of its *Second Report and Order* (*Third Report and Order/Third Further Notice of Proposed Rule Making*), the FCC slightly modified its initial conversion timetable and set the following deadlines: 1996 for HDTV spectrum applications (three years after the ATV system selection); 1999 for the construction of HDTV facilities (after six years); 2000 for the 50 percent simulcasting requirement (after seven years); 2002 for the 100 percent simulcasting requirement (after nine years); and 2008 for full station conversion from NTSC to HDTV (after 15 years) (FCC, 1992c). The revised conversion timetable still infuriated many broadcasters who insisted that such a strict 15-year transition window could sound the death knell for small-market stations that may be unable to absorb the high cost of HDTV equipment and programming (Lambert, 1992d). But by 1995, FCC Chairman Reed Hundt, Congress, and the White House were all considering shortening rather than extending the transition timetable for HDTV terrestrial broadcasting—in order to recover valuable spectrum and schedule auctions as early as possible, much to the dismay of many broadcasters who have stressed that more, not less time, is needed to complete the switch from NTSC to HDTV.

In February 1993, the results of the first round of ATTC tests were released. They revealed that, apart from Narrow MUSE, none of the four all-digital systems was significantly superior to the others. Therefore, there was no clear winner or loser. In light of this dilemma, ACATS suggested two courses of action to the four HDTV finalists: either ready their systems for retesting within the same year or create a "grand alliance" system that would combine the best components of individual systems.[24] The four proponents reached a consensus and formed the Digital HDTV Grand Alliance on May 24, 1993 (see Chapter 7). This consortium comprised seven European and American entities: AT&T, the David Sarnoff Research Center, General Instrument, the Massachusetts Institute of Technology, North American Philips, Thomson Consumer Electronics, and Zenith Electronics. The five subsystems of the Grand Alliance HDTV prototype (scanning parameter, video compression, data transport, audio, and modulation) were selected by February 1994. In December 1994, additional features were described to satisfy multichannel, standard-definition television (SDTV, a digital NTSC system discussed below) demands. But these parameters were never integrated in the hardware to be tested to avoid further delays in the testing schedule and possible dissensions among Grand Alliance members, who were unable to agree on an SDTV implementation plan for competitive reasons. Some members, especially GI and Thomson, were reluctant to cooperate and share information for building an SDTV system, because they were already involved in selling SDTV-type technology in the marketplace (e.g., Thomson's DSS). So although the December 1994 paper suggested that SDTV was theoretically doable, it was *never* tested as part of the Digital HDTV Grand Alliance system (P. Fannon, Advanced Television Test Center, personal communication, March 18, 1996).

As tested, the HDTV system contained seven main characteristics (ACATS, 1995; Fannon, 1994): (1) all-digital audio and video transmission; (2) dual scanning format (1125-line [or 1080-active[25] line] interlaced scanning and 787.5-line [or 720-active line] progressive scanning); (3) MPEG-2[26] compression for video and Dolby AC-3 compression for audio; (4) MPEG-2 encoding and transport scheme; (5) vestigial sideband (8-VSB) modulation;[27] (6) multichannel audio (five channels plus a subwoofer); and (7) 16:9 aspect ratio.[28] The Grand Alliance system featured "interoperability"[29] capabilities, such as dual scanning modes and multiple frame rates (see Table 1.1), to accommodate broadcast as well as nonbroadcast uses and applications (e.g., computer systems, cable television, DBS).

Testing of the Digital HDTV Grand Alliance system at the ATTC began in April 1995 and lasted until August 1995. In April 1995, the full membership of the ATSC endorsed the Grand Alliance system as what is now known as the ATSC Digital Television Standard. Inclusion of SDTV in the standard was approved in September 1995. The role of the ATSC in the HDTV transmission standard-setting process ought not be confused with that of ACATS. These two entities performed different tasks and operated under different rules. On behalf of the Advisory Committee, the ATTC conducted extensive testing on the Grand Alliance system according to a set of product specifications finalized in December 1994. The ATSC then took this blueprint and turned it into a carefully-defined standard—the ATSC standard—by documenting the complete details of the specifications. A metaphorical way to compare the two bodies is to say that the ATTC/ACATS delivered the steel frame and the engine of a car, while the ATSC supplied all the accessories to make the car ride smoothly and comfortably (R. Hopkins, Advanced Television Systems Committee, personal communication, November 17, 1995).

On November 28, 1995, the 25 members of ACATS met for the ninth and last time to sanction the final report and recommended to the FCC the adoption of the ATSC Digital Television Standard as the U.S. standard for ATV broadcasting (ACATS, 1995). In April 1996, public television station WETA-TV, Washington, DC, announced plans to build a broadcast HDTV station by 1998 for an estimated $10-$14 million. Jerry Butler, WETA-TV Vice President of Engineering, Operations, Production and Computer Services, predicted that "90% of the station's equipment will have to be replaced in the project" (Dickson, 1996, p. 60). WRC-TV, an NBC-owned-and-operated station in Washington, DC, made a similar announcement in May, and WRAL-TV, a CBS affiliate in Raleigh, North Carolina, received the first experimental HDTV license (WRAL-HD on channel 32) in June 1996.

On May 9, 1996, the FCC launched a *Fifth Further Notice of Proposed Rule Making* (*NPRM*) to propose the adoption of the ATSC Digital Television Standard, and this despite Chairman Reed Hundt's last-minute misgivings about "whether the government should be in the business of *mandating* standards at all" (FCC, 1996a, p. 6272). On July 23, 1996, WRAL-HD, Raleigh, NC, became the first broadcast station in the United States to transmit HDTV signals.

In its *Sixth Further Notice of Proposed Rule Making* adopted on July 25, 1996, the Commission moved away from the idea of placing virtually all HDTV allotments on the UHF band (FCC, 1992b) and instead proposed to allocate *both* VHF and UHF frequencies for ATV use (FCC, 1996b). According to this new plan, all existing eligible broadcasters would be allotted a digital television (DTV) channel in a core spectrum between channels 7 and 51 (i.e., between 174 and 216 MHz for VHF channels 7-13 and between 470 and 698 MHz for UHF channels 14-51). During the 15-year transition phase, however, some stations would operate on channels 2-6, 52-59, and 60-69, but ultimately they would be "repacked" into the core DTV spectrum. As drafted, this Table of Allotments provided core digital spectrum to 88 percent of the 1578 existing eligible broadcasters (FCC, 1996b).

Despite the quasi-unanimous vote of ACATS on the ATSC standard on November 28, 1995, opposition to the FCC mandating and adopting such a standard grew throughout 1996 to reach a climax in October when all industries but broadcasters and consumer electronics manufacturers objected to it (see Chapter 6). By that time the ATSC digital standard appeared to be in real jeopardy with no consensus in sight. The National Cable Television Association (NCTA) opposed the notion of mandating a standard for advanced television; the American Society of Cinematographers (ASC) and the Directors Guild of America (DGA) advocated a 2:1 aspect ratio instead of 16:9; consumer groups, such as the Consumer Federation of America (CFA), protested the "giveaway" of digital spectrum to broadcasters; and the Computer Industry Coalition on Advanced Television Service (CICATS) argued that the inclusion of interlaced scanning would perpetuate obsolete technology and hamper the convergence of television sets and computers (see McConnell, 1996e). Not only did CICATS "urge the Commission to adopt no more than a minimally necessary DTV standard" (Computer Industry Coalition on Advanced Television Service, 1996, p. 2), but it considered HDTV itself an unnecessary technology that would force broadcasters and consumers to leap beyond standard-definition television (SDTV) and acquire expensive transmitting and receiving equipment. Pressured by FCC Commissioner Susan Ness to settle their differences (Beacham, 1996b), broadcasters, CICATS, and the Consumer Electronics Manufacturers Association (CEMA) reached a compromise on November 26, 1996 by calling upon the FCC to approve the ATSC standard *without* the video format constraints (Table 1.1), that is, the number of active lines, the number of pixels (picture elements) per line, the aspect ratio, the frame rate, and the scanning structure. Concretely, this compromise, which was really an agreement about *removing* a disagreement area, meant that the selection of these

TABLE 1.1 THE 18 VIDEO SCANNING FORMATS OF THE ATSC DIGITAL TELEVISION STANDARD

ACTIVE LINES	HORIZONTAL PIXELS	ASPECT RATIO		FRAME RATE AND SCANNING STRUCTURE			
1080	1920	16:9		30I		30P	24P
720	1280	16:9			60P	30P	24P
480	704	16:9	4:3	30I	60P	30P	24P
480	640		4:3	30I	60P	30P	24P

NOTE: I = interlaced; P = progressive.

parameters would be left to market forces (see Chapter 6). On December 24, 1996, the FCC adopted this modified ATSC standard, referred to as the Digital Television (DTV) Standard (FCC, 1996c). On February 2, 1997, NBC, in cooperation with its owned-and-operated WRC-TV, became the first U.S. broadcast television network to transmit live a program (*Meet the Press*) in HDTV format.

On April 3, 1997, the FCC completed the two remaining ATV rulemakings—general service rules and channel assignments. In the *Fifth Report and Order*, it completely reshuffled the NTSC-to-HDTV transition schedule it had adopted in September 1992 (FCC, 1992c). It required that (1) the affiliates of ABC, CBS, Fox, and NBC in the top 10 markets (30 percent of TV households) build their digital television (DTV) facilities by May 1999 and those in the top 30 markets (53 percent of TV households) by November 1999; (2) all remaining commercial stations construct their DTV facilities by May 2002; (3) all noncommercial stations construct their DTV facilities by May 2003; and (4) DTV licensees simulcast 50 percent of their analog video programming on the DTV channel by April 2003, 75 percent by April 2004, and 100 percent by April 2005. It also shortened the duration of the transition period from 15 to 10 years by setting the target date for the phase-out of NTSC service in 2006. More surprisingly, and contrary to earlier actions, the FCC declined to mandate broadcasters to air a minimum amount of HDTV programming and, instead, left this decision to the discretion of the licensees (FCC, 1997a).

In the *Sixth Report and Order*, the FCC adopted a Table of Allotments for digital television based on use of channels 2-51. Responding to comments, it added channels 2-6 to channels 7-51, which was originally proposed as the core spectrum in the *Sixth Further Notice of Proposed Rule Making* of July 1996 (FCC, 1996b), to determine DTV channel assignments. Upon acceptability of the lower VHF channels 2-6 for

DTV use, the Commission will consider shifting the core spectrum from channels 7-51 to channels 2-46. According to this table, over 93 percent of broadcasters would receive a DTV allotment that reaches at least 95 percent of their existing NTSC service area (FCC, 1997b).

As of June 1997, seven stations were broadcasting experimental HDTV signals: WRAL-HD, Raleigh, NC; WRC-HD, Washington, DC; KOMO-HD, Seattle, WA; KCTS-HD, Seattle, WA; KOPB-HD, Portland, OR; WETA-HD, Washington, DC; and WCBS-HD, New York, NY. The first HDTV receivers are expected to hit U.S. stores by Christmas 1998. In May 1997, Zenith estimated they would cost between $5,000 and $7,000 ("Zenith's First Digital Sets," 1997).

1992: The Failure of the MAC Policy and the Road to Digital Television in Europe

In Europe, MAC, especially D2-MAC, was expected to become the de facto transmission standard for DBS and pave the way for HDTV satellite broadcasting. To achieve this evolutionary approach, the Council of the European Communities adopted a Directive in November 1986 that required all DBS programmers to deliver their services in MAC format (O.J. No. L 311, 1986). Therefore, the fate of MAC was linked to the success of DBS. By mid-1990, however, few DBS services were off the ground due to a combination of bad luck and poor planning (Renaud and Morgan, 1990). The driving forces behind D2-MAC were the national governments (especially France), not the satellite broadcasters, most of whom elected to embrace PAL technology instead. Broadcasters' lack of interest in MAC, PAL's incumbent status for satellite transmission in Europe, and the multiplicity of MAC and encryption standards all contributed to sidetrack the MAC initiative. Therefore, the demand for 4:3 D2-MAC receivers never materialized (see Chapter 4).

It was in this atmosphere of uncertainty that the Commission of the European Community (EC)/European Union (EU) issued its first draft of the second MAC Directive in July 1991 to replace the first one which was scheduled to expire on December 31, 1991. When it initiated this policymaking process, the EU Commission triggered a series of heated debates in which satellite broadcasters and electronics manufacturers clashed, the former decrying this latest example of EU industrial policy and balking at the cost of upgrading to HD-MAC technology and the latter promoting MAC as an evolutionary approach toward HDTV and supporting the creation of a single standard for satellite and cable transmission (Amdur, 1992; "EC Drafts," 1991). Furthermore, some European broadcasters were concerned about technological obsolescence, questioning whether it was even wise to continue supporting an analog HDTV system in light of recent and rapid developments in digital technology that had taken place in the United States (Kaitatzi-Whitlock, 1994). In May 1992, the EU Council of Ministers struck down a compromise that enabled existing satellite broadcasters to continue transmission of their services in PAL, but required all new satellite channels launched in 1995 and later to transmit their signals in D2-MAC (O.J. No. L 137, 1992).

For all intents and purposes, the D2-MAC transmission standard and *a fortiori* HD-MAC was politically "retired" (or more precisely "de-emphasized") in July 1993, at least as a policy initiative, when the Council adopted a resolution calling for a revision of the 1992 MAC Directive and shifting priorities away from MAC technology and the question of standards toward widescreen and digital technology (O.J. No. C 209, 1993). The EU stepped back from the standard-setting process of satellite transmission and left the selection of an appropriate standard to the discretion of programmers. In retrospect, the 1992 Directive faltered for a variety of reasons (see Chapter 4). In addition to those articulated above, the MAC policy fell victim to Council members' political infighting over costs—the United Kingdom's opposition to a proposal ("Action Plan") to help financing the introduction of HDTV being a case in point. But it also failed because its upgrade path (HD-MAC)—the raison d'être of the MAC strategy—had become irrelevant in an increasingly digital television world. It is ironic that the major European consumer electronics manufacturers were still reaffirming their commitment to HD-MAC at the 1992 International Broadcasting Convention, while in another room a group of Scandinavians were demonstrating HD-DIVINE (Digital Video Narrow-band Emission), a digital HDTV system (Fox, 1995). The analog HD-MAC will continue to exist as a broadcast standard, but its practical use, if any, might be only limited to nonbroadcast applications (e.g., transmission of programs to a network of movie theaters) (L. H. de Waal, Advanced Digital Television Technologies, personal communication, March 22, 1995).

With the abandonment of HD-MAC, it is likely that D2-MAC will eventually follow the same path. For now, it is still in operation, although it may be just a matter of time before the inevitable advent of digital television sweeps away all analog-based terrestrial, cable, and satellite transmission systems in Europe. According to *Cable and Satellite Europe,* only 39 out of 239 satellite channels transmitted programs in 4:3 or 16:9 D2-MAC by November 1996. Unlike D2-MAC, the widescreen 16:9 format has generated considerable consumer interest in Europe, especially in France. By the end of 1991, there were only 2,000 16:9 television sets in the European Union all in France (European Audiovisual Observatory, 1994). But by the end of 1995, annual and cumulative sales of 16:9 TV sets in the European Union totaled 220,500 and 482,400 units, respectively (Coopers & Lybrand, 1996).[30] Jean-Luc Renaud (1995b), the editor-in-chief of the newsletter *Advanced Television Markets,* has argued that Europeans' (and Japanese's) growing interest in those new widescreen sets may be more a function of the technology (wider aspect ratio) than programming, which has remained fairly limited so far.[31]

In June 1994, the 22 EUREKA[32] nations approved the creation of a new research project for the development of European digital HDTV. The EUREKA 1187 Advanced Digital Television Technologies (ADTT) project succeeded the EUREKA EU95 project, which started in 1986 with four members and ended in 1994 with more than 80 participants from 13 countries. The new project is expected to cost ECU 250 million ($297 million; $1 = ECU 0.843) over two and a half years and like EUREKA EU95 will be funded primarily by the participants and EUREKA members (i.e., national governments). The ADTT project will develop and test a digital broadcast HDTV system based on a specification adopted by the Digital Video Broadcasting (DVB) group (L. H. de Waal, Advanced Digital Television Technologies, personal communication, March 22, 1995). European officials have stated that the investment made in EUREKA EU95 has been fruitful as most of the research on the analog HD-MAC will be applicable to digital systems (see Homer, 1993). But it is less and less evident that HDTV is still on the industry agenda. "I don't know of one single broadcaster in Europe who wants to do HDTV. The economics simply do not add up," recently declared Rob Van Oostenbrugge, who was involved in HD-MAC development and is now Business Manager for Digital Video Communications Systems at Philips (Homer, 1996, p. 28). Other than investment and bandwidth requirements, European broadcasters and consumer electronics manufacturers may shy away from HDTV because of the recent memory of the HD-MAC fiasco. As the French proverb says, *chat échaudé craint l'eau froide* (in effect, once bitten, twice shy).

In October 1995, the EU Council adopted a Directive on the use of standards for the transmission of television signals to repeal the 1992 MAC Directive (O.J. No. L 281, 1995). It required member states to take appropriate measures for the promotion of advanced television services, including widescreen television, high-definition television, and digital television. It further mandated that terrestrial, cable, and satellite programmers transmit in 16:9 D2-MAC (*if* their service is in widescreen, 625-line format, and not fully digital); or (2) in HD-MAC (*if* their service is not fully digital HDTV); or (3) in a digital format approved by a European standardization body. Therefore, while not totally leaving TV standard matters to market forces, this new Directive displays much greater flexibility than did its predecessor by offering programmers a series of transmission standard options.

This change of direction does not mean, however, that the EU is uninvolved in European ATV policy. Through its 1993 Action Plan (AP), it allocated ECU 94.6 million ($115.7 million) between July 1993 and December 1995 for the European production and remastering of 19,789 program hours in 16:9 format (Commission of the European Communities, 1996). By May 1997, some 50 European stations from 14 countries had broadcast 53,311 16:9 hours with the support of the AP. Despite these encouraging statistics, it remains to be seen whether the AP is able to achieve its ambitious objectives of creating "a critical mass of advanced television services in the 16:9 format" and "a sufficient and increasing volume of programming in the 16:9 format" (O.J. No. L 196, p. 49) by the end of its term in June 1997.

Arguably, the most important technical player in European DTV today is the Digital Video Broadcasting (DVB) group. This entity, created in 1991, is spearheading European efforts to establish technical standards for digital satellite, cable, and terrestrial transmission (see Digital Video Broadcasting [DVB], 1996). In February 1996, its membership counted more than 200 organizations from 25 different countries, representing a wide cross-section of the electronic media industry (e.g., broadcasters, manufacturers, regulatory bodies). It remains fiercely independent of the EU Commission in Brussels and receives no EU subsidies (Fox, 1995). In early 1994, the DVB Steering Board agreed on MPEG-2 based digital standards for cable (DVB-C) and satellite (DVB-S) television. In December 1994, the European Telecommunications Standards Institute (ETSI) ratified the standards. In December 1995, the DVB Steering Board approved a digital television standard (DVB-T) for terrestrial 7-8 MHz channels using OFDM (orthogonal frequency division multiplexing) transmission technology

(DVB, 1996) The DVB-T uses the same generic elements as the DVB-C and DVB-S, but modulation schemes are different. ETSI approved the DVB-T standard in January 1997. In June 1997, in a surprising move, the DVB group announced guidelines for HDTV transmission and reception (DVB-HDTV) that incorporated the Main Profile at High Level (MP@HL) formats from the MPEG-2 family of video coding and data transport standards. These formats are in line with ITU decisions and are similar to those used in the U.S. ATSC digital television standard (P. MacAvock, DVB, personal communication, September 1, 1997).

Several direct-to-home (DTH) digital services were launched between late 1995 and late 1996, and more are expected to follow in 1997 and 1998. For instance, in October 1985, Telepiù, the Italian pay-TV operator, became the first European satellite service to transmit programs in digital format via EUTELSAT II F1. In January 1996, it launched DStv, an 8-channel digital bouquet (i.e., package). Then in April 1996, heralded by an aggressive marketing campaign, French pay-TV operator *CANAL+* introduced *CANAL*SATELLITE *NUMERIQUE,* a digital satellite-delivered package offering basic channels as well as pay-per-view, digital radio, and even a software "teleloading" channel (Alliot, 1996). If the current raging battle between French DTH providers is any indication (e.g., for output deals), the market for digital satellite services in Europe is on its way to becoming as competitive as that for DBS services in the United States.

The years 1997 and 1998 could also be fertile in European digital *terrestrial* television (DTTV) ventures. For instance, the British government is preparing plans to make at least 18 DTTV channels available to 60 to 90 percent of viewers by 1997-98 (see Chapter 4). Over-the-air broadcasters (British Broadcasting Corporation, Channel 3, Channel 4, and Channel 5) would be guaranteed one digital channel in exchange for simulcasting a minimum of 80 percent of their existing PAL programs in digital format ("Coalition," 1995). For the moment then, European broadcasters pursue an SDTV, not HDTV, strategy for satellite and terrestrial delivery.

1994: Japan's Reluctant Reconsideration of Analog HDTV

By 1993, the United States had moved full speed toward the adoption of a digital terrestrial television standard, and Europe had started revamping its R&D HDTV program from analog to digital. Of the triad, Japan was still clinging to its primarily analog MUSE transmission system, broadcasting eight hours of Hi-Vision programming a day from the BS-3b satellite[33] (see Chapter 2). On January 25, 1993, however, the Telecommunications Technology Council of the Ministry of Posts and Telecommunications (MPT) created the ultra-definition television (UDTV) project. The purpose of the UDTV venture was to promote and develop high-resolution and digital image technologies for the 21st century (Y. Sakanaka, Ministry of Posts and Telecommunications, personal communication, January 5, 1996).

Then on February 22, 1994, Akimasa Egawa, Director General of the MPT's Broadcasting Bureau, dropped a bombshell in Japan and overseas by stating at a panel discussion that Japan will begin developing a digital HDTV system that will ultimately replace analog MUSE. "The world trend is digital," he declared to a dumbfounded audience (Choy, 1994, p. 5). Egawa added that holding on to analog technology might endanger Japan's industrial future and that the Ministry might even consider cooperative digital HDTV projects with the United States. Although he backpedaled the next day by saying that "he did not mean to imply the conversion to digital was imminent" (Kageki, 1994, p. 1), his original comment angered NHK and the Japanese consumer electronics industry which have invested an estimated $1.3 billion in developing HDTV (Johnstone, 1993). In particular, NHK, which spent ¥19 billion ($186 million; $1 = ¥102) on HDTV technology (Kageki, 1994), called Egawa's remarks "very regrettable" ("Japanese Official," 1994, p. B11). Tadahiro Sekimoto, NEC President and Chairman of the Electronic Industries Association of Japan, was quoted as saying that "Japan's MUSE system has been complete, and we have no ideas of giving up the HDTV system" (Regelman, 1994, p. 70). But Nicholas Negroponte (1995), the MIT Lab Director, claimed otherwise: "I recall vividly at the time a televised panel of the presidents of the giant consumer electronics companies swearing they were fully behind good old analog Hi-Vision, implying that the deputy minister was off his rocker. I had to bite my digital tongue, because I knew each of them personally, had heard them say the opposite, and had seen their respective digital TV efforts" (p. 40). At any rate it was the first time that a Japanese official publicly acknowledged what some analysts had suspected for some time—a digital system will replace analog HDTV in Japan in a not-too-distant future. It is quite rare for a high-ranking government official in Japan to make such public statements without first consulting the industry and engaging into an elaborate process of consensus-building (known as *ne-mawashi*). Egawa left the MPT in June 1995.

In January 1995, the *Japan Economic Newswire* reported that NHK was planning a double strategy for the early 21st century: to introduce digital broadcasts *as well as* transmit Hi-Vision on its two satellite channels. In March 1995, a private MPT advisory panel, the Study Group on Broadcasting System in the Multimedia Age, recognized that "the digitalization of High Definition TV is a global trend for the near future" ("Outlook," 1995, p. 4). It recommended (1) that low-power communication satellite (CS) broadcasting begin in 1996; (2) that high-power DBS broadcasting satellite (BS) services *either* continue with Hi-Vision for now and introduce digital broadcasts after 2007 *or* consider the introduction of digital broadcasting earlier with the launch of the next BS-4 satellite series around 1999; and (3) that digital terrestrial television be introduced between the years 2000 and 2004. In April 1996, the MPT indicated that it was seriously considering using digital technology for the second BS-4 satellite, although it postponed the decision for a year, apparently to alleviate the pressure from NHK and consumer electronics companies. On October 1, 1996, PERFECTV, a consortium of 28 companies, launched the first of four digital satellite TV (CS) services, offering 57 SDTV-type channels (Johnson and Davies, 1996).

In early 1997, Japan announced two major digital plans, which should enable Japanese broadcasters to catch up with their American and European counterparts by the year 2000. First in February 1997, a study group of the MPT proposed that digital broadcasting satellite (BS) service begin in 2000 using the four channels of the second BS-4 satellite. One of the channels will simulcast the four analog services of the first BS-4 satellite until around the year 2007. The other three channels will deliver digital HDTV signals (Kumada, 1997). On May 30, 1997, the Radio Regulatory Council, an advisory body to the Minister of Posts and Telecommunications, recommended approval of the study group report. Technical standards for digital HDTV satellite broadcasting are expected to be set by the end of 1997.

Then in March 1997, Shuji Kusuda, Director General of the MPT Broadcasting Bureau, disclosed that the introduction of digital terrestrial television (DTTV) in Japan will be moved up before the year 2000, instead of between 2000 and 2005 as originally planned (Pollack, 1997). The Ministry will begin allocating frequencies and testing a DTTV system, which will use the European-based OFDM modulation scheme, by the end of 1998. While recognizing the inevitability of digital terrestrial broadcasting, commercial Japanese broadcasters, not unlike their American counterparts a few years ago, expressed serious concerns about the economic impact of such a policy on their existing station operations ("MPT Speeds," 1997). Meanwhile, full-scale, regular MUSE HDTV broadcasting is still expected to start in 1997.

1995: EDTV in Full Swing

Since the late 1980s, broadcasters in the triad have expressed a growing interest in the capabilities of enhanced-definition (or extended-definition) television (EDTV), whether it is analog or digital. In the United States, several ATV proponents were working on developing EDTV systems in the late 1980s, but by 1990 emphasis had shifted almost exclusively to HDTV systems due to the FCC's *First Report and Order*. Although Press Broadcasting had petitioned the Commission in 1992 to lift restrictions on NTSC-like multichannel broadcasting (Jessell, 1992),[34] it was not until 1994 that the concept of standard-definition television (SDTV), a form of EDTV technology, really attracted the attention of the television industry and formally became part of ACATS's work (McConnell, 1994). In its *Fourth Further Notice of Proposed Rule Making and Third Notice of Inquiry*, the FCC defined SDTV as "a digital television system in which picture quality is approximately equivalent to the current NTSC television system" (FCC, 1995, p. 10541). In effect, SDTV is a digital version of NTSC and, by extension, refers to non-high-definition DTV. In July 1995, the ACATS Technical Group incorporated two SDTV formats into the ATV standard (see Table 1.1). Technically, broadcasters can use up the 19 megabits per second (Mbps) capacity of the 6 MHz broadcast channel with one HDTV program, 4-6 SDTV programs (occupying about 5 Mbps of the 19 Mbps capacity each), or even a series of video games if they so wish. Naturally, the greater the number of SDTV programs that can be delivered within the 6 MHz broadcast channel, the lower the resolution of SDTV one is willing to accept (R. Hopkins, Advanced Television Systems Committee, personal communication, January 1, 1996). So the ATSC Digital Television Standard would allow broadcasters to either transmit HDTV or multiple SDTV channels, although Richard Wiley cautioned that SDTV is technically not as good as HDTV (McConnell and West, 1995).

From a broadcaster's point of view, however, the expected benefits of SDTV over HDTV are self-evident. For some, HDTV is a costly and losing proposition that is unlikely to spawn financial gains, whereas SDTV would allow them to tap into new revenue streams by broadcasting video, audio, and data on multiple channels.[35] Broadcast data channels may be used for product information, credit information, video games, computer software, newspaper information, stock quotes, and audio services. For instance, Preston Padden, then President for Fox Network Distribution, argued that "If broadcasters can begin offering a multiplexed channel selection, they will be able to compete head to head with cable and others" (Stern, 1995a, p. 30). Fox intends to offer HDTV during certain times of the day and multiple programming services at other times.

On the other hand, other broadcasters have contended that endorsing SDTV instead of HDTV does not make much technical sense. As NBC President Robert Wright put it, "Since high definition will be, for an indefinite period of time, the standard for the best picture, it's inconceivable to me that broadcasters, of all people, would want to be in any way left out of complete parity in high definition" (West, 1995a, p. 28). Furthermore, they have claimed that the SDTV strategy could backfire in this day and age when spectrum represents a hot commodity with high resale value. The FCC estimated that the digital TV spectrum alone is worth at least $11 billion, although there is no consensus on the real worth of the spectrum—digital or analog (Stern, 1996b). Receiving "free" spectrum for multiple SDTV channels or services other than HDTV could seem a less legitimate purpose in the eyes of the current budget-cutting Congress, in which some members have indicated support for auctioning second-channel spectrum upfront. The prospect of Congressionally-mandated second-channel spectrum auctions faded away with the departure of Senate Majority Leader Bob Dole, who had promoted the idea, in June 1996 (see Chapter 6).

The United States is not alone in its quest for a terrestrial EDTV system. Unlike the United States, Japan and Europe have never really envisioned delivering HDTV terrestrially. So it is hardly surprising that, to compete better in the marketplace, Japanese and European broadcasters decided to develop an EDTV system for terrestrial use. In Japan, the first generation of NTSC-compatible extended-definition television (EDTV) system, called Clear-Vision or EDTV-I, was introduced in August 1989. It offered an enhanced picture with a 4:3 aspect ratio. In 1995, 115 of the 119 private broadcasters were broadcasting in Clear-Vision (Ninomiya, 1995).[36] But more than Clear-Vision or Hi-Vision—which still cost over $3,000 in June 1997[37]—it has been widescreen television that has enjoyed the greatest success with Japanese consumers in recent years. Widescreen television in Japan is just

NTSC with a 16:9 aspect ratio. Total sales of widescreen receivers soared from 44,000 in 1992 to 8,010,000 by April 1997 (Hi-Vision Promotion Association, http://www.j-entertain.co.jp/hpa/index.html). In June 1997, the price of a 28-inch widescreen receiver averaged about $1,500 (¥180,000).

In early 1996, six private broadcasters in Tokyo were transmitting in EDTV-II, the next generation of EDTV technology, and more were expected to follow suit (Y. Osumi, Ministry of Posts and Telecommunications, personal communication, February 6, 1996). This EDTV system offers a better picture quality than NTSC, as well as a 16:9 aspect ratio. EDTV-II is compatible with the existing NTSC system but requires current owners of widescreen television sets to purchase a decoder for displaying the higher picture resolution (Pollack, 1994). With the recent changes in Japanese broadcast policy, it is unlikely that these analog EDTV systems will survive the digital wave in the long term.

In Europe, a growing number of terrestrial broadcasters are transmitting signals in PALplus format, an EDTV version of 625-line PAL, although the amount of programming remains relatively limited. In 1989, public service broadcasters from Germany, Austria, and Switzerland, in conjunction with European consumer electronics groups, formed a steering group to coordinate research and development on PALplus. In mid-1997, 28-inch PALplus receivers sold for $2,000-$2,500. PALplus is compatible with PAL, but also offers a 16:9 aspect ratio, sharper pictures, and better sound quality (Rawsthorn, 1995). Collectively, over 20 European broadcasters aired about 14,500 hours of programming, mainly films, in PALplus in 1996, a 25 percent increase from the last 18 months (June 1994-December 1995) (E. Matzel, Office PALplus, personal communication, January 5, 1996).[38] Over 20,000 hours are expected for 1997.

1997: And What About the Worldwide Production Standard?

As of this writing, the prospect for the eventual adoption of a single worldwide *production* standard remains unlikely. In fact, it has become virtually a moot point because the ITU-R,[39] the successor of the CCIR, approved a revised Recommendation 709 at the 95 Radiocommunication Assembly that no longer promotes the objective of a single worldwide standard but instead endorses a dual standards (the Japanese 1125/60/2:1 and the European 1250/50/2:1) approach for HDTV production.

In recent years, ITU-R's work on common image format (CIF) and common data rate (CDR) approaches has slowed down because more attention has been given to digital broadcasting. While in the 1980s it was deemed important to set technical specifications for production before transmission because origination is the first step in the television process, priorities have now shifted under the influence of digital transmission systems. In the early 1990s, the HDTV studio format studied by Task Group (TG) 11/1 became somewhat dormant, whereas digital television considered by TG 11/3 (created in March 1992) emerged as a major standardization issue (J. W. Reiser, Federal Communications Commission, personal communication, August 11, 1995). In 1993, the first Radiocommunication Assembly (RA-93) met in Geneva, but little happened in regard to the HDTV production standard because no proposal was presented by Study Group 11 (television broadcasting). In March 1994, Study Group 11 met in Geneva to approve a revision of Recommendation 709 that included, among other things, the dual implementation of the production standards based on the 1250/50 and 1125/60 formats (G. Rossi, International Telecommunication Union, personal communication, April 5, 1995). Also in 1994, TG 11/1 suggested changing the definition of the scanning structure of the 1250/50 system from progressive 1:1 to interlaced 2:1 in Recommendation 709.

In October 1995, the second Radiocommunication Assembly (RA-95) approved the revised Recommendation 709 (709-2), although the U.S. Administration took a reservation on two counts (Radiocommunication Assembly, 1995). First, because the United States endorsed the principle of a single worldwide standard, it opposed language of the Recommendation that supported two separate production standards. While the two standards did have many common parameters, the United States strongly believed that the number of vertical pixels, as well as the number of horizontal pixels, should be common. Second, it rejected the number of active lines per frame proposed by the Japanese (1035) and Europeans (1152) for being inconsistent with the computer age. The United States had proposed 1080 active lines, stressing the importance of square pixels[40] (R. Hopkins, Advanced Television Systems Committee, personal communication, November 17, 1995). At the same meeting, the ITU-R approved both the American Dolby AC-3 and European Musicam technologies for the audio component of digital terrestrial television (DTTV), as recommended by TG 11/3 in October 1994 and by Study Group 11 in June 1995 (J. W. Reiser, Federal

Communications Commission, personal communication, June 3, 1996).

In March 1996, TG 11/1 was dissolved and any future work in the area of HDTV production was referred to the Working Party 11/A on Television Systems and Data Broadcasting. In mid-April 1997, WP 11/A convened to consider several technical modifications to Recommendation 709 (the HDTV studio standard) and agreed to include the U.S. proposal of 1080 active lines in the document. On April 25, 1997, because of this modification, the United States removed its October 1995 reservation. When the U.S. delegate to Study Group 11 proposed the deletion, the audience burst into applause! (J. W. Reiser, Federal Communications Commission, personal communication, May 22, 1997). An important page of international HDTV standardization history was turned at that meeting.

Throughout 1996, TG 11/3 examined two MPEG-2 based digital terrestrial television (DTTV) systems, proposed by North America (the ATSC system) and Europe (the DVB system), and prepared Recommendations for the various elements of DTTV (e.g., data transport, audio and video coding). It completed its work in November 1996, producing, in the words of its chairman, Stanley Baron (1997), "a set of Recommendations and Reports that defines a unique digital terrestrial television broadcasting (DTTB) system" (p. 180), with two compatible subsets (ATSC and DVB). Strictly speaking then, TG 11/3 has not drafted a *single* worldwide DTTV standard, but instead sought to minimize and harmonize the differences between the parameters of the two proposals (e.g., audio and video coding). This situation is not unlike the one faced by IWP 11/6 and its successor, TG 11/1, in the late 1980s and early 1990s with the HDTV production standard. The DTTV Recommendations were approved by Study Group 11 in April 1997 and are expected to be submitted to the Administrations for final adoption through correspondence (i.e., the accelerated approval process) before the Radiocommunication Assembly of November 1997 (J. W. Reiser, Federal Communications Commission, personal communication, May 22, 1997). TG 11/3 was disbanded in April 1997.

Discussion and Conclusions

HDTV, analog or digital, has come a long way since first developed in the NHK laboratories, although it still is only available to consumers in Japan (see Chapter 2). Consequently, this chapter has described the formative years of a technology that undoubtedly will continue to evolve over time in one form or another. HDTV will enter a phase of maturity when broadcasters begin airing regular programs in this format or a variant thereof. This final section reflects on the HDTV international standardization process. Other chapters will provide in-depth discussions of issues that were covered only summarily here, such as NHK's motivations for developing Hi-Vision/MUSE (Chapter 3), the reasons for the MAC debacle (Chapter 4), the role of the FCC and ACATS in the U.S. HDTV standard-setting process (Chapters 6 and 7), and the economic disadvantages of SDTV (Chapter 8).

Perhaps more than anything else the present history of HDTV has taught us another lesson of humility and foresight about the complexities and realities of international television standardization matters. The CCIR delegations came to HDTV standard-setting negotiation tables with great hopes and expectations—that the errors of the past not be repeated and that a single worldwide standard be selected. But by 1986 it was already too late for hammering out a consensus because the political will for doing so was either weak or absent. For many, Dubrovnik 1986 will be remembered as an encore of Oslo 1966. For some, the Dubrovnik rejection is all for the best, because the adoption of a single worldwide *analog* production standard could have delayed or impeded efforts to research digital television systems. The "global TV standard" is now MPEG-2 and future research is needed to chronicle how this standard came about. Yet for other observers, the European victory in Dubrovnik is a Pyrrhic victory, because the Europeans' concerted effort to postpone the adoption of an HDTV Recommendation might have permanently damaged the credibility and existence of the CCIR (and its successor) as an international standard-setting organization (see Savage, 1989).

What both Oslo 1966 and Dubrovnik 1986 suggest is that a world-wide agreement on a television standard bears more far-reaching implications than just a technical stamp of approval. These international television standardization negotiations became exposed to all sorts of pressures—trade issues, corporate and national strategies, and even programming concerns—that had little to do with the goal of setting up common technical parameters. The combination of all these tensions made an international accord on a common standard not just improbable but impossible.

Besen and Farrell (1991) have predicted that the role of the ITU as the preeminent international telecommunications standard-setting body will diminish in importance in the near future due to the emergence of regional standards organizations (RSOs), such as the European Telecommunications Standards Institute (ETSI). The ITU itself recognizes "the mounting pressure from industry, network operators and service suppliers on standards bodies to produce standards in a timely and market-oriented fashion" ("Can the Global," 1996, p. 10) but still affirms the importance of standardization for market stability. Because of their smaller and more homogenous membership, RSOs are able to adopt standards more quickly and efficiently than the ITU. Nowhere is this trend more visible than in the work conducted in a record time by the DVB group to agree on a family of digital video standards. ETSI approved DVB-C and DVB-S in December 1994 and DVB-T in January 1997. Within about three years, the Europeans managed to initiate a complete turnaround from advanced analog television systems (MAC/HD-MAC) to digital television technology and adopt a series of digital standards, a move that could have taken a decade to accomplish at the ITU-R level and without guarantees of success. One can speculate that the next generation of television standards will follow such a two-step international standardization process—they will be first agreed upon in a regional sphere before being submitted for approval to the ITU-R. In fact, this is already the case for the digital terrestrial television broadcasting standard(s).

2   Global HDTV Production

Actors and directors are also influenced by the shape of the television screen. When we started shooting [the film] *Star Trek: Generations* using the actors from the *Star Trek: The Next Generation* television series, I noticed that the actors were leaning very close together, unconsciously, when they were in a two-shot. That comes from years of having to do that for television. I had to go over there and separate them.
—*John Alonzo, director of photography, 1996*

As noted in Chapter 1, the difficulty with establishing any new television production standard is reflected in the classic dilemma. Which should come first—the chicken or the egg? In the case of high-definition television, should program development take priority over disseminating receivers on which viewers can see the programming? The incentive for consumers to invest in a new (and probably expensive) HDTV receiver will be low until there is a substantial amount of programming available. Conversely, television producers have little initial incentive to create programs in HDTV because there are few advanced television sets on which to view these programs.

It is also important to remember that production is only part of the basic television triad of production, transmission, and reception. There can be different standards for production and transmission, while reception technology is linked to the type of transmitter used. Thus, various formats of film and video can be used for television production, but transmitters and receivers need to use the same standard. A PAL receiver or VCR from Europe cannot be used to view or record NTSC programming without some form of transcoding. There is often confusion among casual observers who speak of the Japanese MUSE system as a high-definition production standard, when it is actually a satellite transmission compression technology. The same applies to the new U.S. ATSC advanced television standard which was designed as a terrestrial transmission/reception system. Table 2.1 below provides an overview of current global advanced television production and transmission standards for comparative analysis.

TABLE 2.1 INTERNATIONAL ATV STANDARDS

NAME	SMPTE 240M	MUSE	PALplus	DVB-T	ATSC[a]
PURPOSE	production	transmission	production transmission	production transmission	transmission
REGION	North America Japan	Japan	Europe	Europe	North America
IMAGE CODING	analog	analog	analog	digital	digital
ASPECT RATIO	16:9	16:9	16:9	4:3, 16:9	4:3, 16:9
LINES	1125	1125	625	625 and 525	480, 720, 1080[b]
PIXELS/LINE	NA	NA	NA	varies	640, 704, 1280 1920
SCANNING	2:1 interlaced	2:1 interlaced	2:1 interlaced	2:1 interlaced	1:1 progressive 2:1 interlaced
BANDWIDTH	30 MHz	6-9 MHz	7-8 MHz	7-8 MHz	6 MHz
FRAME RATE	30 fps	30 fps	25 fps	25 fps	24, 30, 60 fps
FIELD RATE	60 Hz	60 Hz	50 Hz	50 Hz	60 Hz
SOUND CODING	digital	digital	digital	digital	digital

NOTE: NA = not applicable
[a]As adopted by the FCC on December 24, 1996, the ATSC image parameters, scanning options, frame rates, and aspect ratios were *not* mandated, but were left to the discretion of display manufacturers and television broadcasters.
[b]The ATSC lines are active lines, while others are scanning lines.

Note that the Society of Motion Picture and Television Engineers' SMPTE 240M standard is the present HDTV production standard in both Japan and the United States. It is also known as the 1125/60 Hi-Vision production system with cameras and recorders manufactured by a number of Japanese companies. In Europe, the original HD-MAC transmission system is being replaced by widescreen Digital Video Broadcasting (DVB) standards. The digital ATSC system in the United States is designed to transmit both film and video source material shot at 24, 30, and 60 frames per second. Each of these systems will be addressed in greater detail below, but first it is important to define our focus.

This chapter will review international HDTV production activities and discuss the role of early ATV production efforts in campaigns by Japanese, European, and American interests to promote their distinctive advanced television technologies. It will also examine aesthetic production factors unique to HDTV. The thesis is that the initial production of HDTV projects had five primary goals:

1. *To provide material that could be used to demonstrate the superior sound and image quality of HDTV display systems, especially in comparison with 35mm film.*

2. *To explore the distinctive aesthetic qualities of HDTV, especially in terms of a wider image aspect ratio, improved color fidelity, multi-track audio, and higher picture resolution.*

3. *To evaluate HDTV production hardware in terms of its relative advantage compared to existing television and film equipment.*

4. *To support the creation of a critical mass of viewers with HDTV sets (in Japan and Europe) that would encourage program producers and distributors to provide additional high-definition program content.*

5. *To demonstrate that country or continent X's HDTV system was superior to any other.*

Unique Attributes of HDTV

HDTV production can be defined as *the electronic processing of video and audio information at a quality level that exceeds that of NTSC, PAL, or SECAM television standards.* It is also necessary to distinguish electronic HDTV production from that of film source material that is transferred to HDTV tape via a telecine. One of the early goals for HDTV was that it would be comparable to 35mm motion picture film when projected on a large screen. Systems developed to date such as Hi-Vision do have dazzling resolution when projected, but I will argue that video has an inherently different look from film that is defined as "telepresence" (Glenn, 1988). The effect is similar to looking out a large rectangular window to a scene happening in real time. A contributory factor is that projected video, unlike projected film, is rock-steady on the screen. Even the best film projectors in the world allow some level of horizontal weave and vertical jitter in the projected image due to the pull-down mechanism of the projector and variations in the sprocket holes on the film. Film and video also vary dramatically in how luminance information (image brightness) is presented, but suffice to say that HDTV images have a different look than projected 35mm film, despite efforts by HDTV promoters to use film as the quality benchmark for high-definition production. It is important to clarify this point since there is a fundamental distinction between programs that are taped in an HDTV production format and transmitted in HDTV and those that are photographed on film and then transferred to HDTV videotape for transmission. This chapter will first examine true electronic production and then consider the implications for film production linked to HDTV post-production and transmission.

A Word About Aspect Ratios

As outlined in Chapter 1, HDTV systems in Japan, Europe, and the United States are typified by a 16:9 aspect ratio. These numbers can be confusing and are easier to comprehend if film rather than television specifications are used for comparison—that is the image height is expressed as a base of 1, with image width as a larger number (see Figure 2.1). Thus 16:9 (16 divided by 9) can be expressed as 1.7777:1 or 1.78:1 in simplified form. An aspect ratio of 2:1, which is common for some widescreen film formats, would indicate that the image is twice as wide as it is high. Traditional NTSC, PAL, and SECAM television systems have a 4:3 aspect ratio or 1.33:1 in comparative form. This television convention is based on the old motion picture standard prior to the advent of widescreen movie formats.

It is interesting to note that the original HDTV aspect ratio proposed by NHK was 5:3 (or 1.67:1), somewhat narrower than the present 1.78:1 standard. In 1984, the Society of Motion Picture and Television Engineers created a Working Group on High Definition Electronic

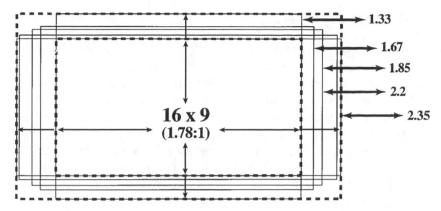

FIGURE 2.1. Television and motion picture aspect ratios (*Source:* Powers 1994).

Production (WGHDEP) to establish technical standards for HDTV. Many of the group members were experienced cinematographers who were familiar with various "shoot-and-protect" schemes that enabled widescreen feature films to be later converted for 4:3 television transmission (Powers, 1994). The SMPTE Working Group proposed a wider alternate HDTV aspect ratio of 1.78:1 (16:9) that would enable the inclusion of "shoot-and-protect" ratios of 1.33:1, 1.65:1, and 1.85:1. Since that time many cinematographers (and the American Society of Cinematographers) have protested the 1.78:1 aspect ratio as being too narrow to accommodate the typical 1.85:1 feature film without some cropping of the image on the sides (see Hora, 1995, for a detailed analysis). Films photographed at wider aspect ratios such as CinemaScope's 2.35:1 would have even more of their edges trimmed for airing on HDTV television channels. This controversy will not go away, but it appears that the likelihood that the HDTV 16:9 aspect ratio will be made wider at this point is close to zero.

The Advanced Television Systems Committee (ATSC) later incorporated the 16:9 ratio as part of the U.S. position on advanced television discussions before the CCIR Study Group 11 of the International Telecommunication Union (Powers, 1994). After the proposed SMPTE 240M standard specified a 16:9 ratio, Japanese and European groups adopted it as a de facto aspect ratio for their HDTV systems (Gordon and REBO Studio, 1996).

HDTV Production in Japan

The chicken-and-egg technology diffusion dilemma was addressed in Japan by encouraging the production of HDTV programming for distribution via direct broadcast satellite (DBS). The Japanese selected DBS transmission due to the bandwidth required by the MUSE transmission system and the mountainous terrain of the Japanese home islands which hampers terrestrial transmission. On June 3, 1989, NHK started airing HDTV programming for one hour a day from the BS-2b satellite in a test of high-definition DBS transmission that spanned three years (Utsumi, Isobe, and Hasegawa, 1995). On November 25, 1991, the Hi-Vision Promotion Association (HPA)[1] began operating the Hi-Vision channel on the BS-3b satellite (BS-9), and the experimental HDTV schedule was expanded to eight hours a day. Programming was supplied by NHK, Japan Satellite Broadcasting (a pay-TV channel), and 11 commercial stations at no charge (Schilling, 1992a). The number of Hi-Vision hours was extended to nine hours a day in January 1994 and to 10 hours a day in April 1994 (from 1:00 to 11:00 p.m.). On November 25, 1994, the third phase of Hi-Vision broadcasting began when the MPT granted licenses to NHK and seven private broadcasters in preparation for full-scale, regular HDTV service[2] which is expected to start in 1997. In April 1995, the number of broadcast Hi-Vision hours was extended to 11 hours a day (from 1:00 p.m. to midnight), with over half of the weekly HDTV schedule being supplied by NHK. As of June 1997, it totaled 14 hours daily (98 hours a week), from 10:00 a.m. to midnight (Y. Ogaki, Nippon Hoso Kyokai, personal communication, July 1, 1997).

Hi-Vision programmers—NHK and seven commercial stations—initially emphasized sporting events broadcast in HDTV. During phase 2 (Nov 1991-Nov. 1994), sports accounted for 34.8 percent of programming time, followed by music (24.8 percent), movies/dramas/ entertainment (20.2 percent), documentaries (18.7 percent), and other types of programs (1.5 percent) (Utsumi et al., 1995). In 1996, the broadcast mix shifted toward entertainment and news programming and away from sports, except during special events such as the Olympics.

While Japanese programmers made an effort to achieve an even balance in program types, a surge in sports programming is obvious during the July 1996 Atlanta summer Olympics in Table 2.2. Another interesting shift in program balance occurred in December when NHK

TABLE 2.2 1996 HDTV PROGRAMMING IN JAPAN AS PERCENT OF TOTAL BROADCAST HOURS

MONTH	SPORTS	MUSIC	MOVIES	NEWS/ DOCMT.	ENTERTAIN- MENT	CULTURE/ EDUCATN.	OTHER	TOTAL HOURS
JAN.	20.7	13.3	24.3	11.5	15.6	13.5	1.1	349
FEB.	13.5	14.0	20.7	16.6	15.2	19.1	.9	320
MAR.	30.9	14.4	17.8	13.3	12.3	10.5	.8	363
APR.	21.3	15.3	17.6	20.6	10.9	12.8	1.5	396
MAY	15.7	14.6	20.9	23.8	8.6	14.7	1.7	411
JUNE	9.4	16.4	20.0	23.1	11.8	17.7	1.6	400
JULY	55.9	11.9	10.6	10.0	3.5	7.7	.4	453
AUG.	51.1	15.1	13.1	9.1	4.1	7.1	.4	462
SEPT.	16.2	22.3	17.4	21.2	9.4	12.5	1.0	400
OCT.	13.5	13.3	17.7	22.8	12.1	19.4	1.2	412
NOV.	14.4	16.3	20.7	23.8	8.6	14.9	1.3	400
DEC.	9.9	21.5	25.6	19.5	9.1	13.0	1.4	438

SOURCE: Hi-Vision Promotion Association.

and the seven commercial broadcasters transmitted 55 feature films in widescreen HDTV as part of a "Hi-Vision Cinema 55" festival designed to showcase motion pictures in an aspect ratio similar to that seen in movie theaters.

No other nation in the world could match the Japanese in terms of the number of hours of HDTV programming transmitted during this experimental broadcasting period. The Japanese philosophy apparently is to stimulate consumer demand for HDTV sets by offering a full broadcast day of high-definition programming. To this end, satellite-delivered HDTV programs were also displayed on large-screen demonstration sets placed in high-traffic public spaces such as airports and commuter train stations where potential consumers could see the broadcasts.

In terms of program production, the Japanese have developed hybrid NTSC-HDTV systems for live simultaneous transmission and recording of sporting events and news. HDTV cameras and signal switchers are interlinked with their NTSC counterparts so that live events can be transmitted in both modes (see Utsumi et al., 1995). This type of simulcasting is similar to what American broadcasters will have to accomplish during the FCC's 10-year transition to digital television broadcasting. The Japanese not only invented the world's first viable HDTV cameras and transmitters, but they also have worked out the practical production details of simulcasting NTSC signals along with their high-definition counterparts.

HDTV Production in Europe

The Italian broadcaster RAI produced the world's first full-length electronic feature in HDTV—*Julia and Julia* in 1987 (Gordon and REBO Studio, 1996). While producers throughout Europe have been active in HDTV, the French led the way in the production and transmission of the widescreen 1250/50 HDTV format. As noted in Chapter 1, this system was developed after the Dubrovnik "rejection" by the European Community of the Japanese Hi-Vision/MUSE system. Like the Japanese, the Europeans also opted for DBS transmission of HDTV signals with the French utilizing the TDF satellites for their first broadcasts. The first live HDTV program was a transmission of speed skating from the Olympic Games at Albertville, France, on February 15, 1991 (Drumare, 1995). During 1991, France 2 transmitted 300 hours of 16:9 programming to 30,000 French households equipped with widescreen sets (Drumare, 1995). Not all of these sets were true HD-MAC televisions as some were 16:9 widescreen D2-MAC models. This is an important point as the Europeans have emphasized a gradual conversion to high-definition satellite broadcasting that places an emphasis on EDTV systems (4:3 and 16:9 D2-MAC) as a bridge to an HDTV standard. In contrast, the American strategy called for a quick transition to an incompatible digital standard, without resorting to any analog intermediate technologies.

French widescreen television production steadily increased from 300 hours of 16:9 programming in 1991, to 1,000 hours in 1992, 1,450 hours in 1993, and 3,300 hours in 1994 (Drumare, 1995). Overall European production in 16:9 video totaled 1,000 hours in 1992 (all French), 3,100 hours in 1993, 10,200 hours in 1994, and 18,900 hours in 1995 as producers in other EU nations started providing programming in widescreen formats including PALplus ("16/9 Broadcasting," 1996). PALplus is an analog EDTV version of the PAL 625-line format that can be viewed in 16:9 aspect ratio on special widescreen sets with stereo sound (see Chapter 1).

Of the 41 European broadcasters transmitting in 16:9 in 1996, 34 broadcast in PALplus using satellite, cable, or terrestrial transmission, and seven used D2-MAC via cable and satellite channels ("16/9 Broadcasting," 1996). D2-MAC is an analog EDTV standard that is likely to be superseded by new Digital Video Broadcasting (DVB) standards for cable and satellite transmission.

The 1993 Action Plan for the Introduction of Advanced Television

TABLE 2.3 16:9 WIDESCREEN PRODUCTION SUPPORTED BY THE EU ACTION PLAN (1993–1996) (BY PROGRAM TYPE EXPRESSED IN HOURS PRODUCED FOR EACH CALL FOR PROPOSALS)

CALL FOR PROPOSALS	SPORTS	CULTURAL EVENTS	FICTION	DOCUMENTARY	CARTOONS	STUDIO	TOTAL HOURS BY PERIOD
1/ 1993	613	230	165	193	0	446	1647
%	37	14	10	11	0	28	100
1/ 1994	471	312	201	464	23	1019	2490
%	19	13	8	19	1	40	100
2/ 1994	1001	402	320	845	19	821	3408
%	29	12	9	25	1	24	100
1/ 1995	446	500	391	650	12	729	2728
%	16	19	14	24	0	27	100
2/ 1995	265	297	256	496	46	714	2074
%	13	15	13	24	0	35	100
1/ 1996	444	338	403	840	30	370	2425
%	18	14	17	35	1	15	100
Hours by type	3240	2079	1736	3488	130	4099	14772
% of total hours	22	14	12	23	1	28	100

SOURCE: Commission of the European Communities (1996), p. 31 (1993–1995 data). Preliminary data for 1996.

Services in Europe stimulated the production of a wide variety of 16:9 programming. Between 1993 and 1995 a total of ECU 95 million was allocated by the Community to cofinance the production of original programming in 16:9 or assist with remastering existing programs to fit the new aspect ratio (Commission of the European Communities [CEC], 1996). There has been a shift in the types of original programs produced as well (see Table 2.3). While sports programming was initially dominant with over 30 percent of the total hours allocated to it in 1993, the production of documentaries has zoomed from 11 percent of the subsidized programming in 1993 to 35 percent of the total hours in 1996 (CEC, 1996). The EU Commission is encouraging the production of what it terms "long-life" programming that includes documentaries, fictional films, and cultural events to provide balance to sports and other live events.

This trend is similar to that in Japan where sports programming was initially dominant but has been gradually superseded by other programming types with a longer shelf-life. These type of productions also have greater marketability around the globe as the success of BBC-produced programming on American public television and cable has demonstrated. Producers estimate that high-quality documentaries have a potential shelf life of 20 years (CEC, 1996).

European governments, manufacturers, and broadcasters are concerned about potential Japanese dominance of the market for 16:9 production and reception hardware, and potential U.S. dominance of programming produced on motion picture film for widescreen electronic distribution on a global basis (Oudin, 1995). Vision 1250, a quasi-official organization, conducts program exchanges, conferences, training seminars and distributes publications to promote European efforts in 1250-line, high-definition production. Its funding comes from member manufacturers and broadcasters and via grants from Directorate-General X (DG X) of the EU Commission ("Vision 1250," 1995).[3] In regard to stimulating HDTV production efforts, Vision 1250 is the European counterpart of the Hi-Vision Promotion Association in Japan.

The DVB group is an organization of mostly European consumer electronics manufacturers and broadcasters that is working on creating standards for digital broadcasting via cable, satellite, and terrestrial transmission. The DVB-T standard for terrestrial transmission is digital, but is technically equivalent to the U.S. standard-definition (SDTV) format since it is only 625 lines and offers both 4:3 and 16:9 aspect ratios (see Table 2.1).

European producers have created thousands of hours of 1250 HD and 16:9 programming since 1991 in every production genre. What presently clouds the production picture in Europe is the lack of any single definitive standard for digital 16:9 widescreen HDTV production. European manufacturers and broadcasters are slowly migrating to an all-digital hardware, but consumers are purchasing 16:9 *analog* sets due to the absence of a common mandated *digital* HDTV format.

HDTV Production in North America

New York video director Barry Rebo was one of the first independent producers in the United States to create HDTV programs. Rebo bought a complete 1125/60 HD production system in September of 1986 and produced two music videos, *Candy* and *Imagine* in October and November of that year (Gordon and REBO Studio, 1996). When Rebo became involved in HDTV production, the aspect ratio of his Sony equipment was still the early 5:3 format. His initial productions were exploratory efforts to determine the unique lighting, composition, and power requirements of early HDTV cameras. Since that time Rebo and his prolific REBO Studio have produced documentaries, short subjects, and feature-length projects in the 1125/60 SMPTE 240M standard (Gordon and REBO Studio, 1996).

In 1994, Rebo formed a partnership with the DBS programming company TVN Entertainment Corporation to create HDLA, a Los Angeles post-production house that can convert film and video material into high-definition programming (Rosenthal, 1995). Forty percent of their work is designed for projected HDTV images for motion simulator rides in amusement parks and science museums, with the balance of their output spread among Japanese broadcasters such as NHK, and U.S. corporate, and medical clients (Rosenthal, 1995). They plan to transmit sports, cultural events, and feature films in HDTV to the 4 million North American households and thousands of sports bars with large C-band satellite dishes. Rebo and TVN are betting that these homes and taverns fit the "early adopter" profile in Rogers' diffusion of innovation model (Rogers, 1995) and that they will be willing to purchase HDTV receivers when first introduced in the U.S. market by the end of 1998 (Gordon and REBO Studio, 1996).

While Rebo and his associates were busy testing early HDTV hardware in New York City, HDTV production was also taking place in Canada. In 1987, the Canadian Broadcasting Corporation (CBC) produced *Chasing Rainbows*, the first television miniseries taped in HDTV (Gordon and REBO Studio, 1996). The Canadians also outfitted a 40-foot production truck with 1125/60 HDTV equipment and used it to record cultural and sporting events for test transmissions across Canada. Canadian production efforts have also been stymied by the lack of a North American HDTV transmission standard. However, neither Canada nor Mexico plans to move as aggressively as the United States in converting to digital broadcasting (Ashworth, 1997).

In the United States other producers have purchased analog Japanese HDTV production gear and gone into business. In addition to the REBO Studio, Randall Dark's HD Vision of Dallas has also been active in HDTV production (see Dickson, 1995b). They have all struggled to make money producing programs that could only be seen by very small specialized audiences. The producers were betting that the United States would create a unique HDTV standard, but underestimated the time that this would take. The original plan called for the FCC's Advisory Committee to recommend such a standard by 1992. The advent of digital technology for high-definition terrestrial transmission delayed this timeline considerably and the issuance of a final FCC order mandating the national conversion to HDTV did not occur until April 1997.

In the interim, there have been very few venues where potential consumers could actually view HDTV images. A working Hitachi HDTV set is on display in the communications exhibit of the Smithsonian's Museum of American History in Washington, DC. Tele-Communications Inc., a major U.S. cable television operator, has created an HDTV theater in the main concourse of the Denver International Airport to promote the image and sound quality of high-definition programming (Dickson, 1995c). Industry professionals have been able to view HDTV productions in special theaters at the National Association of Broadcasters convention held in Las Vegas every spring. Many of these productions were created to promote proprietary technologies developed by companies involved in the FCC's advanced television standards competition. This process is detailed in Chapters 6 and 7.

Sports Production in HDTV

In the summer of 1988, NHK conducted 1125/60 Hi-Vision experimental broadcasts of the Seoul Olympic Games transmitted throughout Japan using both fiber optic and satellite technology (U.S. Congress, 1990). In recent years, the Olympics have presented an ideal opportunity to showcase new media technologies, especially those like HDTV that lend themselves to widescreen coverage of sporting events with a great deal of lateral action. The Olympic opening and closing ceremonies have become mega-spectacles featuring thousands of dancers and musicians and battalions of costumed athletes from almost every nation on earth. Olympic broadcasts attract large audiences, especially in the United States, leading American networks to bid colossal amounts for the rights to transmit the games.

The Olympics are a sporting event that has defined the audience-drawing power of sports television and they have been equally shaped by that television coverage. The games, like the development of high technology, are linked to national pride and are seen as an opportunity to demonstrate the superior talent (athletic and technical) produced under differing economic and political systems. Thus there is a clear parallel between the political spectacle of the Olympic Games and the technical spectacle of the HDTV transmission of the games as a demonstration of the relative merits of Japanese and European high-definition technology. It is not coincidental that the first public transmission of the U.S. Grand Alliance HDTV system was proposed for the 1996 Olympic Games in Atlanta, Georgia, until it became clear that delays in setting a U.S. standard would preclude that possibility.

While 1988 was a test run for NHK, the company was back in full force for the 1992 summer games in Barcelona, Spain, and the winter games earlier that year in Albertville, France. Since the installed base of home receivers was still quite low at the time, NHK utilized its network of large-screen monitors set up in public thoroughfares such as train stations and airports. NHK estimated the Japanese viewing audience at 4,500,000 for the winter Olympics and 12,000,000 for the summer games in Barcelona, but did not explain how these figures were calculated since many viewers saw the programming while passing through transportation terminals (Nakae, 1994).

The 1992 Albertville winter games were the first time that the 1250/50 system was used to broadcast HDTV programming over an extended period (Delesalle, 1994). It was a multinational effort with television crews from many EU nations converging on Albertville to assist with the HDTV telecast. As French director Jean-Pierre Spiero noted, "For the vision, sound, and some of the cameras I had a crew from Thames Television [of the UK]; for the production I worked with my usual French crew. We all spoke the same language because Europe became a total reality at that time; there was a common will to succeed" (Delesalle, 1994, p. 34). So while Europe was merging politically under the new European Union, there was a sense amongst the HDTV crews at Albertville that the 1250/50 advanced television system represented a similar technological merger. It represented a common television system that might supplant the production and program exchange problems induced by the PAL-SECAM standard schism.

In an unusual turnabout, the 1994 winter games in Lillehammer, Norway, were jointly produced in HDTV by NHK and the European 1250 HD group. They used a newly developed standards converter to transcode 1125-line images into 1250-line material for telecast in Europe (Nakae, 1994). The 1250 HD crews covered ice hockey, and slalom racing, while NHK telecast speed and figure skating. Each group separately televised the opening and closing ceremonies in their respective HDTV formats. By sharing high-definition feeds from the Olympic venues, each production team needed fewer crews to cover the games than were necessary at Albertville. It was an example of international cooperation that stands in stark contrast to the acrimonious atmosphere that surrounded the 1986 CCIR meeting in Dubrovnik where many Europeans refused to accept a de facto adoption of the 1125 Hi-Vision system.

The 1996 summer games in Atlanta were jointly covered in a similar manner by crews from NHK and the German broadcaster ZDF ("The Atlanta Olympic Games," 1996). NHK covered the opening ceremony, gymnastics, wrestling, tennis, and the marathon, while ZDF was responsible for coverage of swimming, basketball, hockey, volleyball, and the closing ceremony. The HDTV feeds from event locations were processed at the International Broadcast Center with NHK sending 1125/60 HDTV images back to Japan via satellite. ZDF downconverted the HDTV signal to 625-line images in both 16:9 and 4:3 aspect ratios for transmission to European stations to be broadcast in PALplus and D2-MAC ("The Atlanta Olympic Games," 1996). Japanese and European broadcasters, by working together again during these games, realized significant cost savings in HDTV and EDTV transmissions to their national audiences. The key point in these coproductions was to start with the highest-quality HDTV images and sound possible, then downconvert the signals to 16:9 and 4:3 variants that could be decoded by many types of home receivers. Such simulcasting in varied analog and digital formats will be common during the 15-year global conversion to digital 16:9 television broadcasting.

The Aesthetics of HDTV Sports Coverage

The widescreen image presents unique challenges for certain types of television production. It is ideally suited for television coverage of sporting events, which explains the natural affinity between HDTV and televising the Olympic Games as noted above. The wide aspect ratio is ideal for capturing motion events that move *laterally*. There are many such examples in sports, including football, baseball, hockey, soccer, auto racing, and skiing. Lucky Schmidtleitner (1994), an Austrian who directed the European 1250/50 HDTV coverage of the women's downhill at the 1992 Albertville winter Olympics, noted that:

> The space in the frame gives a better impression of the skier's speed, for in 625 (PAL) the race seemed slowed down by the use of long lenses. With high definition the directors now have to think in advance about how they will construct the frame with the space left and right. There is not just the "middle" like there was before, it is really very different. (p. 35)

Likewise, U.S. director Randall Dark predicts that widescreen HDTV will transform the way American football is telecast:

[With HDTV] there's more action, you're going to see approaching tackles a little earlier. Barry Sanders is an incredible running back and his ability to anticipate is amazing. Often he starts moving before we can see his opponents enter the frame. He can see them, but we can't. With widescreen you will see more action, but it is quite easy to protect (the key action) for 4:3. If you look at virtually all sports this rule applies. (Gordon and REBO Studio, 1996, p. 117)

There is one area where HDTV clearly excels and that is in resolution. Viewers of hockey programs have complained for years that they "can't see the puck" in a long (e.g., wide) shot. In 1996, the Fox Television Network convinced the National Hockey League to use a modified electronic puck that could be made to artificially glow on television screens so it could be seen better by home viewers. Jean-Louis Machut (1994) directed the ice hockey matches in Albertville in 1992 and observed that "the 1,250-line picture is much more precise and allows us to see clearly how the puck moves, which is more difficult in 4:3" (p. 36).

The Aesthetics of HDTV Production and Viewing

To the casual observer in a consumer electronics store, the most readily apparent feature of a high-definition television set is the wider aspect ratio. This 1.78:1 aspect ratio translates into a wider field of view if the screen size is increased proportionally. The viewer will ideally sit closer to an HDTV screen than to a 1.33:1 NTSC screen to take advantage of this wider aspect ratio. The ideal viewing position for an NTSC display is approximately 7 screen heights from the set (yielding an 8 to 11 degree field of view), while the ideal viewing distance for an 1125-line Hi-Vision display is 3.3 screen heights yielding a wider 17 to 28 degree field of view (Glenn, 1988) (see Figure 2.2).

The key to enhancing this wider aspect ratio is a larger screen size, and this will mean either projecting the image in a manner similar to NTSC "big-screen" sets or waiting for the arrival of large flat-panel displays (FPDs). The problem with very large "direct-view" television tube displays is that the mechanics of scanning electron beam systems require a set that is almost as dimensionally deep as it is wide. Thus, a high-definition television that is 48 inches wide would yield a huge 4-foot cube hulking in the living room. On the other hand, a front projection model could be mounted in the ceiling as many classroom systems are today, embedded in the wall as many rear-projection

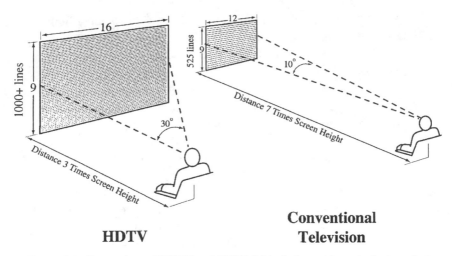

HDTV   **Conventional
                                                       Television**

FIGURE 2.2. Comparison of HDTV and NTSC field of view with equivalent resolution (*Source:* Nippon Hoso Kyokai, *NHK Factsheet*, 1996).

corporate boardroom models are, or might be a "hang-on-the-wall" flat-screen model. Projected television images have grown incrementally larger and brighter as engineering technology has evolved and it is expected that HDTV will accelerate this trend.

The combination of higher image resolution, larger screen size, and closer viewing distances for high-definition television will accentuate the psycho-physical perception of "telepresence" as described by researcher William Glenn (1988). Telepresence is the creation, by the viewer, of a sense of being physically present at an event. If the viewer is sitting close enough to a large screen to fill the central area of their visual field, the results can be as dramatic as they were for the first audiences of the widescreen Cinerama, CinemaScope, or IMAX film formats.

At the 1996 National Association of Broadcasters (NAB) convention in Las Vegas, Nevada, CBS/Westinghouse sponsored a theater showing examples of HDTV programming. The audience was seated near a large 10 x 17-foot screen (with rear projection). Most of the scenes they saw were typical scenic shots of the United States interspersed with American flags waving in slow motion to set the proper patriotic tone for this promotional program (produced with Japanese cameras and recorders). However, one segment was a computer-generated, point-of-view animation of a realistic roller-coaster ride. When the audience witnessed the first precipitous drop from the top of the roller coaster there was a collective gasp as they viscerally experienced telepresence, a rare phenomenon with NTSC narrow-screen televisions. I will argue that, under the proper viewing conditions of large screen size and close viewing distance, observers of HDTV images will have a qualitatively different viewing experience compared with traditional 4:3 television images seen on a narrow screen across the room.

Even the close-up view on a television tube is perceptually different according to those who have worked with HDTV technology in production settings. After telecasting the Norwegian National Holiday parade in HDTV, director John Andreassen recommended that television directors sit closer than normal to their line monitors in the control room. He stated, "in high definition, from my point of view, *close to a monitor where I could see every detail* [emphasis added], I could use medium-long shots and the long shots to pick up whatever detail I wanted to look at" (Andreassen, 1994, p. 30). Andreassen reiterated a common theme amongst directors and producers of HDTV 16:9 programming, namely that the widescreen aspect ratio and larger screen size reduce the need to frequently cut to close-up shots as is the case with 4:3 formats. The narrow aspect ratio and small screen size of conventional television dictates an often-unrealistic intimacy that conforms to the fact that the small image is typically viewed from across the room in home settings. While the 4:3 proportions are in fact based on the pre-CinemaScope 35mm film aspect ratio, those images were enlarged to be 30 feet in height in a darkened cinema. American director of photography John Alonzo noted that aspect ratio also affects the directors and actors who work in television dramatic series:

> When we started shooting [the film] *Star Trek: Generations* using the actors from the *Star Trek: The Next Generation* television series, I noticed that the actors were leaning very close together, unconsciously, when they were in a two-shot. That comes from years of having to do that for television. I had to go over there and separate them. Finally they fell in love with the idea of being able to talk to an actor that's more than eighteen inches away. It's unnatural to talk to someone that close all the time. It's unnatural to perform that way. (Gordon and REBO Studio, 1996, p. 112)

Today it is commonplace for feature films to be broadcast on television within months or even weeks after their theatrical release. Historically, however, one of the primary reasons motion picture aspect ratios widened in the 1950s from 1.33:1 to 1.85:1, and 2.35:1 (Cinema-Scope) was to differentiate film productions from television and to lure patrons back into theaters (Powers, 1994). Motion picture companies and theater owners were desperate to staunch the loss of their audiences to television and adopted widescreen images with spectacular shots such as a roller-coaster point-of-view sequence in films such as *This is Cinerama* (1952). If this sounds much like the HDTV sequence shown for dramatic effect by CBS in Las Vegas in 1996, the coincidence is probably not unintentional. The irony in all this is that feature films are now photographed with eventual conversion to narrowscreen television in mind, but the conversion process is still an art form that often requires panning back and forth across the widescreen image to keep the action and actors in frame. Television directors will face similar compositional quandaries as they record HDTV footage in 1.78:1 that will need to be simulcast in 1.33:1 until the narrowscreen formats (NTSC, PAL, SECAM) are phased out in the next century.

Many American viewers are unaware that when they watch television programs such as the medical drama *ER*, they are seeing the middle portion (1.33:1) of an image that was photographed on widescreen 1.78:1 motion picture film. Since 1993 Warner Brothers, among other Hollywood companies that produce programming for U.S. television networks, has been shooting their television series with widescreen aspect ratios to allow for potential off-network syndication sales after a U.S. HDTV standard is adopted (Cookson, 1995). During the 1994-95 broadcast season, 447 episodes of 22 prime-time series were produced by Warner in widescreen modes, yet no American television viewers saw the full image as recorded (Cookson, 1995). Over 90 percent of Warner's television programming is initially recorded on motion picture film, and this "shoot and protect" system also serves to protect their investment in television series that can be easily converted into several HDTV formats for global resale in coming years. This inventory is very valuable to these multinational companies and they are using widescreen 35mm and 16mm film as a universal high-definition production format that is independent of regional squabbles over global HDTV standards.

The Aesthetics of the Television Close-up

HDTV's higher resolution and larger image size mean that far fewer close-ups are needed to show the viewer what is happening. Director Lucky Schmidtleitner (1994) noted that he could include the entire women's downhill course in a single master shot at Albertville and viewers could still discern individual course gates. Television has often been described as a close-up medium, an attribute driven by the small screen size compared with film and the low resolution of the picture. Many television directors who have worked in HDTV have commented on a reduced need for close-up shots that have typified narrowscreen television (Giustiniani, 1994; Gordon and REBO Studio, 1996; Machut, 1994; Spiero, 1994). High-resolution medium shots and even long shots can now be substituted for close-ups with no loss of viewer comprehension. Fewer close-ups also translates into slower program pacing, a counter trend to ever-faster "MTV" shot-a-second pacing. German director Wim Wenders (1994) sees this as a blessing. "High definition, in replacing low resolution, constitutes an historic opportunity to correct bad visual habits and reveal a way of seeing things that is less terrorist and more congenial, more human" (p. 44).

Wenders translates higher resolution into the ability for the director to stay with a medium shot and to use fewer close-ups than normally required to inform the television audience. While he characterizes the fast intercutting of television as "terrorist," many television directors would argue that the extra close-ups were an aesthetic characteristic of dealing with a small narrow screen. The higher resolving power of HDTV screens may change the directorial pacing of television productions, but given the adoption of very fast "MTV" cutting by feature film directors it may prove to be a moot point.

One significant trend emerges from 10 years of experimentation with HDTV aspect ratios and high resolution—watching high-definition images on a large screen with multichannel audio will be a very different experience from looking across the room at a 19-inch television set with a 3-inch speaker. With all the emphasis placed on image quality in discussions of the future impact of HDTV, the potential improvement in sound quality has been largely overlooked. Until the advent of broadcast stereo systems in the 1980s, television sound had very low fidelity. With the advent of Dolby 6-channel, surround-sound technology as part of the ATSC digital standard in the United States, American television viewers will be able to replicate a theatrical aural experience in the home. This is a neglected area in discussions of the aesthetics of high-definition television, but multichannel sound may be as important to the overall viewing experience as sharper, larger pictures.

Discussion and Conclusions

The link between early HDTV production efforts and the development of telecommunications policy is often overlooked but significant. The production-policy linkage can be best summarized with the five goals of initial HDTV production outlined at the start of this chapter.

1. *Produce material to demonstrate the technical superiority of HDTV compared with conventional television and 35mm film.* The key goal of HDTV developers in Japan, Europe, and the United States was to demonstrate that their systems were superior to existing NTSC, PAL, and SECAM television technologies. Unless this superiority could be conclusively demonstrated to policymakers and broadcast engineers, the prospects for the global adoption of these new technologies would be very slim, especially if they were incompatible with existing broadcast formats. While initial efforts focused on superior image and sound quality, after 1990 the benefits of digital broadcasting assumed greater importance.

HDTV demonstrations for engineers and policymakers were widely reported in the trade press. The 1981 demonstration of the NHK system at the San Francisco SMPTE conference, the first demonstration of the Grand Alliance system for the FCC and Congressional leaders in 1996, and the demonstration theaters set up by HDTV development companies at the annual National Association of Broadcasters conventions are all examples of efforts on the part of system proponents to convince policymakers and engineers of the superiority of high-definition television. At the NAB shows, more often than not, the HDTV images were projected on large screens in darkened minitheaters featuring multichannel audio for maximum visual and aural impact. These demonstrations were an important part of "selling" HDTV technology to those two influential audiences.

2. *Explore the distinctive aesthetic qualities of HDTV* and

3. Evaluate HDTV production hardware in terms of its relative advantage compared to existing television and film equipment.

Early production efforts in Japan, Europe, and the United States yielded important information about shooting with wider aspect ratios and dealing with the influence of higher picture resolution on directing styles. The fundamental concept of telepresence is an aesthetic issue unique to television, especially for HDTV with its closer viewing distance and high resolution. Making a case for HDTV as the logical successor to analog narrowscreen video was easier to make to policymakers if there were distinctive qualities unique to high-definition programming.

In many respects HDTV represents an aesthetic hybrid of television and theatrical motion pictures. It can be easily projected like film (without the image degradation of NTSC video), yet it will have the look or telepresence of television. It can be digitally enhanced to resemble a film image, yet it will allow immediate playback on location—a significant production advantage of video over film. Film directors frequently employ a video "take-off" on motion picture cameras to allow them to view a video version of what was just recorded on film. Aesthetic judgments made on location about the quality of a recorded performance are now often made based on an immediate review of the videotape rather than traditionally waiting for a review of "dailies" (prints of film footage shot the previous day, airshipped to and from a lab).

Another new electronic tool that film and video directors are using on location are digital non-linear editing systems. These computer-based systems permit the director and editor to start assembling sequences of the production before all of the scenes are shot. A rough cut of the film or video can be analyzed in a motorhome on location and scenes can be reshot if necessary before the location sets are struck (see Ohanian and Phillips, 1996). This electronic technology gives the director greater creative control of the production while shooting is underway. This system requires that all film or analog video footage be converted to digital video form for processing in the non-linear computer. After the director and editor are satisfied with the electronic cut of the program, the film negative is cut to conform to the electronic version. This is not to suggest that electronic recording will immediately replace film in the production of motion pictures and television commercials. Thirty-five mm film still yields a projected widescreen image that HDTV cannot match in its present form, and other film formats such as IMAX in 70mm are even more impressive when projected on screens 50 feet high and 50 feet wide.

However, in the future television and film directors are going to be relying on a series of digital electronic tools in the production process to make important aesthetic decisions. Digital HDTV electronic formats, with their widescreen aspect ratios, will be increasingly used as a bridge between film and electronic modes in recording motion media.[4] The ability to see an immediate playback of a performance is one of the most significant technological attributes of electronic recording, an asset to the production team on location that now extends to the editing process.

4. *Support the creation of a critical mass of viewers with HDTV sets in Japan and Europe that would encourage program producers and distributors to provide additional high-definition program content.*

Production efforts in Japan have been designed to stimulate consumer demand for HDTV sets. Given their significant economic investment in system R&D, satellite launch costs, terrestrial production facilities, and actual production costs, the results can only be described as disappointing at best (see Chapter 8). While sales of HDTV models have been meager compared to the total number of sets in use, sales of lower cost 16:9 NTSC sets have been significant, despite the limited amount of widescreen programming available in Japan until 1996. This trend reinforces the point that consumers may eventually buy an HDTV set, but only when they perceive that they are getting a good value for their money.

In Europe, the 1993 Action Plan has been instrumental in stimulating the production of 16:9 widescreen analog programming. As Table 2.3 demonstrates, the production of subsidized "long-life" programming, such as documentaries, has increased significantly since 1993. Total European production in the 16:9 format (both subsidized and independent) had increased six-fold to almost 19,000 hours a year by 1995. Clearly, a great deal of effort was being expended in Europe to resolve the chicken-and-egg technology diffusion dilemma by stimulating the production of 16:9 programming to jump-start widescreen television set sales.

5. *Demonstrate that country or continent X's HDTV system was superior to any other.*

The ultimate goal of companies and governments with a stake in the multibillion dollar development costs of creating unique HDTV systems was to prove the superiority of their technology when compared to the competition. Thus, part of the common mission of the Hi-Vision Promotion Association in Japan and Europe's Vision 1250 consortium is to stimulate the production of HDTV programming that validates the quality claims made by system developers. The image and sound quality of Hi-Vision motivated European and American competitors to develop rival systems that matched or exceeded the Japanese system in quality. A game of international can-you-top-this in television technology has been underway for over 10 years with no end in sight. It is refreshing to see international cooperation in the high-definition telecasts of recent Olympic Games, where production teams from Japan and Europe have demonstrated that multiple formats of HDTV can be used simultaneously to create programming distributed on a global basis. We may never have a single global HDTV standard, but this trend bodes well for the international exchange of high-definition programs.

3

The History of HDTV Development in Japan

with Scott D. Elliott

The real key to understanding the course of HDTV development in Japan
... is not the promotional agencies [MPT and MITI], but NHK.

—*Gregory W. Noble, University of California at Berkeley, 1992*

This chapter retraces the development of Hi-Vision, MUSE, and Clear-Vision in Japan from the mid-1960s to the early 1990s and analyzes the role of five key actors—Nippon Hoso Kyokai (NHK), broadcast electronics manufacturers, the Ministry of Posts and Telecommunications (MPT), the Ministry of International Trade and Industry (MITI), and private broadcasters—in bringing these new television technologies to the Japanese public. What forces motivated them to undertake or support such endeavors? In so doing, this chapter seeks to identify the various influences—social, economic, and regulatory—that shaped the technical evolution of high-definition television (HDTV) and extended-definition television (EDTV) in Japan during these three decades.

Scott Elliott is an information technology policy specialist with the Commonwealth of Pennsylvania's Office for Information Technology. He is completing his doctoral dissertation in communications at the University of Washington. This chapter is a revised version of an article previously published in *Multimedia Review*, 1993, vol. 4, no. 1, pp. 59-72, Scott D. Elliott, Meckler Media: Westport, Conn. Reproduced with permission.

History has shown, indeed, that the development and adoption of television standards involves more than simply technical questions (see Chapter 1). Just as important is consideration of the economic and social ramifications brought on by the introduction of an innovative over-the-air service (Shimizu, 1989). For example, during the Federal Communications Commission (FCC) debates over the advent of U.S. color television in the early 1950s, much attention was given to economic factors, such as the prices anticipated for the first color receivers (see Federal Communications Commission [FCC], 1970). The cheaper cost expected for CBS color sets was, in fact, a primary reason why the Commission gave initial approval to the CBS field sequential system. (This ruling was later reversed in favor of NTSC color[1] when the CBS standard failed to gain a foothold in the marketplace.)

Social factors also played a part in the U.S. color television standardization proceedings. When the FCC decided that any proponent color system had to be compatible with the 6 MHz channelization scheme used for monochrome service, the Commission did so with the intent of ensuring that the higher technical quality of color TV would not limit the diversity of television channels available to the public. Considering the engineering challenge this presented at the time, it is significant that the Commissioners based this "technical" ruling on their desire to advance the social goals of broadcast program diversity and increased competition among stations.

The Role of NHK

The image of a lonesome knight forging through vast, unchartered territories in the quest of an elusive prize is an apt metaphor to characterize the role of Nippon Hoso Kyokai (NHK),[2] the Japan Broadcasting Corporation, as HDTV developer from the mid-1960s to the late 1970s. NHK was essentially on its own for crafting and testing the parameters of Hi-Vision, the Japanese HDTV production standard.[3] It was not until the 1980s that the Ministries, the broadcast equipment manufacturers, and the private broadcasters entered the HDTV scene in earnest. And even then the tremendous influence that NHK had exercised over the design specifications for Hi-Vision as its solitary creator did not end once the program moved into full-scale production. Not only did NHK build many of the first HDTV prototypes, but it also worked with government agencies like the Ministry of Posts and Telecommunications (MPT) to parcel out assignments for the production of various components and full pieces of equipment among selected Japanese manufacturers (Adam, 1990; Fujio, 1985; Schreiber 1991; U.S. Congress, 1990). Clearly, up through the arrival of the first experimental hardware in the early 1980s, NHK practically stood alone as the sole driving force behind the Japanese HDTV revolution.

Groundbreaking Research

It is difficult to establish the exact year when HDTV research started in Japan. One source indicates that it began as early as 1964, following the Tokyo Olympics (U.S. Congress, 1990). Others suggest that the Mexico City Olympics in 1968 provided the impetus that kicked off Japanese HDTV studies (Behrens, 1986; Carbonara, 1990a; Cook, 1990; Tatsuno, 1990). Most Japanese authors, however, point to 1970 as the official starting date for formal investigations into high-definition imaging in Japan (Broadcasting Technology Association [BTA], 1987a; Nakamura, 1988; Sugimoto, 1988).

One is left with the impression that it was through the telecasting of the Olympic Games that the Japanese became convinced that the existing NTSC color TV standard was inadequate and could be greatly improved by making use of technological advances. Another explanation given for the Japanese foray into television imaging is that viewers in that country have smaller homes and, thus, were forced to view NTSC pictures from much closer distances than those experienced by the average American family (Schubin, 1990). As a result, Japanese audiences expressed increasing dissatisfaction with the quality of NTSC color service (Ministry of Posts and Telecommunications [MPT], 1989). For instance, a survey of 796 households conducted by the Nikkei Industry Research Institute in 1985 revealed that only 10 percent of the respondents voiced no significant grievances about the quality of NTSC reception. The most frequent complaints included: "eye strain from prolonged viewing," "poor reception on some channels," and "ghosts" (double images) (Takahashi, 1985).

Keen interest in improving upon the NTSC color system was not the only motivating factor for NHK to undertake HDTV research. Noble (1992) argued that "For NHK, HDTV has above all been part of a strategy to protect and increase revenues in an era of rapid technological change" (p. 4). NHK is commercial-free and relies almost exclusively on a fee system paid by viewers for its funding. Some NHK officials fear that the level of revenue might stagnate and that viewers might balk at paying their monthly fees unless the corporation demonstrates a commitment to offer the latest technology to its audience.

Regardless of when Japanese HDTV research actually began, it is clear that the initial studies were more ambitious than any work that had been done earlier in the United States.[4] The first Japanese tests were concerned with the difficult objective of duplicating the human visual experience through means of electronic imagery (Donow, 1988). The Japanese, at this stage, were interested in producing television pictures of such high quality that viewers would react to them as if they were experiencing the scenes first-hand. Unknown at this time was whether this goal could best be accomplished by means of stereoscopic television, high-resolution television, or some combination of these two systems (Fujio, 1985).

Compatibility with the existing NTSC system was dismissed early on as a requirement that would limit the potential of this project. It was seen as unreasonable to expect that research aimed at imitating human vision should be tied to the outdated technology of the NTSC standard, which had been formulated during the 1940s and 1950s. Instead, this NHK program sought to apply advances in electronics and computer technology to radically improve television images and bypass the problems inherent to NTSC (e.g., interline flicker, line crawl, crosscolor; see FCC, 1987a, for a review of NTSC defects).

The person most associated with this effort was Dr. Takashi Fujio, Director General of NHK Science and Technical Research Laboratories. Created in 1930, these laboratories specialize in research and development of broadcasting and related technologies. It was there that an extensive battery of psychophysiological and perceptual tests were conducted in order to evaluate the operating qualities of the human visual and aural senses. These tests investigated such things as the effect of color on visual response, the eye's ability to follow and reproduce subject motion, and the effect of sound on viewers' perceptions of screen images. Tests were also run to assess how changes in the field of view offered by wider screen sizes affected the psychological impact of projected scenes shown to research subjects (Donow, 1988; MPT, 1989; Poynton, 1990; Sugimoto, 1988). These evaluation results enabled NHK to understand key elements of the home viewing environment and would form the technical foundation on which the Japanese HDTV standard was built.

For instance, NHK determined that the optimum distance for

watching an NTSC receiver was 7.2 times the height of the receiver's screen. At that distance, the viewer's visual angle of the screen along the horizontal plane was 10.7 degrees wide (Fujio et al., 1982). If the subject moved closer to the NTSC set, the scanning lines and artifacts in the TV picture became increasingly objectionable. If the subject moved farther away, the resolution of the set was not good enough to provide a satisfactory picture (see Figure 2.2 in Chapter 2).

NHK also learned that if the limitations of the 525-line NTSC receiver were removed and the resolution of the display was adjusted to allow the viewer to sit at any distance from the screen which felt most comfortable, test subjects invariably wanted to move closer to the picture screen. The most desirable viewing distance was found to be 3 times the screen height (3H). If rapidly moving television images were presented, that distance increased to 4 times the screen height (4H) as viewers found that sitting any closer was straining on the eyes (Fujio, 1985; Fujio et al., 1982).

Initial Parameters of the 1125/60 System

As testing moved from experimentation to implementation, the decision was made that stereoscopic television was not a viable alternative. For one thing, it was difficult to provide high picture quality through a stereoscopic display; for another, the experience was found to be visually taxing for viewers. Emphasis was placed, instead, on achieving a widescreen, high-resolution image which, while not stereoscopic, could still come close to imparting a sense of visual reality to its audience (Fujio, 1985). As a benchmark, the goal of equaling the quality of 35mm motion picture film was selected (Donow, 1988; Schreiber, 1991).

A Decision on Scanning Lines

Subjective tests were carried out by displaying color transparencies on experimental HDTV screens. The intent was to determine the level of picture quality required by a subject sitting at a distance of 3 times the screen height from the test receiver—the value determined earlier as the most preferred viewing distance (Fujio et al., 1982). Employing screens that used interlaced scanning,[5] NHK scientists came to the conclusion that the optimum number of scanning lines to be used should be around 1241—more than twice the number of lines used by NTSC television.

The number of scanning lines ultimately designed into the Japanese HDTV system, however, was 116 less than this optimum value. Fujio (1985) explained: "These tests confirmed that the [high-definition] system with 1125 scanning lines and luminance signal bandwidth of 20 Mhz (aspect ratio 5:3) is *fully acceptable* [emphasis added] at a viewing distance of 3H" (p. 648). NHK's decision to use 1125 scanning lines instead of 1241 serves as an early indication that the Japanese were not merely designing their HDTV system for domestic consumption but for the global marketplace as well (Clark, 1988; McMann, 1982; Nippon Hoso Kyokai [NHK], 1988). Even though the evolving Hi-Vision system—as it came to be called—was not compatible with existing NTSC, PAL, or SECAM color TV standards, it was being structured in a way to make it more easily convertible among these many systems (Carbonara, 1990b; Jurgen, 1988; Mathias and Patterson, 1985). As engineer Charles Poynton (1990) noted:

> The scanning parameters of 1125/60 [as the Japanese production system became known outside Japan] were chosen to be closely related to 525-line [NTSC] and 625-line [PAL and SECAM] systems. The total line counts 1125, 625, and 525 are the odd multiples 45, 25, and 21 respectively of 25. (p. 45)

NHK's decision to downgrade its Hi-Vision system from 1241 scanning lines to 1125 is the first evidence of a technical feature of the equipment being adjusted for economic and social reasons. The economic impetus in this case is obvious and understandable, given NHK's investment of time and resources into this research: The adjustment of the number of scanning lines to be used would not lessen picture quality to any great extent and would greatly enhance the promotion of Hi-Vision as a worldwide standard.

Although the economic benefits of this standard refinement are clear, Japan's desire to develop Hi-Vision in a manner that would aid international communications should not be overlooked. It must be remembered that economic goals and concern for improved world understanding are not necessarily mutually exclusive. Considering the excellent reputation that the Japanese have gained in business for their efficiency and attention to detail—as well as the society's emphasis of such qualities as harmony and conformity in group relationships—it is understandable that the wastefulness and disorder of the three existing global television standards would seem illogical to that culture (Prestowitz, 1989). Therefore, repeated emphasis by the Japanese on Hi-Vision's ability to improve communications among nations must not be dismissed as merely a smoke screen for more basic economic schemes (Fujio, 1985; MPT, 1989; Nakamura, 1988; NHK, 1989; Ono, 1990). As it will become clearer below, these two factors—the goals of worldwide marketing and concern for improved global understanding—worked to influence the specifications of Hi-Vision (and MUSE) in many areas.

The 60 Hz Field Rate

The decision by NHK to employ a 60 Hz field rate[6] in its Hi-Vision prototypes is another example of a technical parameter being manipulated in an attempt to gain international acceptance of a single HDTV standard. NHK research had found that a field frequency as low as 45 Hz (roughly equivalent to the 24 frames per second of feature films) would be sufficient provided that the brightness of the television screen could be maintained at an even level across its surface for the time span of a single field (Fujio, 1985). That, however, is not yet technically possible, because a television picture screen produces a varying degree of luminance as the electron gun scans the front of the tube. Additionally, the 50 Hz field rate common to both PAL and SECAM systems was determined to be too low, resulting in unacceptable large-area flicker[7] within the television picture (Jurgen, 1988). The figure ultimately chosen for Hi-Vision was a 60 Hz field rate, very close to the NTSC field rate of 59.94 and preferable to it.

Aspect Ratio and Line Structure

The topics of aspect ratio[8] and line structure[9] are two areas where the specifications for Hi-Vision equipment were based primarily on NHK's research findings, at least for the first generation of equipment. Regarding aspect ratio, NHK scientists found that the most pleasing dimensions for a video screen were actually 5:3, not the more narrow 4:3 of today's receivers (Fujio, 1985). With viewers sitting 3 times the screen height from the TV display, the 5:3 aspect ratio offered test subjects a 31 degree visual angle of the projected video images along the horizontal plane (see Figure 2.2 in Chapter 2). This, in turn, was found to increase audience involvement because viewers were no longer able to take in the entire scene with a single glance; instead, test participants were found to be looking around various parts of the TV screen much as they would viewing a scene in real life (Cook, 1990; Mathias and Patterson, 1985; Schreiber, 1991). The 5:3 aspect ratio would also conserve the bandwidth needed for a broadcast signal (Mathias and Patterson, 1985). In 1985, it was changed to 5.33:3 (16:9) after the Society of Motion Picture and Television Engineers (SMPTE), and subsequently the Advanced Television Systems Committee (ATSC), in the United States decided that a 16:9 aspect ratio was preferable for film-to-HDTV conversion because it incorporated a variety of film aspect ratios (see Figure 2.1 in Chapter 2).

The specifications established for line structure have remained constant over the life of the Japanese HDTV program. A 2:1 interlaced pattern became the selected design, although multiple interlaced ratios (e.g., 3:1, 5:1) were tested (Fujio, 1985). While slightly increasing vertical resolution, these alternative line display patterns were found to increase annoying artifacts, such as line crawl and interline flicker. During an interview, Fujio explained that progressive scanning[10] had also been considered but was discarded because the improvement in resolution it offered was deemed insufficient in relation to the higher equipment costs that resulted (Roizen, 1986). Additionally, interlaced scanning offered bandwidth efficiencies for transmission and recording (Fujio, 1985).

The Continuing Evolution of the 1125/60 System

The specifications for Hi-Vision were not solidified with the arrival of the first generation of 1125/60 equipment in the early 1980s. Instead, some of the main parameters for Hi-Vision continued to evolve, shaped by debate in industry circles outside of Japan. Much of the foreign influence on Hi-Vision's specifications emanated from HDTV studies conducted by the Society of Motion Picture and Television Engineers (SMPTE) in the United States (see Chapter 2). SMPTE draws its membership from the motion picture and broadcasting industries and is concerned, in great part, with the formation of voluntary national standards for movie and TV equipment. Other influential parties included the Advanced Television Systems Committee (ATSC) and the International Radio Consultative Committee (CCIR) of the International Telecommunication Union (ITU) (see Chapter 1).

In July 1986, SMPTE began standardization procedures for HDTV, concentrating on the early design work completed by NHK (National Association of Broadcasters [NAB], 1989). The Japanese equivalent of the ATSC, the Broadcasting Technology Association (BTA), collaborated with SMPTE in this process (Omura and Sugimoto, 1987). The SMPTE make-over of 1125/60 was finalized in August 1987, and the resulting standard became known as SMPTE 240M. As Sony's Laurence Thorpe (1990b) recounted, the changes to the original 1125/60 standard were substantial, including "a 30 percent increase in bandwidth required for distribution; a wider color gamut; a wider aspect ratio [from the original 5:3 to 5.33:3, or 16:9] … and a narrower horizontal blanking width" (p. 39). The change in aspect ratio was an adjustment also suggested during deliberations within the CCIR (Poynton, 1990).

These changes in the original Hi-Vision standard represented significant concessions made by the Japanese. As the result of this fine tuning, costs for the second generation of Hi-Vision equipment introduced in 1989 were noticeably higher than those for the first-generation gear some five years earlier (Thorpe, 1990a, 1990b). The more expensive price tags for the second generation of Hi-Vision cameras and recorders had been prompted by the higher technical performance demanded by SMPTE 240M changes, and by the retooling which had to be undertaken by Japanese manufacturing plants.

It was, then, through discussions within worldwide standards organizations, as opposed to debate within the standards committees of BTA and the MPT, that the key specifications for Hi-Vision were completed. In fact, BTA held off on the finalization of its HDTV standard in 1988 in order to make adjustments for colorimetry decisions following a meeting of the CCIR's Interim Working Party (IWP) 11/6 (Sugimoto, 1988). The Hi-Vision standard that had been developed by NHK and was subsequently molded by international cooperation finally came to rest at 1125 scanning lines with a field rate of 60 Hz, using interlaced scanning (2:1) and an aspect ratio of 16:9.

Emergence of the MUSE Family

Development of the MUSE[11] *transmission* standard, in all its variations, was a process influenced by a mixture of technical, economic, and social priorities, as had been the case for the Hi-Vision *production* standard. Likewise, the resultant final standards were determined in great part by preferences voiced overseas—in America.

Work on the parent MUSE transmission system began around 1982 (Ninomiya, Ohtsuka, Izumi, Gohshi, and Iwadate, 1987). The decision to focus on a direct broadcast satellite (DBS) configuration for MUSE resulted, in part, from the natural evolution of a program which had already been established at NHK (Donow, 1988; MPT, 1989). The public broadcasting organization was originally incorporated under the Broadcast Law of Japan with the understanding that it would make a sincere effort to provide radio (and then television) service to all of the nation's citizens (Ito, 1978; Kodaira, 1986; Ninomiya et al., 1987). However, because of the mountainous terrain of Japan, as well as the scattered location of more than 1,000 habitable islands, blanket coverage of the entire population with broadcast signals has been difficult and costly to achieve. Also, the continuing erection of high-rise buildings has prevented top-quality reception of terrestrial signals in urban areas (Ito, 1978).

During the 1970s, NHK decided to pursue the delivery of NTSC programming by using broadcast satellites (Ito, 1978). The extension of that directive—to combine satellite broadcasting with Hi-Vision programs—was seen as a logical next step. Additionally, the improved reception to be gained from DBS, in place of terrestrial delivery, was recognized as a fitting complement to the high technical quality anticipated for high-definition programming (NHK, 1988). Not to be overlooked, of course, was the efficiency equated with nationwide satellite telecasting, as opposed to the alternative of trying to provide complete domestic coverage through the construction of interlinked earthbound transmitters (Fujio, 1985; Ito, 1978).

A variety of other factors have been cited as providing explanations for why NHK has pursued satellite broadcasting of HDTV, as opposed to terrestrial delivery which has been mandated in the United States. For example, it was determined that the wider bandwidth required for MUSE delivery of Hi-Vision programming could not be squeezed into the limited spectrum space offered by the existing UHF or VHF channels, whereas newly planned satellites operating in the 12 GHz band could have transponders set aside to handle HDTV delivery (MPT, 1989; NHK, 1989; Shimizu, 1989).

MUSE delivery by satellite was also seen as an opportunity to diversify programming options to the Japanese public without having to take away spectrum space from preexisting broadcast services. As Massachusetts Institute of Technology's William Schreiber (1991) noted, "From the first, the plan was to implement HDTV in Japan as an entirely new service, delivered to viewers by DBS, and intended to supplement, rather than replace, the existing over-the-air (terrestrial) system, which would continue to employ NTSC" (p. 268). It must be added that the delivery of distinctive HDTV programming over newly operational satellite channels—instead of merely simulcasting[12] identical shows in both HDTV and NTSC—fits nicely into the Japanese marketing plan for MUSE as well. This approach supplies the incentive needed to entice consumers into purchasing more expensive HDTV home receivers and 75-cm satellite antennas (Schreiber, 1991; Shimizu, 1989).

A final reason cited by the MPT (1989) for MUSE delivery by satellite—and one that might be less apparent to people living outside Japan—is the protective isolation of orbiting transmitters from the potential damage caused by earthquakes and other natural disasters. Lying as it does along the Pacific Rim of Fire, Japan is the scene of frequent and intense earthquakes. Typhoons are also relatively common there (Ito, 1978). While it is true that satellite earth stations could be knocked out by geological and weather disturbances on the ground, Japanese planners anticipate that mobile backup stations could be quickly put on line to restore national DBS communications.

MUSE development continued throughout the mid-1980s. Following the example set by Hi-Vision's designers, NHK researchers decided that striving for direct compatibility between MUSE and existing home sets would not be desirable since it would tie the new transmission standard to limitations set by the outmoded NTSC technology. As it had been for Hi-Vision, the benchmark established for MUSE quality was that it attempt to satisfy the full dynamic range of the human visual system (Ninomiya et al., 1987). As the result of this clearly established engineering goal, Japanese scientists came to emphasize convertibility rather than compatibility in their research approach:

> The MUSE system is not directly compatible with the current [NTSC] television systems. If we wanted to have direct compatibility, the picture quality of the received HDTV signal would have to be sacrificed. What we aimed at was to develop a system with which we can transmit the best HDTV picture with a limited bandwidth. In spite of this, some consideration has been paid to designing the MUSE system for easy conversion to current systems. (Ninomiya et al., 1987, p. 158)

It should be noted that, originally, only one MUSE standard was developed—the standard designed for DBS service, which has since become known as MUSE-E. However, as research on MUSE progressed, an entire family of MUSE standards was assembled, largely at the request of organizations in the United States.

At the urging of the National Association of Broadcasters (NAB) and the Association of Maximum Service Telecasters (MST), NHK was encouraged to produce variations on the MUSE theme that would be more friendly to conditions in America by either being compatible with the current NTSC system or at least compatible with the 6 MHz channelization scheme established there (Carter, 1989; Donow, 1988; NHK, 1988; Schreiber, 1991). Further work by NHK resulted in the creation of NTSC MUSE-6, which is both receiver and 6 MHz compatible; NTSC MUSE-9, which is receiver compatible but requires a 9 MHz transmission channel; and Narrow MUSE, an adaptation of the MUSE bandwidth compression algorithm that fits into a 6 MHz channel but is not receiver compatible.

This situation involving the evolution of the MUSE family of transmission standards bears resemblance to the development of the final Hi-Vision standard; both were greatly influenced by international scientific and political considerations. In the case of MUSE, however, the original MUSE-E standard to be used in Japan was not changed, which was not the case with the studio standard finally accepted by BTA and the MPT.

NHK initiated the experimental distribution of NTSC programming by satellite in 1987. In June 1989, this DBS testing came to fruition with the establishment of two regular channels of NTSC service (both supplied by NHK) over satellite BS-2b. At the same time, NHK began airing one hour a day of MUSE programming. An experimental broadcasting schedule of MUSE programs (eight hours daily) on BS-3b began on November 25, 1991, and was expanded in subsequent years (see Chapter 2; see also Adam, 1990; CasaBianca, 1992; Jurgen, 1989a; MPT, 1989; NHK, 1995; Thurber, 1990; U.S. Congress, 1990). While BS-3a, launched in 1990, transmits the two NHK DBS channels (Channel One and Channel Two), BS-3b, launched in 1991, delivers Hi-Vision and WOWOW, a pay-TV service offered by Japan Satellite Broadcasting (JSB), a consortium of Japanese commercial broadcasters.[13]

The Role of Broadcast Electronics Manufacturers

The impact of broadcast equipment companies, such as Sony, Panasonic, and Ikegami, on HDTV development is not to be understated, although it must be understood, at least initially, in connection to NHK's HDTV endeavors. It is true that NHK involved private electronics firms as early as 1970 to manufacture equipment, but it did so selectively to preserve its autonomy and control over the project. In addition, according to NHK, many manufacturers were hesitant to participate in the process of manufacturing prototypes which were so different from their existing product lines and for which there might never be any market (Noble, 1992). The major exception was Sony, which built the first-generation HDTV system in April 1981 as part of a long-term economic strategy. Sony, a consumer electronics firm, had just entered the broadcast market, and Akio Morita, then its chairman, had envisioned the eventual replacement of 35mm cinematography with *electronic* cinematography. Winston (1989) explains:

> NHK research opened the door to a wonderful commercial possibility. Effectively exploited, HDTV could give Sony a commanding lead in the professional field. The film studios were moving towards video anyway. Only the mismatched quality of 35mm and existing television remained an obstacle to the electronic penetration of that market. (p. 128)

So it was not until the late 1980s, when the specifications of the Hi-Vision standard were virtually set, that broadcast equipment manufacturers became a major force in HDTV development and diffusion by offering professional users a vast range of HDTV production hardware. At the 1983 International Television Symposium and Technical Exhibition in Montreux, Switzerland, the Sony HDVS was the only HDTV system available on display ("The Mood," 1983). In contrast, at the 1989 Montreux symposium, no fewer than 25 companies showcased equipment that dealt with every aspect of HDTV studio production ("High Definition," 1989).

The Role of MPT and MITI

Starting in the early 1980s with the emergence of the first working Hi-Vision equipment, groups outside of NHK began to have an increasing impact on Hi-Vision's ultimate specifications and on the marketing of the 1125/60 standard at home and abroad. During that decade, the Ministry of Posts and Telecommunications (MPT) grew more involved in HDTV policy by coordinating standard-setting and promotional activities. Consistent with its mission, the Ministry of International Trade and Industry (MITI) confined its role to HDTV promotion. This section elucidates the HDTV standard-setting and

promotion mechanisms used in Japan and identifies the key committees involved in those activities.

HDTV Standard-Setting Activities

By the mid-1980s, more and more domestic governmental and
broadcasting groups brought their influence to bear on the HDTV
standard-setting process (see Figure 3.1). As part of its regulatory powers, the MPT aggressively pursued the rapid formalization of standards for Hi-Vision, in recognition of the boost that definitive standards could have by encouraging economies of scale for manufacturers and lessening confusion for consumers (Tatsuno, 1990). Although
this Ministry is responsible for approving telecommunication standards, it also seeks to exchange ideas with and promote coordination
among interested parties in the early stages of the standard-setting
process (see MPT, 1996). As political scientist John Campbell (1984)
points out, "Japanese tend to see a policy 'decision' as emerging from
a complex network of relationships among individuals or institutions
with conflicting interests" (p. 325). The MPT's quest for consensus-
building (known as *ne-mawashi* in Japan) is consistent with the frequently noted collectivist mentality that permeates the Japanese society and economy.[14]

The most significant force behind HDTV standardization in Japan
has been the Broadcasting Technology Association (BTA) (renamed the
Association of Radio Industries and Businesses [ARIB] in May 1995).[15]
Created in September 1985, BTA was an industry-led organization similar to the ATSC in the United States, consisting of broadcasters and
electronics manufacturers. It considered two different approaches for
bringing advanced television (ATV)[16] technology to the Japanese people: HDTV and EDTV, which will be discussed later in this chapter. In
April 1986, as part of its Satellite Broadcasting Group, BTA established
a High Definition Television Committee to investigate HDTV studio
and transmission standards, as well as program production technology (see Figure 3.1).[17] The committee members participating in these
studies came from NHK, commercial broadcasting stations, and
equipment manufacturers.[18] "The committee [drew] up the plan for
HDTV development with the approval of the Ministry of Posts and
Telecommunications, and the member companies and organs [carried]
out their own research or joint development based on this plan"
(Sugimoto, 1987, pp. 5.1.5-5.1.6).

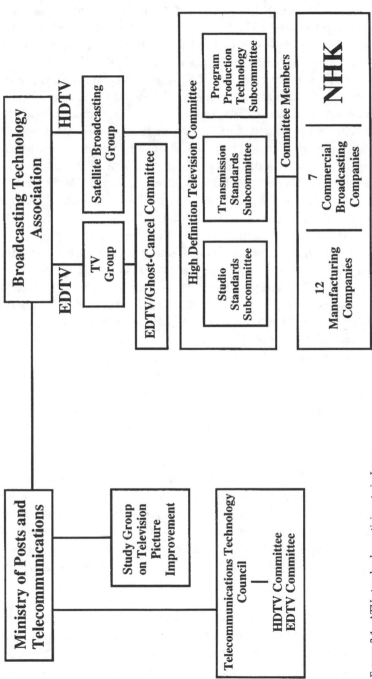

FIGURE 3.1. ATV standards participants in Japan (*Source:* Broadcasting Technology Association and Ministry of Posts and Telecommunications).

Therefore, the High Definition Television Committee (not to be confused with the MPT's HDTV Committee) was the central hub within which industry debate over HDTV standards took place. It is significant to note that, since the BTA's inception, no changes were suggested for the primary elements of NHK's Hi-Vision standard by this HDTV committee. In fact, adjustments to key parameters of this standard have been influenced more often by foreign standards bodies—primarily from the United States—than by this Japanese organization. Also, while private broadcasters were able to participate in the formulation of HDTV standards through BTA, so much of the character of both Hi-Vision and MUSE had already been decided by NHK, and the involvement of commercial telecasters was more of a superficial exercise than an act of actual substance (Sanger, 1989).

From the outside, it appears that the role played by the High Definition Television Committee was one of facilitating information exchange and consensus-building, and not as a forum for heated squabbles over Hi-Vision's technical specifications. This is easily explained, since NHK had already worked for 15 years to perfect its HDTV "champion" before BTA even began its deliberations. As previously noted, NHK's extensive testing had left little room for doubts as to its performance capabilities. Also, considering NHK's status as Japan's public broadcasting giant—combined with the cultural emphasis on teamwork and cooperation commonly identified with Japan—it is not surprising that Hi-Vision passed through this committee with its basic character left unchanged (see Prestowitz, 1989).

No impediments to the existing specifications of the Hi-Vision standard were offered by the MPT either. The MPT is responsible for overseeing Japan's broadcasting infrastructure and, much like the FCC in the United States, handles the allocation of television frequencies for NHK and the stations comprising the country's five commercial networks (Ito, 1978; Senitt, 1988). A major difference between the MPT and the FCC, however, is illustrated by the Japanese Ministry's enthusiastic promotion of Hi-Vision within the broadcasting industry and to Japanese consumers (Shimizu, 1989).

BTA approved Hi-Vision in August 1987, and the Telecommunications Technology Council (TTC) sanctioned it tentatively in June 1988 (interim report). The TTC, a permanent consultative organ of the MPT, began advising the minister on technical requirements for HDTV as early as June 1984. It established an ad hoc committee, the HDTV Committee, in April 1985 for HDTV standards consultation (S.

Takizawa, Ministry of Posts and Telecommunications, personal communication, December 25, 1996). Following the CCIR Plenary Assembly in May 1990 and a final TTC recommendation in June 1990, the MPT ratified the BTA Hi-Vision system as the national studio production standard in March 1991.

The standardization process of MUSE, on the other hand, was more controversial. Noble (1992) explains:

> This time, ... NHK faced competition. A professor ... from Nagoya University approached the Telecommunications Technology Council with an alternative proposal, though he never submitted it formally. More important was an application by NTT [Nippon Telegraph and Telephone], potentially a deadly rival of NHK, since fiber optic cables could eventually obviate the need for 'broadcasting' as we (and NHK) now know it. (p. 18)

Ultimately, the MPT approved MUSE in March 1991 as a national broadcast standard, because it was "more complete and more cost-effective" (Noble, 1992, p. 18). What really happened internally remains a mystery because TTC records are not open to the public. One can surmise, though, that the fact that NHK had worked so long on MUSE technology influenced the MPT's decision to a significant extent. As Japan's subsequent reaction to digital television indicates (see Chapter 1), it is not a Japanese habit to abandon a technology, even an inferior one, when so much time and money has been invested in its development.

Promotional Policies

The rejection of the 1125/60 system at the 1986 CCIR Plenary Assembly in Dubrovnik, Yugoslavia, prompted the MPT and MITI to consider promotional policies for HDTV (Noble, 1992). These actions began in earnest in 1987, and by 1989, the number of study groups and organizations involved in HDTV promotion reached staggering proportions. While it is true that the MPT was an active participant in standards discussions, much of its energy concerning Hi-Vision has been spent administering the promotional agenda established by its own Round-Table Conference group. As to MITI, it was solely involved in the promotional aspect of HDTV because its primary mission was, and still is, to stimulate the growth of Japanese industries domestically and internationally (see Johnson, 1982; Okimoto, 1989).

Both the MPT and MITI have used their influence to coordinate Hi-

Vision public relations campaigns of dramatic proportions (Jurgen, 1989b; Kumabe, 1988; Nakatani, 1988; Prentiss, 1990; Tannas, 1989; Tatsuno, 1990; Udagawa, 1988; U.S. Congress, 1990). For example, Hi-Vision displays were featured prominently during the World Fashion Fair held in Osaka in 1989 and during the Exposition of Flowers and Greenery the following year. High-definition programming has also been shown in restaurant settings, intended as a sort of dinner enter-tainment that has come to be known as "Western Theater." Both also set up advisory committees, created internal offices, founded HDTV promotion organizations, supported the extension of HDTV to urban settings, and even coordinated R&D consortia for display technology. In sum, the two Ministries have rivaled each other, initiative for initia-tive (Grout, 1988; Noble, 1992). Although dysfunctional and counter-productive for those private companies caught in the middle of a min-isterial feud, this kind of policy overlap is not unusual in Japan. Campbell (1984) explains:

> While such overlaps occur in all governments, they appear particular-ly frequent in Japan, and moreover in other countries usually some sort of liaison committee is established for at least minimal coordination among the agencies involved. In Japan, often the agencies do not com-municate at all—once a Welfare Ministry official asked me, as a foreign scholar, to go over to the Construction Ministry and find out what the officials there were thinking about housing for old people. (p. 320)

The infighting between the MPT and MITI over HDTV promotion pol-icy is not an exception as documented in the case of the so-called value-added network (VAN) war in the early 1980s (see Muramatsu, 1991).

Among the cornucopia of MPT initiatives, four deserve special mention. In March 1987, the MPT established the Round-Table Conference Concerning HDTV Promotion, an advisory committee to the Minister, to discuss and propose measures how to best promote HDTV in Japan (Okai, 1987). Second, in May 1987, it established its own internal office within the Broadcasting Bureau, the Hi-Vision Promotion Office, to prepare policies for the promotion of HDTV. Third, in September 1987, the Ministry created the Hi-Vision Promotion Council (incorporated into the Hi-Vision Promotion Association in October 1991)[19] to recommend specific measures for encouraging the diffusion of HDTV. This affiliated organization was composed of broadcasters, electronics manufacturers, and MPT staff. For instance, it actively planned and coordinated the HDTV demonstrations during the 1988 Seoul Olympics, which were seen by an estimated 3.7 million people, by installing 208 Hi-Vision receivers in 81 locations nationwide (MPT, 1989). It selected November 25 as "Hi-Vision Day" (because this date—11/25—designates the number of scanning lines used in HDTV) to hold special HDTV events every year.

Finally, with a vision surpassing that of the European Community and the United States, the Japanese government was making a long-term commitment to Hi-Vision—recognizing the technology as more than an extension of today's method of broadcasting and, instead, as an element essential to the operation of tomorrow's information-based society (see BTA, 1987a; Carbonara, 1990b; Jurgen, 1989b; Sugimoto, 1988). Nowhere was this more apparent than in the MPT's "Hi-Vision Cities" program introduced in 1988 (MPT, 1989; Nakatani, 1988; Tatsuno, 1990; U.S. Congress, 1990). The thrust of this ongoing initiative was to create "a city-wide video information network providing high-quality images serving as the core, which aims at 'creating a charming and intellectually stimulating city'" ("Hamamatsu City," 1994, p. 3). The Ministry reasoned that advanced visual media will be critical to the formation of the information society in the next century and that HDTV will contribute to it by fostering social, economic, educational, and cultural goals. Specifically, the MPT (1989) envisioned that Hi-Vision Cities would:

- encourage community formation and interpersonal communication
- promote health and welfare
- support education and culture and raise the quality of life
- make the urban environment safer and more comfortable
- promote the creation of attractive resorts
- expand leisure activities and encourage family togetherness
- stimulate a new visual information industry
- strengthen the local economy
- promote urban internationalization.

Out of 71 applications submitted to the MPT in 1989, 14 cities were chosen to participate in the program. Using a regional approach to the development of nationwide information systems, they were encouraged to apply high-definition imaging tools in manufacturing, education, electronic publishing, video cataloging, and a variety of other information-intensive uses. As of August 1996, there were 39 Hi-Vision Cities.

In addition to being somewhat futuristic in scope, the MPT's Hi-Vision City concept has served as a practical approach for creating diversified markets that can use high-definition imaging hardware. Also, unlike the situation surrounding international broadcasting, these innovative markets for Hi-Vision equipment are much less restricted by global agreement on a single set of operating specifications. Therefore, if the Japanese were able to use their head start in HDTV to establish 1125/60 as the de facto standard for use by medical or educational institutions, for example, they could effectively block any challenge that might come from other formats that have arrived on the scene more recently (Jurgen, 1989b).

Compared to other sources of funding, MPT's direct financial support for the domestic and international promotion of HDTV has been minimal. For instance, the Ministry contributed ¥111 million ($815,000) between 1989 and 1992: ¥28 million ($202,896; $1 = ¥138) in 1989; ¥32 million ($220,690; $1 = ¥145) in 1990; ¥25 million ($186,567; $1 = ¥134) in 1991; and ¥26 million ($204,724; $1 = ¥127) in 1992. The MPT also offered tax breaks in 1989 and 1990, although the amounts were not revealed. Most of the funding for Hi-Vision Cities came from the Japan Development Bank (JDB) in the form of no-interest loans (up to ¥150 billion or about $1.1 billion in 1989) and low-interest loans (up to ¥473 billion or about $3.4 billion in 1989) (Noble, 1992).

Like the MPT, the Ministry of International Trade and Industry (MITI) established an advisory committee, Future Prospects for HDTV (in January 1987), a ministerial office in charge of HDTV promotion, the New Visual Industry Office (in January 1988), and an affiliated promotion organization, the Hi-Vision Promotion Center (in July 1988) (Noble, 1992). This center is primarily concerned with HDTV promotion in fields other than broadcasting (e.g., multimedia). MITI also matched the MPT's Hi-Vision Cities program with its own "Hi-Vision Communities" initiative. The purpose of these advanced communities was similar, albeit more specific, than their MPT counterparts: to "'create an international ceramics information center,' 'introduce Hi-Vision into the Gifu Museum [of Art],' 'help summer tourists (to a small town in Hokkaido) call forth the winter ice skating season' and so forth" (Noble, 1992, pp. 12-13). MITI has contributed little direct financial assistance to the "Hi-Vision Communities" program (¥62 million [about $457,000] from 1989 to 1992). Tax breaks were also offered during those years, but no estimates are available. Again, no-interest loans (up to ¥70 billion or about $510 million in 1989) and low-interest loans (up to ¥530 billion or about $3.8 billion in 1989) from JDB, Hokkaido Tohoku Development Bank, Small and Medium Size Enterprise Bank, and the Bank of Commerce and Industry made up the bulk of the support for the program (Noble, 1992; see also Grout, 1989).

The Role of Private Broadcasters

By the early 1980s, the Europeans were not the only ones to feel threatened by the potential worldwide adoption of NHK's Hi-Vision and MUSE technologies. Japanese commercial broadcasters[20] were growing distressed over the competitive advantage NHK could gain should these two technologies be adopted as domestic standards. They opposed Hi-Vision because it required significant investment in new equipment, because it could only be transmitted via satellite, creating difficulties for local terrestrial broadcasters, but mostly because it was dominated by their archrival, NHK (Noble, 1992).

This fear led private broadcasters to begin work in 1983 on an alternative ATV design, an NTSC-compatible, extended-definition television (EDTV)[21] system, with the expectation that such an improved system would undermine interest in MUSE and challenge the hegemony of NHK. In June 1985, the MPT responded to their concerns, albeit unenthusiastically (see Noble, 1992), by establishing a Study Group on Television Picture Improvement to determine the feasibility of improving the picture quality of existing NTSC television and creating an EDTV system (MPT, 1989). In early 1986, BTA began looking into ways of improving the quality of terrestrial broadcasting, emphasizing the goals of maintaining compatibility with NTSC receivers and keeping down the transition costs demanded by any new system (Figure 3.1;[22] BTA, 1987a; Jurgen, 1988). These criteria were quite different from those which were used to lead the development of Hi-Vision and MUSE.

After reviewing 25 proposals and selecting 10, the Study Group submitted a first report to the MPT in September 1987 (M. Wakao, Association of Radio Industries and Businesses, personal communication, January 8, 1997). The same month, the TTC established the EDTV Committee. In February 1989, BTA submitted to the TTC its EDTV System Study Report describing the parameters of EDTV-I. This pro-

posed standard, also known as Clear-Vision, used ghost cancelling cir-
cuitry and progressive scanning in studio cameras and home receivers
to improve the final display's horizontal and vertical resolution as well
as the system's signal-to-noise ratio (BTA, 1987a; Carter, 1989). In
March 1989, the TTC issued an interim report to the Minister recommending the BTA EDTV system, as well as a two-stage approach for
EDTV in Japan (EDTV-I and EDTV-II). EDTV broadcasts debuted in
August 1989 (see Chapter 1).

The very fact that commercial broadcasters were motivated to pursue the development of Clear-Vision illustrates the threat they felt
from the advent of HDTV by NHK ("Does Japan," 1989; "MITI and
Trade," 1989; Tatsuno, 1990; U.S. Congress, 1990). The commercial networks had determined that, if they did not upgrade the quality of their
own TV images, the possibility existed that their stations would be
looked upon by viewers as providing a second-rate service in comparison to the image quality promised by NHK's Hi-Vision.

This development of a second round of standards activities highlighted some of the dynamics impacting the growth of Japanese ATV.
First, it illustrated how the country's commercial broadcasters had
come to perceive that the establishment of Hi-Vision was not necessarily a factor in their favor. It also reinforced the point of how small
an impact the commercial broadcasters had on the establishment of
Hi-Vision and MUSE standards.

Discussion and Conclusions

Since their initial efforts to promote Hi-Vision and MUSE as potential, global high-definition TV standards, the Japanese have encountered a wall of international obstacles (see Chapter 1). This situation
illustrates a difficult lesson of the global marketplace: Leading the
innovation of a new technology does not always ensure one of capturing a leadership position. The United States has learned this lesson as
well, when it introduced NTSC color technology 10 years ahead of
PAL and SECAM standards. Whereas a decade ago the Hi-Vision standard was being hailed by most observers as the heir-apparent to
NTSC, PAL, and SECAM, today it has fallen victim to technological
obsolescence with the advent of digital television systems. Ironically, it
was precisely this type of situation that NHK sought to avoid when it
launched its HDTV research program. Although somewhat reluctant-

ly, the Japanese have now reached the conclusion that a digital standard offers the best performance potential for 21st century television (see Chapter 1).

By taking a historical approach, this chapter has identified various technical, economic, and social priorities that have affected the development of Hi-Vision, MUSE, and Clear-Vision. It becomes increasingly important to recognize society's varied influences on the character of any such innovations. It must be remembered that new technologies do not simply appear on the scene by divine intervention but, rather, are called forth and given form by the cultures in which they are ultimately employed (Slack, 1984). In other words, while technologies obviously influence the societies in which they operate, they are also equally influenced during their formation by the societies that conceived them.

In retrospect, NHK decided to undertake HDTV research for a variety of technical, economic, and social reasons. Soon after beginning color TV transmission, the Corporation set out to ameliorate television technology and eliminate the perceived NTSC defects using the human visual experience (and later 35mm film) as a baseline. As the HDTV effort moved towards the building of the first prototypes, the specifications to be used were increasingly shaped by input from foreign organizations, especially SMPTE and the ATSC in the United States. One commentator, in fact, has gone so far as to suggest that the 1125/60 standard is as much an American creation as it is a Japanese design (Schubin, 1990). It is also clear that NHK initiated HDTV research to protect its revenue base and justify higher viewer fees. NHK revenues rose by almost 500 percent between 1970 and 1991 and over 100 percent between 1980 and 1991 (MPT, 1992), although more direct evidence is needed before we can claim that HDTV development induced an increase, even partial, in the amount of the receiving fee.

Perhaps less highly visible than the technical and economic considerations which have molded the profile of Japanese HDTV, social priorities have had their impact as well. As an NHK official points out, "NHK has a task to contribute to the culture and welfare of the public and is always interested in new technologies which are thought to contribute to society" (Y. Onuki, Nippon Hoso Kyokai, personal communication, May 21, 1996). By developing Hi-Vision and MUSE technologies, NHK was complying with Article 9(2) of the 1950 Broadcast Law as amended, requiring it to "conduct researches and investigations necessary for the improvement and development of broadcasting and the reception thereof" (*Japanese Legislation*, 1991, p. 10).

This chapter also describes how the leading parameters of field frequency, aspect ratio, and the choice for the optimum number of scanning lines were all affected by consideration of the ways that Hi-Vision might best be promoted as a unifying champion of global television standards. This appears to have been done partially out of social concern that the next generation of video equipment will introduce a worldwide "television community" (Fujio, 1985, p. 646), improving the chances for better international communications. Realistically though, a leading reason for this manipulation of NHK's evolving ATV standards must be recognized as the economic coup that would have resulted from approval of Japanese-designed HDTV specifications as the agreed-upon global format.

A more concrete social goal of the Japanese has been the development and promotion of Hi-Vision as the imaging tool that will tie together each workplace and home and, in so doing, establish an information network of far-reaching capability. Admittedly, this will have economic implications as well. However, as an interconnected information and entertainment resource offering a new freedom of participatory involvement in the handling of multimedia data sources, the dynamic social implications inherent in the advent of such an information-based culture should not be overlooked (Dertouzos, 1991; U.S. Congress, 1990).

Other social priorities are evident through the decision to distribute MUSE transmissions by DBS. Such a plan will ensure that HDTV program channels are available to rural and urban dwellers alike, and that it will be accomplished efficiently and at the lowest possible cost. On the other hand, the choice of DBS for MUSE transmissions may also reflect a lack of emphasis on the value of local programming, since satellite operations tend to strengthen the role of centralized network uplinks.

Finally, the chapter highlights some of the similarities and differences that exist between Japanese and American approaches to standardization and promotion of broadcast technologies. Like the Americans, the Japanese have a regulatory agency—the MPT—that is expected to handle the allocation of radio frequencies and other such spectrum management duties. In both countries, the amount of responsibility left to the broadcasting industry, especially in standard-setting, is notable. In the situations concerning both HDTV standard-ization—led by NHK—and EDTV standardization—debated within BTA—the framework for broadcast specifications was determined overwhelmingly by industry participants. As recently demonstrated in the U.S. HDTV transmission standard-setting process (see Chapter 6), the involvement of the industry in broadcast standardization matters is also a staple in U.S. communications policy. One senses, however, that this cooperative process is a cultural factor in Japan and that stan-dard-setting there is less often disrupted by interindustry wrangling and positioning.

Even more dramatic in its uniqueness is another role stressed by the MPT in addition to its standardization responsibilities. As evi-denced by the creation of ad hoc bodies, part of the Ministry's mission has dealt with the promotion of ATV use as opposed to just the regu-lation of ATV standards. Judging by the ambitious promotional poli-cies carried out since the late 1980s, both the MPT and MITI have assumed a very active role in the coordinated manufacture and imple-mentation of ATV products within Japan. In the United States, the responsibility for promoting HDTV falls primarily with the electronic media and consumer electronics industries, not with the FCC.

In conclusion, one has to be impressed with the depth of commit-ment that has been shown by the Japanese government and broad-casting industry to the advancement of HDTV as part of a carefully planned national agenda. This is in marked contrast to the situation in the United States, where ATV technologies have been dropped as the centerpiece of a coordinated, national industrial policy (see Chapter 5; Passell, 1989; Pollack, 1989). That the Japanese value highly the impor-tance of ATV to their domestic economic and social vitality is evident by the comments of one Japanese broadcaster who has suggested that the HDTV acronym stands not for high-definition television but, instead, for "higher devotion to television" (Kumabe, 1988, p. 7)—a devotion which has played itself out through over 30 years of advanced television research and development.

4 HDTV and Standardization Policymaking in Europe

by Sophia Kaitatzi-Whitlock

HD-MAC reached obsolescence as it reached technical maturity.
—*Jean-Luc Renaud, editor-in-chief,* Advanced Television Markets, *1995*

The rise and fall of Europe's high-definition television (HDTV) project and its associated common transmission standards policy, as well as the subsequent advanced television policies of the European Union (EU),[1] offer a case study of the impact of global capital integration on the political economy of the region. This case affords valuable insights into the interplay between the relatively new political and legal institutions of the EU and market forces in the audiovisual field. It highlights the nature of political power structures and decisionmaking processes in a period of rapid technological development and economic restructuring.

In 1986, the EU set out its grand advanced television technology strategy with the aim of conquering new-generation product markets and strengthening the competitive stance of the European consumer electronics industry. It intended to use the muscle of the European market to fore-

Sophia Kaitatzi-Whitlock is lecturer in the Faculty of Liberal Arts at City College, Thessaloniki, Greece, and a researcher for the Institute of Audiovisual Media in Athens. She holds a Ph.D. in communication from Westminster University, London.

stall existing Japanese and prospective American endeavors to impose advanced television transmission standards worldwide. The key to this strategy was the EUREKA 95 (EU-95) project, whose purpose was to develop a European HDTV system based on the MAC (multiplexed analog components) standard for the 1990 CCIR (International Radio Consultative Committee) Plenary Assembly.[2]

But in 1993, the EU officially abandoned its controversial standards strategy and its exclusive support of MAC, leaving the widescreen 16:9 format as the only remaining common factor. Concurrently it gave up its competitive rhetoric and entered into an explicit policy alliance with its former rivals, the United States and Japan.[3] In 1995, a third Directive enshrined the neutrality of transmission standards and looked to the multimedia promise and interactive options of digital technology. So five years after the Rhodes Summit of 1988 all the premises of this strategy were turned upside down. Neutrality prevailed over the objective of an EU standardization, HD-MAC plans were effectively abandoned by all, former enemies became friends, and there emerged a new language of objectives related to multimedia and the global "information society."

Meanwhile, crucial shifts had taken place in the international power structure, giving rise to a multilaterally liberalized regime in trade in services. Technological innovation performed quantum leaps, and further advances in international capital integration[4] and consolidation were recorded. On the internal front the revision of the "constitution" of the EU (Treaty of the European Union) put its mark on policymaking procedures and economic goals.

During the decade, an intense techno-nationalism and grave policy errors worked against the MAC agenda. Emergent all-digital systems compounded the damage, obliging the EU to relax its standards policy. More importantly, the EU was forced to abandon its proactive style of policymaking that had sought to foster regional initiative and independence. So the D2-MAC and HD-MAC models, nurtured until 1992, were not the only casualties of this techno-economic and political war. The defeat of the standards policy also swept away accepted principles and policy practices. The policy mix of an aggressive neo-mercantilism and the liberalization trends of the 1980s, largely in competition with each other, was succeeded by accommodation and cooperation with ex-competitors. The organizing system of standardization, in line with the 1987 Single European Act (SEA), was replaced by the new "doctrine" of neutrality in transmission standards. Clumsy attempts at interventionism succumbed to market-led developments.

This chapter examines a decade of European transmission technology policymaking[5] in two phases (see Appendix). The first phase (1986-1992) retraces the evolution of the D2-MAC/HD-MAC policy and the resultant policymaking deadlock, concentrating on the 1986 and 1992 Directives plus the intended MAC Action Plan (AP) and the Memorandum of Understanding (MoU). The second phase (1993-1995) covers the emergence of a new organized power center, the Digital Video Broadcasting (DVB) group, whose role has been seminal since 1991, and the conditions under which the transition from supporting analog to embracing digital technology have taken place. This policy U-turn occurred through the concurrent adoption of the Resolution and the neutral Action Plan of 1993, followed by the Resolution of 1994 and the new Directive of 1995. The chapter ends with a discussion of Dahlian and Lukesian power models and analyzes whether they may apply to the EU HDTV decisionmaking process.

The Political and Economic Climate of the Decade

The consumer electronics market was showing clear signs of saturation in the 1980s. Stagnation and recession were affecting many sectors of the economy, destabilizing trade balances and leading to economic decline. The race for HDTV between Europe, Japan, and the United States coincided with a period of rapid technological development, capital concentration, and fierce international and interregional competition for global market share. This conjuncture was further marked by the technological/economic convergence between previously independent sectors. The strategy for a world HDTV system was the means of capturing not only the estimated 750 million TV set market but also a wide range of consumer electronics and services markets: semiconductors, studio and production equipment, satellite transmission hardware, and nonbroadcast HDTV services. Control of this technology would secure a range of network and non-network-based[6] economies of scope and scale according to prevailing information economy visions (Joosten, 1992). Thus HDTV was expected to create demand pull in several neighboring sectors.

Ever since the NHK 1125-line/60-Hz system was proposed as the world production standard at the 1986 CCIR Plenary Assembly in Dubrovnik, Yugoslavia (see Renaud and Schilling, 1993), the EU feared that the Japanese would use HDTV to control the consumer electronics industry. This fear sparked off a European neo-mercantilist offensive. The EU first coordinated efforts to fend off the standard backed by Japan and America at the 1986 CCIR Plenary Assembly, effectively putting off adoption of a single worldwide HDTV production standard until 1990. The successful attack against the 1125/60 system hinged on its incompatibility with incumbent PAL, SECAM, and NTSC television systems (see Chapter 1). It is precisely this element that provided the impetus for developing the European D2-MAC and HD-MAC standards.

EU-95 was one of over 200 collaborative European-wide research and development (R&D) projects that supplemented ongoing research projects in a market-friendly and flexible way. It was a "unique" cooperative venture between 30 research centers from 19 European countries aiming to define a world production standard for HDTV by the 1990 CCIR Plenary Assembly and to strengthen the competitive position of European products and services. Thus, CCIR planning schedules became crucially linked to the implementation and timing of EU-95. The originators and most important backers of the project were: Philips (Netherlands), holding the presidency of EU-95, Thomson (France), vice presidency, Thorn EMI (United Kingdom), and Bosch (Germany). These firms proposed their project and had it adopted at the London EUREKA Ministerial Conference of June 30, 1986 (EUREKA EU95, 1988).

"Mission-oriented industrial policies such as the Eureka HDTV project would be unimaginable without the development of the internal market as a backdrop" (Peterson, 1993, p. 497). As a result the EU-95 project was welcomed as a paradigm of intergovernmental cooperation in Europe. This in turn fortified the rationale of the Single European Market (SEM), which sought to harmonize legal instruments and regulatory provisions. With the SEM in view it was felt and believed that the EU could take up the challenge of global competition. The objective of creating economies of scale, perceived as crucial for competitiveness, was also high on the EU agenda. Toward that end, standardization was embraced as a means of (1) creating, solidifying, and delimiting the European market; (2) fending off competitors; and (3) creating worldwide standards of European origin.[7] This *dirigiste* approach (i.e., heavy-handed governmental interventionism) at the "supranational" level was further vindicated by the threats perceived in the covertly hostile communications policies of EU competitors, including the U.S. Department of Defense.[8] So the EU launched its standards offensive against the threat of a global cartel by its major competitors.

In the mid-1980s, the EU Commission was gaining in recognition, and policy momentum was at its peak. DG XIII (Directorate-General for Telecommunications, Information Market and Exploitation of Research)[9] policy proposals were quickly adopted. In 1986, six documents were approved by the Council on common strategies in the information technology (IT) sector, including the first MAC Directive. At the Rhodes Summit of December 1988, EU heads reconfirmed the high priority of the standards policy on the wider EU agenda. The DG XIII estimated that world TV set markets would grow to 100 billion of European Currency Units (ECU) by 1992 and that world production of HDTV sets would reach one billion units by 1995 (Commission of the European Communities [CEC], 1992a), thereby promising to revive both the industrial and the services sectors. Demand for integrated circuits, semiconductors, and color cathode ray tubes would create broader market pull, as an HDTV set requires five times as many semiconductors as a conventional TV set (Joosten, 1992).

The 1986 MAC Directive

The first Directive on the MAC/packet family of standards for direct broadcast satellite (DBS), adopted in November 1986 for five years, was shaped by the climate of thrust towards the SEM and the hype over DBS television fostered particularly by the Franco-German axis.[10] "Much kudos was also associated with the joint Eurosatellite manufacturing venture between France and West Germany in the construction of Europe's first DBS satellites—TDF-1 and TV-SAT" (Hughes, 1988, p. 50). "A conflict of interests became abundantly evident in the summer of 1985, when Schwartz-Schilling [then West Germany's Telecommunications Minister] signed a controversial agreement with his French counterpart, Louis Mexandreau, to introduce a new television transmission standard for DBS, called D2-MAC" (Humphreys, 1988, p. 134). This agreement for the introduction of the D2-MAC transmission standard spawned reactions at various levels, first within these countries, notably between broadcasting and telecommunications authorities, such as the conflict between TDF

(Télédiffusion de France) and DGT (Direction Générale des Télécommunications), and secondly between member states, such as Luxembourg and the United Kingdom, which coveted the satellite services' markets for their own industries, and the Franco-German axis. Most notably it became apparent that Luxembourg deployed its ambitious domestic communications strategy by setting up SES/Astra, while concurrently "adopting" the Franco-German and wider European project (see Stüdeman, 1996). But the complex disputes that had evolved before and during the CCIR 1986 Plenary Assembly between EU partners were hard to overcome. The C-MAC, preferred by the British, had the advantage of double sound channel capability but the Franco-Dutch D2-MAC was finally chosen after a decisive display of industrial muscle and political dexterity. The French carried the day after threatening to withdraw from the conference (Hills and Papathanasopoulos, 1991).[11]

The MAC standard was developed in 1980 as a rival to PAL and SECAM in view of satellite transmissions. The three variants of MAC (C-MAC, D-MAC, and D2-MAC) use the same video format (625 lines) and aspect ratio (4:3) but have different data rates, number of audio channels, and applications. While C-MAC was devised for satellite broadcasting given its high sound and data capacity requirements, D-MAC was intended to be the C-MAC version for cable distribution. D2-MAC, on the other hand, fit into 7 or 8 MHz channels and could be used for either cable or satellite transmission (*D2-MAC*, 1988; Long and Stenger, 1986). D2-MAC was designed to be an intermediary phase before HD-MAC[12] (forward compatibility) and yet be viewable on PAL/SECAM screens using set-top converters (backward compatibility). Compatibility was thought to be a key advantage as it would enable the gradual introduction of the new system without rendering obsolete the incumbent technology.

But the evolutionary nature of D2-MAC and the compromise it represented between the various European transmission systems affected all phases of its overall market performance. It contributed to diminishing viability of programs by splitting economies of scale since two types of dishes (MAC/HD-MAC) had to be produced (Hills and Papathanasopoulos, 1991). Most crucially, this split delayed production of intricate integrated circuits indispensable for MAC receivers. Then, in a key action, Rupert Murdoch announced in June 1988 that he would broadcast in the PAL format, thus circumventing the 1986

Directive and the intent of this European policy (Fox, 1995; Peterson, 1994; Peterson, 1993).

Although the EU promoted MAC by adopting this Directive, it concurrently created a cause for its demise. It regulated only the broadcasting satellite services (BSS), leaving a loophole for fixed satellite services (FSS).[13] "Direct broadcasting by satellite means a service [i.e., BSS] ... using channels assigned to member states in the 11.7-12.5 GHz band ... " (O.J. No. L 311, 1986, p. 28). It was a grave mistake. Brown, Cave, Sharma, Shurmer, and Carse (1992) argued that "the Commission *consciously* [emphasis added] excluded low and medium power satellites [i.e., FSS] ... from the requirement because of opposition from the Luxembourg company SES [Société Européenne de Satellites] which was operating Astra as a medium-powered satellite" (p. 27). Negrine (1988) concurred with this assessment: "By 1985, technological advances had ensured that the same desired effect—individual reception on small dishes—could be achieved by medium power satellites and that the dishes could be even smaller than 0.9 metres" (p. 11). Technological progress facilitating broadcasting from medium-power (FSS) satellites with increasingly smaller parabolic antennas was fatal for MAC. Those wanting to circumvent the Directive transmitted through satellites like Astra using the FSS band. Quite simply this option made the best business sense, so satellite and telecommunications operators such as British Telecom (BT) were keen to exploit these competitive advantages. "Hence Eutelsat's distress at British Telecom International's involvement with a rival medium power satellite carrier, SES/Astra" (Hughes, 1988, p. 55). The initial reluctance toward MAC, compounded by the poor performance of the DBS industry, gradually turned into an outright rejection.

In the late 1980s, British Satellite Broadcasting (BSB), German TV-SAT 2, and French TDF 1 were the main channels that volunteered to operate in the MAC format (CEC, COM(88)299, 1988), although none achieved significant market penetration. BSB, the only commercial user, folded within months after "being forced into a merger with its rival," Sky Television (Brown et al., 1992, p. 29). The British Independent Broadcasting Authority (IBA) revoked BSB's MAC transmission license after its merger with Sky, a PAL adherent and a strong anti-MAC lobbyist. In Germany, the public service broadcasters ARD and ZDF abandoned MAC to defect to the rival FSS Astra system, after suffering huge losses. In France, the delay of the DBS satellite TDF 1 in launching its channels resulted in a less-than-spectacular promised MAC debut and did not bode well for the fledgling European DBS industry (Slaa, 1991). Ex-MAC users were deterred and won over to the unregulated FSS band haven.

In view of identified technological advances enabling the transmission of DBS via medium-power satellites using PAL and SECAM standards, the Commission, as it should, pointed out that the loophole could produce a boomerang effect due to inequality between the regulated DBS and the unregulated FSS band (CEC, COM(88)299, 1988). It called for a mid-term correction that would free the BSS band from its competitive disadvantage vis-à-vis the FSS band, but to no avail. The proponents of the anti-MAC agenda, led by SES/Astra, BT, RTL, and Sat 1, continued to disregard and undermine the European global HDTV system. This reluctance to take corrective actions suggests if it does not prove that the loophole of the 1986 Directive was intentional and probably the result of trade-offs within the Council and between Council members and leading entrepreneurs.

In spite of the reluctance to make corrective incremental actions and the persistent introduction problems of the MAC/packet, a "global strategy" was laid out again at the 1988 Rhodes Summit and in a 1989 Council Decision on HDTV. This initiative set out five main objectives for the introduction of HDTV in Europe:

(1) to ensure that the European industry develops in time all the technology, components and equipment required for the launching of HDTV services.

(2) to promote the adoption of the European proposal based on the 1250 lines, 50 complete frames per second progressive scanning parameters, as a single world standard for the origination and exchange of HDTV programme material.

(3) to promote the widest use of the European HDTV system throughout the world.

(4) to promote the introduction, as soon as possible—and in accordance with a suitable timetable from 1992—of HDTV services in Europe.

(5) to ensure that the European film and television production industry achieves the capability, experience and dimension required to occupy a competitive position on the HDTV world market. (O.J. No. L 142, 1989, p. 2)

With regard to the last target no significant efforts were made either nationally or "supranationally." The Commission did not incorporate HDTV production projects either in its *Television Without Frontiers* (TWF) Directive of 1989 or in its MEDIA program.[14] In December 1989, yet another supportive Decision was adopted, commanding, as usual, common action for its worldwide adoption (O.J. No. L 363, 1989).

In late 1989, the Commission established a European Economic Interest Grouping (EEIG)[15] for HDTV called Vision 1250 to fulfill the HDTV objectives in Europe and foster greater cooperation between member states. Two of the main goals of Vision 1250 were to make HDTV mobile equipment available to audiovisual professionals and companies at a pan-European level and demonstrate the achievements of EUREKA 95, particularly at the Olympic Games of 1992 (see Brown et al., 1992). This program received funding from national governments and corporations, as well as from the Commission (which put up a sum of ECU 2.5 million or about $3 million). Still broadcasters were reluctant to incorporate MAC in their satellite TV initiatives. This situation was due to five factors: (1) the costly investments required (both for the new technology and for simulcasting); (2) the preferential regulatory treatment of medium-power band of satellite transmissions; (3) the widely perceived technical reliability of the existing PAL/SECAM transmission standards; (4) the lack of available programming in the new format; and (5) the foreseeable advent of digital technology.

Animosity against MAC gained momentum in the interim period between the first and the second Directives (December 1991-May 1992). "HD-MAC will never be the HDTV standard in Europe," declared Pierre Meyrait, Director General of the SES, on January 8, 1992, adding that "in the post-Directive era the agenda would be dictated by purely commercial considerations" (Laroche, 1992, p. 3). A month later, at the World Administrative Radio Conference (WARC) of the International Telecommunication Union (ITU) in Torremolinos, Spain, a European Broadcasting Union (EBU) official statement resounded equally ominous for the prospects of MAC: "The television of the future will be digital … will be high-definition … will exploit the wide-band transmission technologies … will have digital multichannel sound and enhanced teletext. And the highest-quality services will be delivered by satellite" (Renaud, 1992, p. 1). With the speedy growth of the pro-digital lobby, the anti-MAC block grew proportionately.

The Turning Point

At the Dusseldorf Plenary Assembly of the CCIR in May 1990, the Administrations approved 23 technical parameters for specifying a single worldwide HDTV production standard (CEC, 1992b). These corresponded to European proposals and among them figured the 16:9 wide aspect ratio format. The Assembly, however, failed to approve the European production standard as the winner of the world HDTV race. This was a critical moment, because it meant that the HDTV system would be from then on confined to the regional space plus possibly to EU-dependent areas. A number of scholars and experts have argued that the window of opportunity for developing worldwide HDTV production or transmission standards was already closed by the late 1980s (see Brown et al., 1992). After the Atlanta and the Dusseldorf CCIR meetings of 1990, more experts came to adhere to this view (Joosten, 1992; Solomon, 1990). At these meetings world regulators abstained from "locking high resolution imaging into the past" (Solomon, 1990, p. 59) by adopting either the Japanese 1125/60 or the European 1250/50 proposals entirely.

In February 1991, an alarming report leaked out of the French Ministry of Foreign Affairs which assessed that the D2-MAC system would be made redundant because an all-digital system was being developed by General Instrument in the United States. This situation, moreover, attracted the two "Euro-champions" (Thomson and Philips) over to the rival camp of digital technology (Peterson, 1993), which meant that MAC was being undermined by its own proponents. As could be expected, by mid-1991 signs of ambivalence in the promotion of the European HDTV transmission standard could not be concealed. From December 1991 to May 11, 1992, when the new Directive was adopted, there was an interim vacuum of valid standards for MAC signals. This event signaled the policy paralysis and indecisiveness troubling the EU Commission and Council. Moreover, it reinforced and rewarded de facto standard-setting trends and circumventions of EU regulation. Although there had always been opposition to MAC, attacks on the fledgling standard intensified in response to the updating of the Directive in 1991, clearly demonstrating the EU's inability to orchestrate its act and have its proposal accepted by the main market forces. Underlying this opposition were industrial intersectoral (e.g., broadcasters were antagonizing hardware manufacturers) and intra-sectoral (e.g., Scandinavian manufacturer Nokia, a proponent of HD-

DIVINE,[16] was battling with the manufacturers of analog hardware) controversies and corresponding divergent national interests.

Since consumer demand for broadcast HDTV was anything but secure, HD-MAC strategies were obviously based on uncertain ground. It was argued, however, that there was a definitive short- and medium-term niche market option in professional uses of HDTV such as electronic cinematography rather than in broadcast consumer services (Slaa, 1991; Solomon, 1990). HD-MAC proponents did not concentrate on such entrepreneurial targets in any notable way. EU competitors did not ignore these possibilities, though. The Japanese, particularly after their 1986 debacle, retaliated by focusing on nonbroadcast, high added-value HDTV applications such as teleconferencing, medical uses, and teleshopping. They also targeted home video—the nonbroadcast introduction of HDTV programming via HD VCRs and HD laser disc players (Peterson 1993; Renaud and Schilling, 1993).

The final blow to HD-MAC was the development of digitalization techniques, namely digital bandwidth compression, which offered a sudden window of opportunity for diversifying business strategies at the professional (e.g., medical imaging) and consumer (e.g., interactive services) levels. These incentives were too strong not to steer the market away from a crippled MAC standard and straight towards digitalization in the converged broader information service sectors.

The 1992 MAC Directive

In the early 1990s, the EU's MAC/HDTV policy consisted of three interlinked parts: (1) a second Directive on standards for satellite broadcasting to replace the 1986 Directive scheduled to expire on December 31, 1991; (2) an Action Plan to fund MAC equipment, which was never approved as originally intended; and (3) a Memorandum of Understanding (MoU) to spell out the obligations of the different parties in D2-MAC development, which was supplanted by a toothless Declaration of Intention. So the novel element of this policy initiative was that it attempted, albeit unsuccessfully, to bring about an actively cooperative (with market forces) style of policymaking. It aspired to coordinate regulators and industry forces through the MoU.

The Commission issued its first draft of the updated MAC Directive on July 9, 1991. But despite considerable concessions from the pro-MAC group, allowing PAL operators to transmit till the end of the century, and despite the efforts of DG XIII Commissioner Filipo

Pandolfi, the Commission and the industry had not reached a compromise during consultations leading to this first draft Directive. The anti-MAC camp, which grew thanks to the fatal loophole of the first Directive, represented by the Association of Commercial Television (ACT), but especially by British Sky Broadcasting (BSkyB) and SES, had rejected any deadlines for the phasing-out of PAL services. They had even rebuffed the proposal that large-screen receivers should carry in-built MAC chips ("Down to the Wire," 1991), although this was indispensable for backward and forward compatibility and for simulcasting (i.e., concurrent transmission in several formats). So the MAC strategists were trapped by the loophole regarding transmission from the FSS band but also by their own argument for compatibility.

The first official inside attack on the EU analog standard was made in the European Parliament (EP) in November 1991. The EP rapporteur Gérard Caudron admitted that 90 percent of the members of the European Parliament (MEPs) expressed reservations about the draft Directive (Peterson, 1993). The MEPs approved it with the proviso that it did not exclude the potential all-digital U.S. system ("HDTV Competitor," 1992). Thus the "European" standard did not achieve unequivocal support inside the EP. Notwithstanding the bellicose rhetoric and the hype about EU competitiveness within the triad, this stance revealed crucial divergences in the EU's positioning towards the United States. Was the United States a competitor or an ally? While certain Europeans appeared to fear—and were therefore poised to challenge—U.S. competition, others embraced U.S. projects, forging industrial alliances in America. Yet another category of Europeans, it seems, chose to combine both attitudes. This, however, militates against a strategy of "global competition." On these very ambivalent premises the Commission attempted to build its newly acquired "corporativist" approach to policymaking through the MoU.

The Commission released its second, much watered-down draft Directive on December 5, 1991. After this time lapse, in view of the EP pressures but also in view of the WARC of 1992 in Torremolinos (Spain), the advent of digital technology could no longer be ignored and was incorporated both in the rationale of the preamble and in Articles 1 and 2, in the form of exceptions or reservations to HD-MAC. EU policy drafters repeated the mistake of 1986: They enlarged the existing loophole that had threatened the very survival of HD-MAC, thus succumbing to unworkable EU-type compromises.

In February 1992, the Council approved a Common Position (i.e., first-reading agreement between member states' ministers). The document insisted on avoiding a dual standards regime (MAC/non-MAC). Furthermore, it stated that:

(1) there should be complete agreement between broadcasters, satellite operators, manufacturers and cable operators on the introduction, as soon as possible, of 16:9 D2-MAC services in conformity with the objectives set out in Decision 89/337/EEC;

(2) such agreement might be reached by means of a MoU;

(3) the MoU would set out the obligations of the respective parties for development and promotion of 16:9 D2-MAC services in accordance with the provisions of the Directive, and that it would constitute an integral part of the overall strategy for the introduction of HDTV; and

(4) it was vital to ensure that audiovisual programs adapted to the new 16:9 format were available in sufficient quantity and quality. (Common Position, 1992)

Even though the Council eventually adopted the second MAC Directive on May 11, 1992, many contentious and unresolved issues remained. The updated Directive was characterized by the *compatibility-led* profile of its predecessor. This again would "double the cost of broadcasting and halve the number of transmitters in Europe" (Peterson, 1993, p. 507). The only way to get through with such a policy was to promise subsidies—to cover the cost difference for conversion and simulcasting and thus secure the project's viability. MAC policy entailed extra costs both to broadcasters—which is why most satellite programmers "would willingly see MAC wither away" ("Down to the Wire," 1991, p. 16)—and to consumers. Paradoxically, while aiming to serve consumers by not making old equipment obsolete, the Directive imposed the extra cost of conversion and simulcasting onto consumers, who would have to pay more for acquiring their TV sets. The Economic and Social Committee (ESC) had advocated that only receivers of DBS services should be obliged to carry decoders and suggested that market forces should be allowed to "act freely" (O.J. No. C 40, 1991), thus pointing to the peculiar blend of *dirigisme* and liberalization which was being applied.

Competitiveness was allegedly served by the Directive, which promoted a "single" new standard. But in fact, there were two transmission standards being endorsed: the transitory 16:9 D2-MAC and the final 1250/50 HD-MAC. Moreover, PAL, SECAM, PALplus, and the prospective digital systems were not out of the picture. On the contrary, the Directive fit them all in a heterogeneous mix (Articles 3, 4, and 5, O.J. No. L 137, 1992). This confusion was indicative of the EU's frustration of attempting to regulate the electronic media at the "supranational" level. How could the "new market" and competitiveness be achieved in this widely fragmented and segmented framework? Cost efficiency, consumer demand, and accessible pricing, all of which would presumably ensure successful introduction and adoption of HDTV, were not in evidence.

The ESC had pointed out that HDTV should be the only available option for advanced television and that it ought to be of *European* conception and ownership (O.J. No. C 40, 1992). But in fact, the chief architects of HDTV R&D, Philips and Thomson, were multinational companies with transcontinental strategic alliances, obligations, and interests. These two Euro-champions were involved in the development of the competing American digital HDTV system in cooperation with American-based firms, as leaked out in February 1991. They became members of the Advanced Television Research Consortium (ATRC), which comprised North American Philips, Thomson Consumer Electronics, NBC, the David Sarnoff Research Center, and Compression Labs Inc. They are now part of the Grand Alliance consortium that designed the U.S. digital terrestrial HDTV system (see Chapter 1). The two firms' participation in the development of the rival U.S. digital HDTV system raises the question of whether these mutually exclusive business strategies were not directly undermining the existence of HD-MAC in Europe. It made perfect business sense for broadcasters and satellite operators to refuse to co-sign the MoU with those who were pursuing mutually exclusive policies. Could Thomson, a state-owned French giant, and Philips both sign the European MoU and be bound by rival transatlantic obligations simultaneously? Why should broadcasters and consumers risk subsidizing Philips and Thomson's EU-95, when these same firms were banking on a rival contender across the Atlantic? Logically an effective *dirigiste* policy should be able to bind such firms to European interests. But as noted below, the concept of *dirigisme*, though descriptive of intentions, is in fact misleading when assessing policies in liberalized, open markets.

Soon after the adoption of the May 1992 Directive, the Commission reckoned that unless funding was given to accelerate the introduction of HDTV, the transition from conventional to advanced technology would occur very slowly (O.J. No. C 139, 1992). These estimates were criticized as grounded on a self-fulfilling prophecy type of survey by UK consultants KPMG. "In each case, then, the body known to have an interest in conclusions reached has had a hand in the reports' preparation" (Flynn, 1992b, p. 2). Anyhow, in accordance with the February Common Position, both the preamble and Article 8 of the Directive stated that financial support should be given but would be dealt with by a parallel instrument, the notorious Action Plan (O.J. No. L 137, 1992). However, the proposal for an AP on the introduction of advanced television services of May 5, 1992 became an immediate source of discord between the 12 European partners. Discrepancy between prior commitment and decisional acts was manifested particularly by the UK government. So time-consuming delays cancelled out the implementation of HDTV "at the earliest possible moment." Member states whose interests were not directly served ignored their original pledges and used these delays to sabotage a possible, medium-term breakthrough for HD-MAC.

Since the February 1992 Common Position, the Council had urged the Commission to intensify contacts with the prospective MoU signatories so that both the Memorandum and the Directive could be signed simultaneously. Negotiations between EU policymakers and broadcasters focused on bartering financial support for development of HDTV hardware through the Action Plan, in exchange for a promise to jump-start the new system and create consumer demand. That is, the aim of the MoU was to condition commitment to advanced television on the basis of the Action Plan. In all, 38 companies signed the Commission's "new and much diluted MoU" on July 20, 1992 (Brown et al., 1992, p. 30). Signatories consisted of satellite operators, broadcasters, and manufacturers. But the ink was not even dry on the agreement when André Rousselet, the owner of the pay-TV service CANAL+, reneged it. "D2-MAC is dead in the short term ... better an association with the Americans today than being forced to manufacture under American license tomorrow," he stated (Flynn, 1992a, p. 2). So the signing of the MoU was never enforced and it was instead supplanted by a nonbinding Declaration of Intention.

The decisionmaking void from May 1992 onwards simply increased the number of those who declared D2-MAC depassé and unmarketable. This situation made suggestions for subsidizing all-digital systems and declaring EU-95 a commercial failure all too plausible. Those on the losing side blamed political inconsistency, inertia, and

sabotage as well as bad publicity for HD-MAC for the unfortunate outcome.[17] During the deadlock, Philips reported a 56 percent drop in
profits for the second quarter of the year, while its share value plummeted ("Philips Profits," 1992). Thomson-CSF for its part was reported to be the last among the 10 worst performing companies of France
at the end of 1992 (*The European*, 1992).

The will to meet prior pledges had vanished. Despite the Common
Position of 1992, the UK Department of Trade and Industry (DTI)
Minister Edward Leigh, chairing the Council, blocked the funding of
the Action Plan on repeated occasions during the second half of the
year 1992. The United Kingdom did not stand to gain much from the
Action Plan (AP). As a program rather than a hardware exporter it
only stood to gain a portion of the 25 percent slice (of the ECU 850 million subsidy) to support software activities rather than from the more
substantial 75 percent slice aimed to fund hardware. It was, moreover,
still equivocal on the European Union. It was Britain along with
Denmark that were delaying both the ratification of the Maastricht
Treaty and the AP. With a possibly earlier than anticipated emergence
of a fully digital HD system and with valuable time running out, the
next option was a compromise for a face-saving reduced MAC subsidy.

The Background to the 1993 Policy Shift

The Council was unable to swiftly replace and reinforce the 1986
MAC Directive with its successor in 1991. When it did adopt it in May
1992, it failed to concurrently approve the Action Plan which was seen
as its indispensable complement. The subsequent series of vetoes
against MAC funding by the EU presidency (held by the United
Kingdom) during the remainder of 1992 deteriorated things further.
Confusion and speculation, thriving on policy paralysis, peaked in the
course of the first six months of 1993.

This climate produced a cascade of perplexing and sometimes contradictory statements. Although MAC was effectively abandoned by
its proponents, attempts were made to present a situation under control. "Neither the U.S. nor European alliances would force the
Community to drop its current support for the analogue based MAC
HDTV standard. Nor would the looming U.S. digital standard become
automatically adopted in the EU," said a DG XIII spokesperson (Fuller,
1993, p. 10). Prior to this, DG III Industry Commissioner Martin

Bangemann stated that "a digital standard doesn't need to be set by Europe" first because "global standards are always the best solution" (Hill, 1993a, p. 16) and secondly because "European companies [are] already bidding to produce the U.S. system" (Hill, 1993c, p. 3). At any rate, "digital HDTV might not arrive until 2005," argued Dominique Strauss-Khan, the French Industry Minister (Hill, 1993b, p. 3). Martin Bangemann further amplified that sentiment by stating that "despite the promise of digital technologies, D2-MAC and HD-MAC are still judged by experts to have excellent performance and they are available *today* [emphasis added]" (Kehoe, 1993, p. 16).

The succeeding Danish presidency in 1993 undertook the brave initiative to open the way to digital systems in the EU rather than just stalling HD-MAC. Danish DTI minister, Helge Israelson, warned of "chaos" if swift action were not taken to set digital HDTV standards. Apart from breaking the impasse this agenda-setting initiative opened up opportunities for the promising Scandinavian digital HDTV system, HD-DIVINE, and liberated the EU from a failed strategy (Hill, 1993c).

This long sought-after policy shift was precipitated by technological, market, and political moves. First, digital spectrum bandwidth compression allowed as many as eight channels to fit into a bandwidth previously required for one. This proliferation was commercially profitable instantly and heralded the quicker advent of digital HDTV. Not only did digital bandwidth compression offer eight-fold enabling multimedia, but it was also the catalyst that made MAC redundant because of digital technology's absolute superiority on all counts. A second event was the creation of the European Launch Group (ELG) which soon evolved into what is now the Digital Video Broadcasting (DVB) group. Finally, in the face of irresistible digital prospects and of the political deadlock of 1991-1993, MAC proponents, Philips and Thomson, abandoned their HD-MAC goals and opted for widescreen 16:9 TV sets with lower definition. A year after the abortive attempt at establishing the MoU, Philips admitted that the failure of MAC was due to the absence of economic advantages for programmers and satellite operators, thereby acknowledging a mistaken business strategy and stating that its role in developing EU standards for satellite HDTV would diminish in the future ("Philips Opts," 1993). In early 1993, Philips and Thomson joined the ELG-DVB forces together with such prominent anti-MAC players as the British DTI and the German public broadcasters ARD and ZDF (Fuller, 1993).

In the light of mounting corporate and political initiatives the EU had no choice but to retreat from its standards strategy. The anti-MAC lobbying could no longer be contained, and the dice was cast against MAC. Meanwhile, Richard Wiley, Chairman of the Federal Communications Commission (FCC)'s Advisory Committee on Advanced Television Service, called for a corporate/political entente at the Montreux MIP-TV of 1993 (Doyle, 1993). After the 1991-1993 policy deadlock, widescreen 16:9 was the only common vehicle upon which the industry could ride. "There is a general agreement that the future television will be widescreen in the 16:9 format regardless of the transmission medium, technology used or standard. This will be the foundation for a range of advanced services in the decades to come" ("Commission Spells Out," 1993, p. 4). The 16:9 format, close to the golden mean and compatible with the cinema format, was commercially interesting for professional applications of digital technologies (e.g., substitution of traditional cinema by electronic cinema).

Into the Digital Era: The Rise of the DVB Group

The birth of the Digital Video Broadcasting (DVB) group, coupled with the standardization developments for a digital terrestrial TV system in the United States (see Chapter 1), mobilized European entrepreneurs into facing the all-digital challenge. Attempts to build up a forum for concerted action started in 1991. European media interest groups from the converging information industries gradually came together to form the European Launching Group (ELG). They then drafted a Memorandum of Understanding, establishing the rules for their collective action. "The concept of the MoU was a departure into unexplored territory, and meant that commercial competitors needed to appreciate their common needs and agendas" (Digital Video Broadcasting [DVB], 1995, p. 4). In September 1993, ELG Secretary, Armin Silbernhorn, announced that the prerequisite 12 signatories to the MoU had been attained (Flynn, 1993b). So the ELG was renamed the Digital Video Broadcasting (DVB) group. It soon attracted all those interested in developing an all-digital TV system that would be grounded on a unified approach. DVB was broad and diversified enough to be considered a "milestone": a virtual convergence of interests in action (Flynn, 1993b).

DVB then rapidly proceeded to create its organizational structure.

A General Assembly, the ultimate decisionmaking body, was to meet once a year, and a Steering Board overseen by it would control day-to-day work. There were four "modules" accountable to the Steering Board and focusing on specific activities: (1) the Terrestrial Commercial Module; (2) the Commercial Satellite/Cable Module; (3) the Interactive Services Commercial Module; and (4) the Technical Module (DVB, 1995). As of mid-December 1995, DVB membership counted more than 195 organizations from 23 countries, representing a wide cross section of the electronic media industry (e.g., broadcasters, manufacturers, regulatory bodies).

The DVB group aspired to advance the standardization of all major technologies used in digital TV and HDTV transmission, and to coordinate all initiatives that would culminate in a digital Action Plan to be presented to the EU Commission. This aspiration was backed by a wide range of entrepreneurs. Satellite and terrestrial operators anticipated standardization of the channel coding and modulation technologies in accordance with the MPEG-2[18] standard proposal, considering that stable standards have positive implications for the cost and the availability of equipment (Healy, 1993). Indeed, the DVB group adopted MPEG-2 in the autumn of 1993.

With standardization as one of its declared objectives, DVB proceeded with the testing and approval of specifications for satellite (DVB-S) and for cable (DVB-C) transmission systems. In December 1995, the DVB Steering Board approved a digital television standard (DVB-T) for terrestrial television using OFDM (orthogonal frequency division multiplexing) modulation technology. This specification uses the same generic elements as the DVB-S and the DVB-C systems ("Editorial," 1995). The DVB-T standard was approved by the European Telecommunications Standards Institute (ETSI) in January 1997, and it is estimated that digital terrestrial services could begin as early as 1997. Because the major part of the digital standardization process has been completed, DVB members are now gearing up for the launch of new products and services.

The thrust to set digital standards by market forces demonstrates DVB's determination to self-regulate, thereby assuming a direct politico-economic role. This was encouraged by the EU Council which explicitly welcomed the role of market agents in policymaking. It declared that "the preferred way to achieve the objective of harmonious market development ... would be by means of a consensus process involving all relevant economic agents including broadcasting organizations, and that it looks forward, with great interest, to any voluntary agreements which may be made by such agents in this regard" under the auspices of the European Digital Video Broadcasting Project (O.J. No. C 181, 1994, p. 4).

The Transition from a Single Standard to a Plurality of Standards

After several months of intense internal controversy the EU member states agreed on abandoning preferential treatment of the MAC system, thereby burying the neo-mercantilist delusions that had nurtured it. This bold step broke a prolonged policymaking paralysis and promised a commitment to digital technology. The momentum of an open-ended economy and wider, if not global, capital integration, combined with the potential of digital technology, had finally defeated EU analog plans. Once policy deadlock was overcome the transition was swift. It involved four policy initiatives: the Resolutions of 1993 and 1994; the Action Plan of 1993; and the Directive on the use of standards of 1995. These four policy instruments founded the structural coordinates of the new era. We shall examine each of these in turn.

The Council Resolution of July 22, 1993 essentially decided to replace the 1992 MAC Directive and to establish "timely mechanisms" and implementation measures for the new digital era (O.J. No. C 209, 1993). It targeted six "possible needs":

(1) to expand the scope to allow other standards, in addition to D2-MAC, to be used for the broadcast of not completely digital 625 line television services in the 16:9 format;

(2) to expand the scope to cover standards for terrestrial transmission and cable distribution;

(3) to limit the number of different standards as far as possible;

(4) [to establish] a European non-proprietary encryption/conditional access system serving a number of competing service providers;

(5) [to standardize] all new television transmission and encryption systems to be used in the Community by the competent European standardization bodies;

(6) to change other Articles of the Directive to ensure consistency following any changes introduced under the above provisions. (O.J. No. C 209, 1993, pp. 1-2)

The problem is that some of these options are mutually exclusive and cannot be concurrently pursued. Furthermore, by allowing a multiplicity of transmission standards, including D2-MAC, the Council still remains ambivalent in its standards policy. This "openness" or lack of clear direction is symptomatic of the malaise that had crippled the Council's HDTV decisionmaking process since the adoption of the second MAC Directive.

Yet, an acknowledgment of the shortcomings of previous policymaking can be "read" in the four calls, in the short space of the preamble, for "coherence" in the new policy. The 1993 Resolution unequivocally put digital television at the top of the future EU agenda. It provided for the single common element among the many competing technologies: the 16:9 screen format. In its second provision, the Council Resolution required the Commission to consider:

(1) the early agreement on common Community perspectives for the digital television market, most notably the feasibility for a single (or family of) standard(s) and encryption system(s);

(2) a timetable for development, specification, implementation, and evaluation of the new digital system; and

(3) the possible need for Community funding of new projects. (O.J. No. C 209, 1993)

Because of the new wisdom that "digital technology is essential for future television systems," the review of the second MAC Directive was precipitated so as to bring it into line with prevailing market and technological conditions. The Council therefore required a new Directive proposal reflecting both the necessity for a "flexible and workable regulatory framework" and the needs of the market in light of current technological developments before October 1993. It was stated that Europe needed a "coherent global approach" to the development of technology and standards for new digital systems (O.J. No. C 209, 1993, p. 1).

The Action Plan of 1993: Minimal and Neutral

In accordance with the concurrently adopted Resolution, the Action Plan applied exclusively to the promotion of the 16:9 format (625 or 1250 lines) irrespective of standards and of broadcasting modes employed (terrestrial, satellite, or cable). "The Action Plan should facilitate the uptake of all technologies, including fully digital technology" (O.J. No. L 196, 1993, p. 48). Notwithstanding the strategic preference for digital technology in the Resolution, the Action Plan celebrated the newly found doctrine of "neutrality." The allocation of funding did not favor any particular transmission standard or mode. This would allow MAC manufacturers to recover some of their losses.

This time the stated objective was modest, primarily fostering a critical mass of programming in an advanced TV format (widescreen) by 1997. This was a good policy cocktail from the point of view of manufacturers and broadcasters alike. ECU 405 million was thought to be a sufficient amount for the moderate objectives of the new AP. The EU was to allocate ECU 228 million to broadcasters and producers, who themselves would put up the remainder (ECU 177 million). This EU contribution would cover just under 50 percent of the cost difference between production or broadcasting in the ordinary 4:3 format and the 16:9 format. It corresponded to just over a quarter of the ECU 850 million proposed for the original Action Plan while no funding would go to manufacturers (O.J. No. L 196, 1993). This reversal of the funding terms was linked to the "new wisdom" that the sector was software-driven. Manufacturers would benefit indirectly.

The uptake of AP funding for the 1993 call was reportedly excellent. Eleven broadcasters received funding. Together they promised to offer 100,000 hours of TV programs in widescreen format over the four years of the AP's duration. The acceleration of the migration from the 4:3 to 16:9 format was thus on course. This migration was also expected to bridge the passage from analog to digital technology (Brown, 1994). Despite this success, the EP decided, within its budgetary controlling prerogatives, to halve AP funds in June 1995 (proposal of MEP Roy Perry). The rationale was that the market had responded by giving priority to "obsolete technologies" and that the 16:9 lowest common denominator was not sufficient to push either the electronics or the audiovisual industries in Europe (L. Castellina, President of the EP Media Committee, personal communication, November 4, 1995). Indeed, out of 25 broadcasters transmitting programs in widescreen format in 1995, 20 used PALplus and 5 used D2-MAC (Renaud, 1995a). MEP Gérard Caudron later argued that the EU still lacked the technological basis for the introduction of digital HDTV and that the wider aspect ratio in itself was not a sufficient enticement to stimulate consumer demand in advanced and more expensive television sets (Leclercq, 1994).

The Information Society: New Rules for a New Game

"The information society is on its way. A digital revolution is triggering structural changes comparable to last century's industrial revolution with the corresponding high economic stakes. This process cannot be stopped and will lead eventually to a knowledge based on economy" (CEC, COM(94)347, 1994, p. 1b). This introductory sentence of the information society framework of the EU maps concrete politico-economic transmutations while reflecting a technological determinism.

On June 24-25, 1994, the Corfu European Summit unanimously adopted the report and the recommendations of the high-level group set up by Commissioner Martin Bangemann. On this basis the Commission designed, and announced a month later, its framework called the *Action Plan on Europe's Way to the Information Society*. This document provided that "Public authorities [Member States] will have to set new 'rules of the game,' control their implementation and launch public interest initiatives." The aim of this exercise would be "to back up this development by giving a political impetus, creating a clear and stable regulatory framework and by setting an example in areas of their own responsibility" (CEC, COM(94)347, 1994, p. 1b). The need for an acceleration of the liberalization process was by far the most prominent new rule of the Action Plan which, moreover, required that "the deployment and financing of an information infrastructure will be the primarily [sic] responsibility of the private sector" (p. 1b). Apart from the acceleration of liberalization, the Action Plan guidelines sought a "global coherence," encompassing cooperation with global partners such as the United States and Japan, and introduced a climate of trust and urgency. It stated, moreover, that "emphasis would be placed on action at the EU level, but collective effort would also be needed involving all interests in the Member States" (CEC, COM(94)347, 1994, p. 12). This delineation of responsibilities between private and public actors is virtually a political manifesto for less state and more laissez-faire. In addition, the Action Plan acknowledges the role of the DVB group, particularly with regard to standardization and issues of electronic legal protection.

The 1994 Framework for Digital Video Broadcasting

In the wake of the Bangemann group elaborations which asked for "clear and stable" rules, the Commission submitted a draft Resolution on a framework for EU policy on digital video broadcasting in November 1993 (CEC, COM(93)557, 1993). This draft repeatedly invokes the need for an "orderly regime." Yet the qualifier "orderly" was substituted in the adopted document in three occurrences by the word "harmonious," and even deleted in another occurrence (O.J. No. C 181, 1994). This semantic intervention signals perhaps a departure from conceptions of orderliness and the associated difficulties with regard to regulation. Certainly it is easier to ask for "harmony" than to impose "orderly market conditions."

Essentially the 1994 Framework Resolution provided for officially incorporating market forces, notably the DVB group, in the processes of policy formulation, standardization, and implementation. It stated that "the preferred way to achieve the objective of harmonious market developments ... would be by means of a consensus process involving all relevant economic agents, including broadcasting organizations" and that "it looks forward, with great interest, to any voluntary agreements which may be made by such agents in this regard" (O.J. No. C 181, 1994, p. 4). The Council, however, was willing to introduce regulatory measures if market forces failed to do so consensually or if the requirements of fair and open competition, protection of consumers, and the public interest required it (O.J. No. C 181, 1994; see also Leclercq, 1994).

The second major element of the Resolution was the endorsement of the new "global" strategy of cooperation. The Council welcomed the Commission's intention to "dialogue with third parties including the United States of America and Japan" (O.J. No. C 181, 1994, p. 4), aiming at identifying and agreeing on common terms for worldwide TV systems in the future. This shift is justified by the new wisdom that digital television technology is a world phenomenon and that "the search for the highest degree of compatibility between the various regions of the world in this area is a desirable objective" (O.J. No. C 181, 1994, p. 3). To that effect minimal regulation, smoothness, and flexibility summed up the goals of the new approach.

The 1995 TV Standards Directive

The October 1995 Directive on the use of standards for the transmission of television signals effectively repealed the 1992 MAC Directive. With the exception of the provisions on conditional access, it essentially introduced no new policy. It was a celebration of de facto standard-setting.

Although this Directive was grounded on the defeat of its predecessor, on the heralded arrival of fully digital television services, and on the "chaos" that would ensue if the framework for the arrival of digital technology were not duly prepared, this Directive remains equivocal in its choice between analog and a fully digital HDTV technology thanks to the doctrine of "neutrality." In response to concurrent market demands the actual issue of the agenda changed. From an issue about HDTV and single transmission standards it became an issue on the management of digitally-enhanced and multiple delivery systems, notably subscription services as opposed to universal access services. A proprietary access system, SimulCrypt (which is liaised with PALplus), was selected by default of decisionmaking ability both in the DVB and in the institutions of the Community. This decision indicated a gravitation towards PALplus, despite an official policy of neutrality. Hence, the outcome of regulation so far tended to favor traditional broadcasters and standards in the pay-per-view (PPV) services. This is undoubtedly a triumphant vindication for the non-fully digital and anti-MAC forces which coincidentally are the leading forces of the Association of Commercial Television (ACT).

Technological neutrality—obtained by killing the objective of the single (standard) market—raised other types of political and economic problems. It actively favored fragmentation and non-tariff barriers to trade in TV services, which did not suit broadcasters. "We may need to broadcast in several different digital multiplexes [a multiplex is a group of frequencies/channels] in order to reach across all Europe— even less standardization and a less common market in television than we have today in PAL, SECAM and D2-MAC and we would lose the transponder cost savings promised by digital compression" (Morgan, 1995, p. 2).

Articles 1 and 2 regarding the scope of the old Directive and the MAC standard were repealed by the new Directive. This was broadened so as to include all operating transmission standards and all modes of transmission—satellite, terrestrial, and cable. In the hypothetical event that a broadcaster would venture to launch analog HDTV services these ought to be in the HD-MAC standard. With regard to TV programs, Article 1 admonished the governments to promote the transition to advanced and digital television services. Article 2 stipulated that:

> All television services transmitted to viewers in the Community whether by cable, satellite or terrestrial means shall: (a) if they are in wide-screen format and 625 lines, and are not fully digital, use the 16:9 D2-MAC transmission system, or a 16:9 transmission system which is fully compatible with PAL or SECAM. ... The 16:9 format is the reference format for wide-format television services; (b) if they are in high definition, and are not fully digital, use the HD-MAC transmission system; (c) if they are fully digital, use a transmission system which has been standardized by a recognized European standardization body. (O.J. No. L 281, 1995, pp. 52-53)

Article 3 required that television sets larger than 42 cm (diagonal) sold or rented on the Community market "be fitted with at least one open interface socket (as standardized by a recognized European standardization body) permitting simple connection of peripherals, especially additional decoders and digital receivers" (O.J. No. L 281, 1995, p. 53).

Article 4, which was as long as all the other articles combined, addressed the thorny issue of conditional access (CA). Since the Council had, in putative terms only, asked the Commission to consider "the possible need for a European non-proprietary encryption/conditional access system serving a number of competing service providers" (O.J. No. C 209, 1993, p. 1), the Commission in its Directive proposal shunned this issue that could have preempted decisions in the DVB group. This position was congruent with the prior Community invitation to DVB to resolve such contentious issues on a voluntary and consensual basis.[19] But as noted by the Commission (CEC, COM(94)455, 1994), the DVB group was unable to hammer out a compromise on conditional access.[20]

The first EP opinion on the proposed standards Directive came on April 19, 1994 and was published in the *Official Journal* in May 1994 (O.J. No. C 128, 1994), when DVB had not yet issued its final position on conditional access. Meanwhile, fears had been expressed that major pay-TV operators could abuse their dominant positions if a CA system were only optionally linked with receiver equipment. "Leaving the interface out could be a condition of purchase or subsidy" (Leclercq, 1994, p. 4). Thus it was suggested that, so long as inclusion of a decoding interface remained *optional*, that is, not specifically regulated, proprietary CA systems' operators could make deals with TV set manufacturers to include interfaces exclusively compatible with their system (de facto standard-setting). The EP amendment dictated that interfaces in TV set equipment be mandatory. Still, CA regulation was left at the decoding level, that is, at the reception end, but restrictive control for access can particularly occur at the encoding level (see Kaitatzi-Whitlock, 1997). "All consumer equipment ... shall possess capability to allow descrambling of such signals," and manufacturers should include conditional access systems "on fair, reasonable and non-discriminatory terms" (CEC, COM(94)455, 1994, Article 3bis iii). Interestingly then, the EP avoided to preempt DVB and did not favor the establishment of alternative conditional access systems, such as the open-accessed MultiCrypt, that would have allowed CA at the encoding level (Article 3bis).

The EP stressed, moreover, that fully digital TV systems should use a common transmission standard "recognized" by a European standardization body and be defined prior to their introduction (CEC, COM(94)455, 1994). The EP did not want DVB to replace EU standardization bodies even though it praised its role in policy formulation (Leclercq, 1994).

The second-reading EP amendments of June 13, 1995 focused again on the controversial issue of conditional access owing to DVB's inability to vote on a single system. The EP amendments embodied in its added Article 3bis were eventually incorporated in Article 4 of the final Directive. These mandated: (1) that consumer equipment for descrambling digital television signals display them transmitted in the clear (Article 4(a)); (2) that conditional access service providers offer all broadcasters, on a fair, nondiscriminatory basis, technical services enabling the broadcasters' digitally transmitted services to be received by viewers by means of decoders; (3) that separate accounts be kept for CA businesses; (4) that holders of industrial property rights to CA systems be prevented from prohibiting, deterring, or discouraging manufacturers from including a common interface allowing connections with several other access systems; and (5) that an appropriate dispute resolution mechanism be set up to deal with potential conflicts between manufacturers and property-rights holders of CA systems. Thus, it was thought that while endorsing the proprietary SimulCrypt system, the EP alterations conferred satisfactory elements of fairness, transparency, and dispute resolution to this aspect of the policy (M. Wagner, Member of the EBU Legal Committee, personal communication, October 12, 1995).[21]

It may be asked, then, what really changed with the new standards Directive? With a unitary, rational, and incremental policy actor we should expect a rationale for each crucial strategy and policy shift. The Directive preamble makes reference to the importance of establishing "common standards for the digital transmission of television signals ... as an enabling element for effective free market competition" (O.J. No. L 281, 1995, p. 52). Two sentences in the preamble to the new Directive are revealing. "Whereas, in order ... to contribute to the proper functioning of the internal market ... it is necessary to take steps to adopt a *common* [emphasis added] format for wide-screen transmissions" (O.J. No. L 281, 1995, p. 51). But an instrument whose most crucial contribution is to enshrine and augment the fragmentation (through neutrality) of the market cannot invoke the principle of the Single European Market (SEM). Or else, given the fact that the International Telecommunication Union (ITU) has endorsed the widescreen format, the SEM could theoretically extend globally, wherever widescreen is adopted. But obviously this is not a sufficient condition for the SEM. The second sentence, contradicting the first, states that: "Whereas the Presidency's conclusions at the G7 Conference on the Information Society ... highlighted the need for a regulatory framework ensuring open access networks and respect for competition rules" (O.J. No. L 281, 1995, p. 52). Common sense suggests that open systems do not constitute internal markets. Yet the EU builds its standards policies on this blatant contradiction, failing once again to achieve coherence in its policymaking.

Digital But Not High-Definition Television

In the process of amending the 1995 Directive the EP reiterated the importance of setting up a European body in charge of the allocation of digital frequencies and the rationalization of the EU satellite capacity (Leclercq, 1994). In a fragmented European territory such an initiative could have strengthened European integration and led to coherent and rational management of these "strategic" resources. Frequency allocation is handled nationally and the communications sector is therefore not the best paradigm of a "common policy."

In practical terms this entails the already observable trend towards regional economic disintegration. Some member states are ahead in digitalization while others are left behind, unable to attract global investments. International companies are poised to introduce digital services and derive the advantages accruing from multimedia technologies and from being first-comers in the most appealing markets of the Union. The UK government, an active proponent of this trend, has already drafted its final plans for digital frequency allocations and bidding terms. Eighteen new digital channels are to be distributed, and during a transitional period digital service providers will need to simulcast in the current analog systems. These new channels are to be shared out among all incumbent broadcasters, with 3½ being open for bidding. In June 1997, the Independent Television Commission, the UK television regulatory body, awarded all three commercial multiplexes (groups of about five frequencies/channels) to British Digital Broadcasting, a joint venture of the Granada Group and Carlton Communications (both ITV companies) (McCormick, 1997). Once diffusion of digital television sets reaches 50 percent penetration, simulcasting is to be phased out. Like its U.S. counterpart, UK terrestrial analog spectrum will then be devoted to other lucrative uses ("Digital Era," 1995).

Unlike the United Kingdom, Italy is banking on *satellite* digital services. In late 1995, Telepiù conducted experimental digital satellite television services via EUTELSAT II F1. International media consortia there, with Nethold[22] in the forefront, now shareholders of Telepiù, are busy building up the commercial pay-television model of the future. The chosen path to success is simulcasting with a terrestrial transmitter during the initial phase. On offer to consumers is a bouquet (i.e., package) of thematic channels including CNN, TNT, and MTV (Muscarà and Causin, 1995) (see Chapter 1).

What about digital HDTV? Though the digital fever has taken Europe by storm, HDTV service projects, notably digital, are at present nowhere to be seen. This absence is due first to an unfavorable cost/demand relationship, and secondly to a lack of programming. Optimistic market forecasts had suggested that HDTV receivers would be available by the end of the century. In any case, both HDTV-only or simulcast HDTV services are conditional on substantial initial investments in production and transmission.

> Significantly, with regard to introducing digital TV and HDTV ser-
> vices, there are more immediate plans for introducing the former
> rather than the latter through all industry groups. 35 percent of cable
> operators, 60 percent of satellite operators and 14 percent of terrestrial
> broadcasters said they plan to introduce digital TV before 1998. As
> regards the introduction of digital HDTV proper, only cable operators,
> 31 percent of them, plan to do so before 1998. (Healy, 1993, p. 12)

These predictions suggest that the introduction of HDTV requires either a higher level of vertical integration of potential entrepreneurs or long-term state intervention. Long-term and large-scale projects, which moreover extend over a number of economic branches or even related converging sectors, involve exceptionally high financial risks in terms of sunk costs and upfront investments. This type of risk can only be undertaken with the financial backing of public authorities that can afford slow and long-term amortizing, or by an adequately consolidated consortium holding effective (via vertical integration) or legal (via political assignment) monopoly guaranteeing the return of profits on its vast initial investment. Given, however, that this indus-try requires by definition a pluralism of information sources and out-lets, private as well as public, at least in the democratic Western states this solution would be extremely unorthodox.

Notwithstanding the lack of large-scale business projects, digital HDTV R&D is plodding along. The EUREKA 1187 Advanced Digital Television Technologies (ADTT 1187) project that has been established in the wake of EU-95 aspires to keep Europe abreast of technical exper-tise and make the European industry an active participant in the devel-opments of digital broadcast and nonbroadcast systems. ADTT 1187 is in effect the inheritor of EU-95, benefitting from the achievements of its predecessor. So it too covers production, studio, transmission, and reception phases of the TV chain and builds on technical principles and specifications adopted by the DVB group.

European MEPs have suggested that a new regulatory framework for fully digital systems be drawn up as soon as the Directive had been revised (Leclercq, 1994). As pointed out by MEP Gérard Caudron there is a conflict between quantity of channels and quality (definition) of picture. Worse still there is a disproportionate relationship between channel proliferation and a lack of "meaningful" programming. The EU that aspired recently to a coherent set of communications policies (information society Action Plan) now has to address these conflicts.

So far it has failed to structurally support the programming end of the industry, restricting itself to inopportune and transitory support programs of questionable value. It is perhaps for this reason that entrepreneurs do not foresee high demand among their subscribers for digital services, even with the same subscription cost (Healy, 1993). This forecast confirms Gérard Caudron's doubts on the ability of widescreen alone to achieve a commercial breakthrough. The "short-sightedness" of which he accuses the Commission is not ungrounded. The EU, in spite of its recent digital fever, still has "no basis for the introduction of digital HDTV" (Leclercq, 1994, p. 4).

Discussion and Conclusions

The European HDTV industrial policy project and its associated common transmission standards approach—as part of the SEM project—were placed at the highest political level of the EU. This policy ensemble aimed at realizing an overarching strategy that encompassed regulatory, technological, and cultural objectives. More specifically, the HDTV and standards initiatives were one part of a European audiovisual policy triptych (HDTV, the TWF Directive of 1989, and the MEDIA programs of 1988, 1992, and 1995). But in spite of sustained rhetoric to the contrary this part has not been adequately coordinated with either of the two other parts of the EU audiovisual policy.

Moreover, apart from anything else, the root problem of this project was that it was dominated by the hardware satellite and electronics industry and by the member states which had set those priorities. This bias in favor of the manufacturing sector and the corresponding lack of consideration for program making and for all non-DBS broadcasters fostered furious reactions against MAC which took the form of both techno-nationalism (rival national overt or covert strategies by member states such as Luxembourg and the United Kingdom) (see the 1986 MAC Directive discussed earlier), and intra- and intersectoral rivalries.

The condition of competing interests and approaches was most clearly manifested in the loophole left in the first MAC Directive. To reach a minimum agreement, member states left an unregulated loophole by exempting the FSS frequency band from MAC-specific regulation. Such a policy blunder clearly demonstrated the existence of fierce techno-nationalism and fierce inter- and intrasectoral competition and tensions in the face of coveted market shares in the European and global markets. Ever since that first policy "omission" two parallel and competing advanced television policies were on course in the EU: the official EC policy versus its opponent embodied by commercial media moguls and the rival national policies of the United Kingdom and Luxembourg.

Hence, two competing policy approaches were tested in this case, the "neo-mercantilism" of the Franco-German axis and the "neo-liberal Atlanticism or globalism" of the British and the Luxemburgers. Both drew Europe away from a potential policy course that could have welded the region together by serving the common European interest. Technological innovation compounded the damage at a later stage. Therefore, this European strategic failure was only secondarily due to the fact that "HD-MAC reached obsolescence as it reached technical maturity" (J. L. Renaud, editor-in-chief of the newsletter *Advanced Television Markets*, personal communication, September 25, 1995).

In this policy case we are witnessing the discarding of a policy-making methodology: the use of standards for integrating the regional European market. This is evidenced in the new "doctrine" of neutrality in transmission standards, which blessed the current heterogeneous multistandard regime. The latest Directive adds nothing new on transmission and limits its focus to a small but growing segment of the broadcasting market: pay-per-view and subscription services. The significant shift is that standards are no longer politically- but market-developed. The DVB group, an industry-led organization, created and adopted a family of digital standards (e.g., DVB-C, DVB-S, DVB-T) subsequently submitted to the European Telecommunications Standards Institute (ETSI) merely for technical approval and codification. But if standards remain similar to those obtained in July 1993, what "chaos" did the EU forestall or deter with its technology policy since the Danish presidency of the first semester of 1993? And what has actually changed in the "new" digital era with respect to the objectives of the Delors White Paper on growth, competitiveness, and employment? None of these objectives could be served with the neutrality of standards and the resultant fragmentation of the market. Such market heterogeneity entailed the wasting of resources (by splitting the scale of the market and by multiple tentative investments) rather than securing conditions for the promotion of competitiveness and growth.

Europe's HDTV project failed. But it tells us much about the political power structuring in the fledgling superpower called the

European Union, whose political institutions are, it must be remembered, in their infancy. To further probe into the power relations between the various players involved in the HDTV controversy and flesh out a possible theoretical framework for analyzing how HDTV policy outcomes came about in the Union, the following paragraphs discuss the application of two power models to the EU HDTV decisionmaking.

According to the pluralist model of decisionmaking, of which Robert Dahl is the principal proponent, competing groups of interests in society struggle to influence the outcomes of government policy. These groups are viewed largely as being equally influential and possessing similar or the same means for influencing policy results. This model then focuses on "behavioral" or agency factors only, in analyzing the competition for specific policy results, thereby ignoring significant economic-structural divisions and constraints and associated dominant positions. Economic power as an unequal relationship, which other than reflecting also reinforces inequality, is disregarded in this conception.

Specifically, Dahl (1961) defines power as what enables **A** to make **B** do what **B** would not do otherwise. In our case study, **A** includes MAC proponents, such as interested member states (France, Germany, the Netherlands), the proponent manufacturers (Philips, Thomson, and Nokia), and the EU Commission, notably the DG XIII. **B** comprises the anti-MAC camp, including SES/Astra, Sky Television (later BSkyB), CANAL+, RTL, SAT 1, ACT, and the British, Luxembourg, and Danish governments representing divergent national interests. But as our analysis pointed out, **A** unsuccessfully attempted to make **B** do what **B** would not otherwise have done. In fact, a reversal of positions occurred in terms of agenda setting and policy outcomes, because **B** made **A** do what **A** would not otherwise have done. **B** managed to make **A**: (1) give up its HDTV project; (2) repeal its common MAC standards policy; (3) drop transmission standards-setting as a (regional and SEM) policy tool; and (4) concede parts of its policy prerogatives (e.g., standards-setting) to market and "global" forces. Thus, Dahl's model is not well suited here because it is too restrictive. By focusing on behavior (e.g., Community decisions and actions) only, it ignores crucial structural inequalities and in-built biases of the competing forces in the policy process (e.g., the fact (1) that MAC opponents transmitting in PAL were incumbent and well entrenched in the market; (2) that British Telecom decided to join Astra; and (3) that the

Independent Broadcasting Authority (IBA) did not grant renewal of the BSB license to transmit in MAC).

Unlike Dahl, Lukes (1974) stipulates that **A** exercises power over **B** (1) when **A** affects **B** in a manner contrary to **B**'s interests, or (2) when **B** remains unaware of the conflict of interests and the nature of the power relationship between **A** and **B**. The purpose of Lukes' three-dimensional model of power and decisionmaking is to demonstrate that power relations are not merely "behavioral," and to highlight unequal economic and political power relations which, while shaping outcomes, are not readily tangible and are not attributable to one set of transparent and negotiable factors. Specifically, Lukes looks at the *structural in-built biases between the competing forces* in the policy process. As stated in the paragraph above, one example of in-built bias in our case study was the 1986 decision of British Telecom (BT) to join Astra, a rival consortium of the MAC project, and to abandon EUTEL-SAT. This was a counteracting move outside the policy scope of the formal and common EU policymaking procedures. As a result, the vested interests of BT were directly linked to those of the anti-MAC camp, which might explain why the UK government in EU Council meetings resisted MAC from that moment onwards. Moreover, Lukes' model focuses on *control of the decisionmaking agenda*, either directly (e.g., the FSS loophole dictated by the anti-MAC camp) or indirectly— not necessarily through formal policy decisions—(e.g., BT's joining the Astra group and Murdoch's key choice to transmit in PAL). Finally, it factors in *potential issues* (e.g., the Community's failure to orchestrate all three prongs of its audiovisual policy triptych in such a way as to make the final outcome mutually supportive rather than internally biasing or counteracting). For instance, the potential issue of setting up HDTV program production measures in the TWF Directive and/or in the MEDIA programs was not addressed, precisely because of the restrictive EU decisionmaking structure. The "brake" of the principle of unanimity precluded agreement on such issues as support for contents. Quite reasonably then, lack of available programming in the new format was the most often invoked pretext by broadcasters for not adhering to the MAC standard. The same reason accounted for the missing consumer demand of advanced TV services. A lack of consumer demand is also linked to another potential issue: the cost/benefit ratio for the consumer. This would seek avoiding the dual MAC/HD-MAC standards that could have split the market and created additional costs (for decoders) for the consumer. Similarly, potential issues derive from MAC's too intimate relationship with the DBS satellite transmission mode.

Lukes also examines *observable (overt or covert) and latent conflicts* and the notions of *subjective and real interests*. The case of the two Euro-champions (Philips and Thomson) joining forces with the rival digital HDTV project of the United States constitutes a case of a double agenda (which was also a hidden agenda until leaked out) and covert business strategy which went against the real interests of the European taxpayer who was asked to subsidize the production of HDTV hardware.

The controversial conditional access system issue, which eventually evolved quite differently from what was originally intended, constitutes yet another case of originally ignoring real interests in the Lukesian sense, at two levels: first the real business interests of competing service suppliers and secondly the real interests of the viewers whose choice would be drastically reduced by the establishment of a proprietary monopoly supplier system (the SimulCrypt CA system), not subject to internal pluralism regulations.

The political dilemma facing the EU was succinctly put forward by the Commission to the Council and Parliament: "The Community is thus faced with a major political choice: can this transition [to digital technology] be left to the market alone, where time scales for return on investment are typically much shorter than the time needed to make this transition in an optimum fashion, or is it appropriate for the Community to develop a longer-term vision of the future global networks and encourage market parties to make their investments and market offerings in ways which are compatible with such a long term vision?" (CEC, COM(93)557, 1993, p. 2). Policy developments in the second phase (1992-1995) suggest that the EU has been led to the first course of action, without offering a coherent rationale for it. This choice demonstrates the operation of structural and agency constraints beyond the control of the "supranational," intergovernmental policy actor, which counteract the EU's framework of regional integration. Moreover, in Lukesian terms, this choice constitutes a case of ignoring the potential and latent conflicts arising from a hands-off policymaking style which progressively makes the European economy dependent on imported hardware and software. HDTV policymaking in Brussels is thus marked by a "dialogue" with market forces and with former rivals from abroad.

Although Lukes' model of power relations and decisionmaking is a useful tool to uncover complex relations and influences on the course of policy formation, it is too static to capture certain policy aspects specific to rapidly changing developments in our particular historical case. This model fails to capture the *dynamic* element of timing as a strategic tool to influence and to produce policy outcomes. Indeed the tool of timing has been used by a number of policy actors in different moments of our case. Three typical instances can be recalled: first, the decision of Luxembourg to set up SES/Astra while concurrently promising to go along the Franco-German axis on the DBS project; second, the timing-wise strategically accurate decision by Rupert Murdoch to transmit in PAL; and third, the use of the timing factor by the British government (and presidency of the EU) in its consecutive vetoing of the MAC Action Plan.

In addition, the Lukesian policy analysis model does not cover aspects of implementation. But our case study clearly demonstrates that special attention to implementation of decisions is necessary. First, we are not dealing here with a unitary government with cohesive power and unquestionable legitimacy, but with a challenged "supranational" actor or regime, which, moreover, is still developing. Second, the historical case under examination takes place in the era of deregulation, of safe havens, and of calculated illegality. Hence, only stringent examination of implementation of decisions can illuminate the degree to which the EU commands legitimacy in this particular sector. It is undeniable that in terms of implementation the first two Directives did not command the required legitimacy and application.

Even at the time when the SEM framework objectives were at their highest political validity, European "supranational" policymaking failed to impose a single and common advanced television transmission standard for the new media. Since then there has been a steady decline in commanding regulatory legitimacy in the EU. The loophole which was left for medium-power transmission satellites and the reluctance to close this loophole and take due corrective measures promptly contributed to establishing a parallel anti-MAC agenda right from the inception of the "European global standards strategy." Consequently this policy cannot be considered *dirigiste* in the traditional sense.

Had the EU promptly corrected this failure which had been identified by the Commission, it could have been argued that EU policymaking was incremental and that EC institutions reasonably controlled the agenda and the policy outcome. Developments suggest the opposite. Moreover, evidence indicates that EU HDTV/standards policy was not a coherent strategy but the outcome of structural and agency constraints (intra-EC and international). Companies such as the SES, *CANAL+*, and BSkyB depended directly on the "rival" USA for their provision of both hardware and software (Hughes Aerospace and the Motion Picture Association of America, respectively). Conversely, had a European MAC standard been established unequivocally, it would have secured the domestic industrial and cultural domain, notably the audiovisual programming production, from the incursions by extra-European exporters. Yet, paradoxically, these anti-MAC companies managed first to gain "deregulation rights" through the 1986 loophole and secondly to maintain the same rights by the non-timely correction of this constraint. In this way they prevailed as the self-interested arbiters of the success or failure of the European HDTV and associated standards project. This illuminates why the initial phase (1986-1992) of the EU HDTV/standards policy was neither purely interventionist nor deregulatory but was a fierce competition of these two styles, and in places an unorthodox "coexistence" of both.

APPENDIX:
Chronology of Key EU-Related Advanced Television Activities and Policies

1980 Development of the MAC standard as rival to PAL and SECAM.

1984 June
Television Without Frontiers Green Paper foreshadows a common satellite television transmission standards policy based on the MAC standard.

1986 November
The first MAC Directive requires the use of the MAC transmission standard on high-power satellites (BSS frequency band) only, excluding regulation of medium-power satellites (FSS frequency band).

1988 Development of a successful HD-MAC prototype by EUREKA 95.

1988 June
Rupert Murdoch announces satellite transmission in PAL on the FSS band, thereby circumventing the MAC Directive and setting up the anti-MAC camp.

1990 The Dusseldorf Plenary Assembly of the CCIR does not approve the European 1250/50 proposal as the world HDTV production standard.

1990 The start and stop of BSB (the first commercial satellite) transmission in MAC; BSB merges with Sky to form BSkyB and is denied license to continue transmitting in MAC.

1991 Broadcasters (ACT) oppose making the new MAC Directive legally binding.

1991 Creation of the European Launching Group (ELG) to develop digital television in Europe.

1991 February
Thomson and Philips reportedly cooperate with the rival digital camp in the United States, thus revealing the adoption of a double (and hidden) agenda.

1991 December
The first MAC Directive expires. A regulatory void follows lasting until May 1992.

1992 May
Text of the watered-down second MAC Directive is adopted.

1992 July
The Memorandum of Understanding or "Gentlemen's Agreement," integral to the second MAC Directive, is signed. It is not, however, binding.

1992-1993 The Action Plan, integral to the MAC Directive, is vetoed by the UK.

1992 Europe-wide demonstrations of HD-MAC at Winter and Summer Olympic Games.

1992 October
CANAL+ and News Corporation launch a joint research venture on digitally compressed scrambled signals.

1993 Spring
The (Danish) EU Presidency urges the adoption of digital standards.

1993 July
Council Resolution on the development of technology and standards in the field of advanced (digital) television services.

1993 July
A revamped Action Plan for advanced television services in any transmission standard is agreed, marking the capitulation of the "policy of standards."

1993 September
ELG evolves into the DVB Group after the endorsement of its MoU by most players; DVB agrees on digital standards including the MPEG-2 standard.

1993 November
Commission proposals (1) for a Directive on the use of standards for the transmission of television signals, including the repeal of the MAC Directive and (2) for a framework Community policy on digital video broadcasting.

1994 June
Council Resolution on a framework policy for digital video broadcasting.

1994 July
Commission's Action Plan on *Europe's Way to Information Society* is adopted.

1995 June
EP introduces amendments to the draft Directive on conditional access.

1995 October
Council Directive on the use of standards for the transmission of television signals and repeal of the MAC Directive of 1992.

5 The Rise and Fall of HDTV Industrial Policy in the United States

We cannot approach HDTV by asking Uncle Sugar [the U.S. government] ... to put up the money, tempting as that may be.
—*Robert Mosbacher, U.S. Secretary of Commerce, 1989*

If a researcher were to read U.S. congressional documents, he or she would be mystified by the huge surge in interest in high-definition television (HDTV) in 1988 and 1989 followed by a dramatic reduction in HDTV references in 1990 and later years. According to the *Congressional Information Service*, only four HDTV hearings took place between 1990 and 1994. In contrast, House and Senate Committees and Subcommittees held no less than eight main hearings between October 1987 and September 1989 (six between March and May 1989!) to examine HDTV standardization and manufacturing matters. At that time, many policymakers and industry leaders voiced serious concerns about the United States' lack of competitiveness and involvement in the potentially huge HDTV consumer market in light of the faltering state of the U.S. consumer electronics industry. For instance, the American Electronics Association (AEA), a

Parts of this chapter were previously published in *The Information Society*, 1990, vol. 7, pp. 53-76, Michel Dupagne, Taylor & Francis, Inc., Washington, DC. Reproduced with permission. All rights reserved.

trade association of some 3,500 U.S.-owned companies and 45 engineering universities, predicted that U.S. absence from the manufacturing of HDTV products would greatly affect the growth of related electronics industries. If U.S. companies captured 10 percent or less of the HDTV market, the AEA warned, manufacturers of personal computers and semiconductors would lose 50 percent of their current worldwide market share (American Electronics Association [AEA], 1988). In 1988, Japanese producers had increased their share of the world's semiconductor market to 50 percent, leaving U.S. chip makers behind with a 37 percent share (U.S. House of Representatives, 1989c). The AEA reported that only 6 percent of the semiconductor production in the United States was devoted to consumer electronics, compared to 50 percent in Japan (AEA, 1988). In view of the evolution of the Japanese consumer electronics industry, the AEA contended, these estimates portended a serious and legitimate warning to U.S. electronics manufacturers. The industry reasoned that HDTV receivers would require high-capacity memory chips, such as dynamic random access memory chips, and if those were unavailable in the United States, foreign-owned companies would be able to control the totality of the $40-billion U.S. consumer electronics market.

Therefore, in the mind of many, HDTV offered an ideal opportunity to reclaim the U.S. consumer electronics market and prevent further siphoning of it by foreign multinationals. Plans to salvage the U.S. consumer electronics industry and organize a government-led riposte were outlined in numerous bills and hearings between 1988 and early 1989. This period of intense Congressional scrutiny characterized the rise of attempted HDTV industrial policy in the United States. But then soon after U.S. Secretary of Commerce Robert Mosbacher testified at March and May 1989 hearings, things began to go downhill (U.S. House of Representatives, 1989a; U.S. Senate, 1989a). By mid-1989, the Bush Administration effectively blacklisted the word "HDTV," and none of the HDTV bills under consideration in Congress would ever become law. What events transpired in 1989 and 1990 that prompted such a negative response from the White House and the subsequent decline in interest from Congress? Who, if anyone, is to blame for derailing HDTV industrial policy initiatives in the United States? How did the Democrat-dominated Congress and the Republican Bush Administration differ in their approaches for encouraging the creation of an HDTV manufacturing industry?

This chapter seeks to address these issues by analyzing the rise and fall of U.S. HDTV industrial policy initiatives between 1988 and 1990.

First, to understand the circumstances that motivated Congress to propose HDTV legislation, it is indispensable to examine the main reasons for the decline of the U.S. consumer electronics industry. Consequently, the first section briefly describes six problems that contributed to the downfall of this sector between the late 1960s and the late 1980s: lack of managerial leadership, lack of R&D investments, high labor costs, decreasing productivity, shortage of engineers, and Japanese anticompetitive trade practices. The second section begins by reviewing the provisions of Congressional HDTV-related bills, introduced in the late 1980s, that were designed to promote U.S. HDTV development and manufacturing. Then it chronicles the events that led to the demise of these bills and any HDTV industrial policy plans, and explores the role of the five protagonists involved in this controversy (White House, Congress, Secretary of Commerce Mosbacher, the AEA, and DARPA Director Craig Fields). In so doing, it offers a fascinating behind-the-scenes look at the way policy is created behind closed doors in executive offices by powerful individuals inside the government. Finally, the chapter discusses the reasons for the failure of HDTV industrial policy and assesses whether it would have been a viable option in the United States.

The Decline of the U.S. Consumer Electronics Industry

From the late 1800s to the late 1950s, American inventors and firms managed to position their country as an uncontested world leader in electrical and electronic technology. Thomas Edison's phonograph (1887), Lee de Forest's three-element vacuum tube "audion" (1906), Vladimir Zworykin's iconoscope (1923), Edwin Armstrong's frequency modulation technology (1933), AT&T and Bell Laboratories' transistor (1948), and Ampex's video tape recorder (1956) are just a few examples of these innovations. Not only were American companies competitive on the domestic market, but they also exported their business skills and technological expertise abroad. By the mid-1960s, RCA and General Electric were still actively involved in color television manufacturing in the United Kingdom and West Germany through joint ventures and overseas subsidiaries ("A Rosy Hue," 1966).

But since the late 1960s, under pressure from Japanese imports, this tradition of industrial leadership has progressively faded. In 1964, 94 percent of color TV sets sold in the United States were produced by U.S. firms. This figure fell to 67 percent in 1975 and to 43 percent in 1986 (Staelin et al., 1988). Of the 27 U.S.-owned companies that manufactured television sets in 1960, only Zenith remained a major contender by 1990 (U.S. Congress, 1992). Zenith's share of the U.S. color TV market plummeted by 50 percent from 1975 (24 percent) to 1990 (12 percent) (see Table 5.1; *Television Digest*, 1995). According to BIS Mackintosh data, six Japanese firms ranked among the world's top 13

TABLE 5.1 THE U.S. COLOR TELEVISION MARKET IN 1990

COMPANY	COUNTRY	BRAND	SHARE (%)
THOMSON	France	RCA (16.60%) GE (5.65%)	22.25
NORTH AMERICAN PHILIPS	Netherlands	Magnavox (7.75%) Sylvania (3.20%) Philco (0.65%) Philips (0.60%)	12.20
ZENITH	United States	Zenith	11.65
SANYO	Japan	Sears (4.90%)[a] Sanyo (1.50%) Fisher (0.30%)	6.70
MATSUSHITA	Japan	Panasonic (3.20%) Quasar (1.85%) JVC (1.50%)	6.55
SONY	Japan	Sony	6.50
SHARP	Japan	Sharp	5.00
TOSHIBA	Japan	Toshiba	4.00
EMERSON	United States	Emerson[b]	3.80
MITSUBISHI	Japan	Mitsubishi	3.50
HITACHI	Japan	Hitachi	2.50
MOBIL	United States	Montgomery Ward[c]	2.40
GOLDSTAR	South Korea	Goldstar	2.00
SAMSUNG	South Korea	Samsung	1.80
OTHERS	—	—	9.15

SOURCE: *Television Digest with Consumer Electronics*, 6 August 1990, p. 12. Copyright 1990 by Warren Publishing. Reprinted with permission.

NOTE: [a]Sears sets were manufactured mainly by Sanyo. [b]Emerson sets were manufactured by Orion Electric America. [c]Montgomery Ward sets were manufactured by Zenith and Sharp.

TABLE 5.2 TOP WORLDWIDE COLOR TV MANUFACTURERS IN 1988 (ESTIMATED PRODUCTION CAPACITY IN MILLIONS OF UNITS AND GROSS SHARE IN PERCENT)

COMPANY	COUNTRY	PRODUCTION	GROSS SHARE
PHILIPS	Netherlands	6.5-7.0	7.2-7.8
MATSUSHITA	Japan	6.5-7.0	7.2-7.8
THOMSON	France	6.5-7.0	7.2-7.8
SAMSUNG	South Korea	5.5	6.1
GOLDSTAR	South Korea	5.5	6.1
SONY	Japan	4.5	5.0
SHARP	Japan	3.5-4.0	3.9-4.4
TOSHIBA	Japan	3.5-4.0	3.9-4.4
SANYO	Japan	3.0-3.5	3.3-3.9
HITACHI	Japan	3.0-3.5	3.3-3.9
GRUNDIG	Germany	2.2	2.4
NOKIA	Finland	2.2	2.4
ZENITH	United States	2.1	2.3
TOTAL	—	90.0	100.0

SOURCE: BIS Mackintosh, Consumer Electronics Information Service (Basis: all end-equipment manufacturing plants). Reprinted with permission.

color TV manufacturers in 1988 with 28 percent of the production capacity, followed by the Europeans with 20 percent and the South Koreans with 12 percent (Table 5.2). Besides Zenith with 2.1 million units, U.S. presence in color TV manufacturing became insignificant.

The decline of the U.S. consumer electronics industry is inextricably linked to the rise of its Japanese counterpart. The growth of the Japanese consumer electronics industry and its irresistible ascension to international preeminence is due to a variety of factors, such as availability of basic patents from the United States, ability to adapt imported technology, lower production costs, prompt acceptance of Japanese products overseas, government stimulation, aggressive pricing, and R&D investments (see Burton and Saelens, 1987a; "L'Industrie Electronique," 1985; Okimoto, 1986). In four decades, Japanese consumer electronics companies have propelled themselves from being mere technology imitators to undisputed technology leaders. Given the favorable conditions of the U.S. market (i.e., large size, modest tariffs, and a common television standard), Japanese firms first marketed and exported their products to the United States. From the 1960s onward, U.S. consumer electronics manufacturers have urged authorities to act upon dumping complaints and to press charges against Japanese manufacturers but to little avail. Dumping refers to the illegal practice of selling goods in a foreign country below cost or at a lower price than in the home country. Similarly, due to protracted litigation, they achieved little success in suing Japanese companies and their U.S. importers for antitrust violations. In 1977, the U.S. government negotiated a three-year orderly marketing agreement (OMA) with its Japanese counterpart to curtail the number of consumer electronics imports. An OMA refers to a bilateral agreement between the exporting and importing countries that limits the number of exports to the importing country (see Matsushita, 1988). To circumvent OMA and antidumping actions, Japanese companies began building assembly plants in the United States. While the installation of foreign plants in the United States has provided some jobs to Americans in the consumer electronics sector and generated revenues for the federal and state governments, the bulk of capital income has ultimately returned to foreign owners and shareholders, aggravating the trade deficit of the United States.

Economic statistics are even more revealing than policies and lawsuits. In 1955, U.S. production in consumer electronics totaled $1.5 billion, while Japanese output only amounted to $70 million (Rosenbloom and Cusumano, 1987). In 1989, estimated U.S. exports in consumer electronics amounted to $1.3 billion, but imports, primarily from Japan, exceeded $12.7 billion (U.S. Department of Commerce, 1990). This $11.4 billion imbalance represented 10.5 percent of the $109 billion foreign trade deficit in 1989 (*Survey of Current Business*, 1990). Total U.S. employment in consumer electronics fell by 35 percent from 1972 (114,500) to 1982 (74,400), and the number of production workers dropped by 41 percent, from 84,730 to 49,848 (U.S. Congress, 1983). In 1990, there were 22,500 production workers in the entire U.S. consumer electronics industry. By comparison, Japan employed 290,000 workers in that sector (Zampetti, 1994).

There is no shortage of explanations for the fall of the U.S. consumer electronics industry. It has been argued that this decline is primarily due to competitiveness-related factors, such as an overemphasis on profit margins, lack of vision, and low R&D investments. Industrial competitiveness refers to the ability of domestic companies to manufacture products that are higher in quality and lower in costs than those of international competitors (U.S. House of Representatives, 1992). It assumes that these firms operate under "free and fair market conditions" (The President's Commission on Industrial Competitiveness, 1985, p. 6). But U.S. manufacturers are quick to point out that there has hardly been a level playing field in the consumer electronics sector. They blame foreign dumping and the U.S. government's reluctance to take action against it for the demise of the industry. Although, as noted by Clyde Prestowitz (1989), former counselor for Japanese Affairs to the Secretary of Commerce and a leading critic of U.S.-Japan trade imbalances, "The Japanese did a superb job, and the Americans made some mistakes" (p. 352), there is also considerable evidence that practices of U.S. importers of television sets, Japanese protectionism, price collusion, and U.S. government policies have all contributed to prevent U.S. manufacturers from competing in the U.S. and Japanese markets on free and fair terms.

Lack of Managerial Leadership

In the 1950s, economies of scale and a strong post-war economy helped the United States establish their competitive edge in world electronics markets, and the U.S. management style was hailed as the wave of the future (see Prestowitz, 1989). But as international competition grew tense and Japanese imports flooded the U.S. market, U.S. consumer electronics companies failed to respond to their foreign competitors promptly by curtailing production costs, adopting long-term strategies, conveying a sense of corporate vision, and championing new products. Examples of U.S. shortsighted planning abound. Although Motorola pioneered solid-state color television technology in 1967, it was Hitachi that first introduced it commercially in 1969. RCA and Zenith did not supply complete solid-state lines until 1973 (Prestowitz, 1989). "Zenith was the last of the leaders to move to printed circuit boards, one of the last to automate plants, the last to go to overseas production for the domestic market, the last to diversify, one of the leaders in the antidumping legal actions, and quick to sacrifice profit margins for market share" (Staelin et al., 1988, p. 60). After the first demonstration of the Ampex video tape recorder (VTR) in the mid-1950s, some engineers at RCA and Ampex recognized the potential of a consumer VTR, but the top management at both companies preferred to concentrate their efforts on the broadcast equipment niche (Rosenbloom and Cusumano, 1987). Generally, U.S. firms have focused their R&D activities on improving existing products, while their Japanese competitors have stressed new product development and process innovations (see LaFrance, 1985). The paucity of corporate visionaries and product champions has contributed in great part to the decline of the U.S. consumer electronics industry.

In contrast, Japanese management philosophy was founded on long-term vision, cooperation between labor and management, and an unconditional commitment to high-quality products (Morita, Reingold, and Shimomura, 1986; Staelin et al., 1988; *The Competitive Status*, 1984). Consistent with the spirit of Japanese corporate culture, each individual at every stage of the production process participated in stringent quality control checks. In 1977, the number of faults found in the production of a television set averaged 0.01-0.03 in Japan, as opposed to 1.4-2 in the United States (Staelin et al., 1988). As an anonymous television manufacturing expert observed, "The focus in the 1960s was on the number of sets produced, with less emphasis on quality. We became complacent about product quality and our new Japanese competitors were not" (Seel, 1993, p. 13). This sense of corporate vision and emphasis on innovative production practices explains to a large extent why the Japanese firms have become so successful in the global electronics marketplace.

Lack of R&D Investments

Because U.S. companies are under constant pressure by owners, shareholders, and the economic system to perform and improve profit margins, they have been more inclined to maintain a short-term, bottom-line orientation and sacrifice research and development investments. This is also the case in the U.S. consumer electronics sector. For instance, U.S. television makers' R&D expenditures averaged 3.1 percent of sales in 1977 and 1978, while Sony's stood at 5.6 percent in 1977 and Hitachi's at 5.8 percent in 1978 (LaFrance, 1985). Between 1984 and 1990, Zenith's R&D expenditures decreased by more than 20 percent and were lower than those of its foreign competitors (see Table 5.3). Akio Morita, then Chairman of Sony Corporation, argued that too often Western consumer electronics companies would rather abandon a promising project than invest more in R&D (Morita et al., 1986). From the outset, Japanese consumer electronics companies had no choice but to commit substantial funds to R&D to survive in the highly competitive domestic market. Morita et al. (1986) reminisced that at a certain time there were 40 domestic television set manufacturers vying for a share of the Japanese market. But by the mid-1980s only six major companies had survived. A study of the shakeout of the U.S. consumer electronics industry between 1967 and 1976 revealed that company size, market share, and low-cost products did not necessarily guarantee the successful survival of consumer electronics compa-

TABLE 5.3 RESEARCH AND DEVELOPMENT EXPENDITURES OF SELECTED CONSUMER ELECTRONICS COMPANIES AS PERCENT OF (NET) SALES FROM 1984 TO 1990

COMPANY	1984	1985	1986	1987	1988	1989	1990
ZENITH[a]	5.1	5.9	5.3	4.4	3.7	3.3	4.0
RCA[b]	2.8	2.8	—	—	—	—	—
PHILIPS[a]	6.7	6.7	7.6	8.3	8.2	8.0	7.9
THOMSON[b]	NA	6.0	6.7	6.3	5.6	5.4	6.0
SONY[a]	7.9	7.8	8.4	8.4	8.3	6.6	5.7
MATSUSHITA[b]	4.2	4.8	5.5	6.2	5.6	5.8	5.8

SOURCE: Annual reports of Zenith Electronics Corporation, RCA, General Electric Company, N.V. Philips, Thomson S.A., Sony Corporation, Matsushita Electric Industrial Co., Ltd., 1984-1990.
[a]Net sales equal gross sales minus returns, allowances, and discounts.
[b]Sales.
NA = not available.

nies. On the other hand, companies who had invested their resources in R&D and produced high-quality color receivers were more likely to be survivors (Willard and Cooper, 1985).

High Labor Costs

Historically, lower wages have bolstered the competitiveness of Japanese consumer electronics firms (see Table 5.4), although by the early 1980s differences in labor costs between Japan and the United States had leveled off, due in part to the soaring Japanese currency. In 1963, the average hourly earnings of U.S. television receiver production workers were almost six times as high as those of Japanese electronic machinery, equipment, and supplies production workers. But by 1977, this wage gap had diminished to one and a half times (LaFrance, 1985). To remain competitive, U.S. producers, such as Zenith and RCA, began moving manufacturing to Taiwan and the Mexican *maquiladoras*. "This was facilitated by the U.S. government, which quietly encouraged these countries to create packages of tax holidays and tariff measures designed to entice U.S. investment" (Prestowitz, 1989, p. 358). The regulatory *maquiladora* regime has allowed U.S.-owned manufacturers to assemble television sets from U.S.-imported parts (e.g., picture tubes) in these border factories and export them to the United States essentially duty-free (see Seel, 1993; U.S. Congress, 1992). In 1990, the average hourly wage in the Mexican electronics *maquiladoras* ($0.92) was 10 times lower than that in the U.S. consumer electronics industry ($10.58) (U.S. Congress, 1992).

Table 5.4 Monthly Manufacturing Wages in Japan and the United States from 1960 to 1990

Country	1960	1965	1970	1975	1980	1985	1990
Japan	¥22,630	¥36,106	¥71,447	¥163,729	¥244,753	¥299,531	¥370,169
dollar equivalents	$63	$100	$198	$552	$1,079	$1,256	$2,561
United States	$323	$380	$533	$763	$1,154	$1,545	$1,767

Source: Various editions of the *Yearbook of Labour Statistics* and the January 1992 *Employment and Earnings*; Exchange rates from *Historical Statistics 1960-1986* and the *1990 OECD Economic Outlook*.
Note: Exchange rates are calculated by averaging the daily rates of the year.

Decreasing Productivity

Industrial competitiveness is often measured by labor productivity rate. Walter Kunerth, Siemens Executive Vice-President, argued that "If Rolls-Royce had achieved the same productivity gains as the TV industry, a Rolls would cost the same today as a bicycle" (Jackson, 1993, p. 19). Based on July 1996 statistics from the U.S. Department of Labor (see Table 5.5), annual average manufacturing productivity rates[1] (as measured by output per hour) increased only by 1.8 percent in the United States between 1977 and 1990, compared to 4.2 percent in Japan, 3.4 percent in France, 2.3 percent in Germany, 3.6 percent in the Netherlands, and 4.0 percent in the United Kingdom. The consultancy firm McKinsey estimated in 1993 that Japanese workers in the consumer electronics industry were 18 percent more productive than their American counterparts (Zampetti, 1994). Robert Galvin, Chairman of Motorola, attributed "some of the problem to differences in the educational level and stability of the two labor forces," but most of it "to the U.S. approach to manufacturing, which concentrates too heavily on high output and reduction of specific costs without considering the entire process" (Prestowitz, 1989, pp. 365-366).

Shortage of Engineers

Although the U.S. consumer electronics industry cannot be held responsible for this national problem, it has been directly affected by it. In 1969, 16,200 and 11,800 engineers graduated from American and Japanese universities, respectively (Humbert, 1984). And even then, unlike their Japanese counterparts, only a minority of the new U.S. electronics engineering graduates chose to enter the consumer elec-

TABLE 5.5 OUTPUT PER HOUR IN MANUFACTURING FOR SELECTED OECD COUNTRIES FROM 1960 TO 1990 (1992 = 100)

COUNTRY	1960	1965	1970	1977	1980	1985	1990
UNITED STATES	NA	NA	NA	75.5	72.8	83.7	95.7
JAPAN	14.0	21.2	38.0	55.9	63.9	77.3	95.4
FRANCE	23.0	31.8	45.5	64.1	70.5	83.8	99.1
GERMANY	29.0	39.2	51.9	72.7	77.1	88.7	98.2
NETHERLANDS	19.9	25.8	39.6	61.9	70.0	89.8	98.6
UNITED KINGDOM	30.5	36.1	43.7	54.4	54.9	71.9	90.1

SOURCE: U.S. Department of Labor, Bureau of Labor Statistics, July 1996.
NOTE: NA = not available.

tronics field (LaFrance, 1985). In 1981, 75,000 engineers graduated in Japan, compared to 63,000 in the United States, although the population of Japan was about half that of the United States (U.S. Congress, 1983) "The thousands of engineers pouring out of Japanese universities today are the result of policies that deliberately encourage students to pursue engineering studies" (Prestowitz, 1989, p. 253). Since the 1960s, the Ministry of International Trade and Industry (MITI) has emphasized the technological aspect of the Japanese industry. Therefore, it is no surprise that such a mentality has been embedded in Japanese universities and students (see Okimoto, 1989). On the other hand, perhaps for cultural reasons, U.S. students have concentrated their attention on professional careers, such as business administration and law. In addition, they have performed poorly on international achievement tests in scientific disciplines. For instance, the International Association for Evaluation of Educational Achievement found that U.S. 12th graders ranked almost last on mathematics and science achievement tests, compared to students from other countries (The Task Force on Women, Minorities, and the Handicapped in Science and Technology, 1988). This finding suggests that high school students in the United States are not as well prepared academically as their peers abroad to study engineering in college.

Japanese Anticompetitive Trade Practices

With the recent work of Prestowitz (1989), Choate (1990), and Curtis (1994), there is ample evidence that Japanese consumer electronics manufacturers, with the consent of MITI, engaged deliberately in dumping actions and price-fixing schemes in the 1960s and 1970s to control the U.S. market. There is also support for the claim that they used their influence to oppose distribution of American television products in Japan. These authors have documented how the U.S. government (Treasury, Customs Service, Justice, Commerce, State, International Trade Commission) failed to act in an effective and timely manner to end these violations of U.S. trade laws, thereby indirectly contributing to the "destabilizing" of the U.S. consumer electronics industry. For instance, in March 1968, the Import Committee of the U.S. Electronics Industries Association filed a complaint with the Treasury Department alleging that Japanese manufacturers were dumping television sets on the U.S. market. But it was not until March 1971, three years later, that the U.S. government formally *acknowledged* that dumping was being practiced by the Japanese (Choate, 1990). Choate (1990) and Curtis (1994) went even further by accusing the U.S. government of selling out its own consumer electronics industry in the name of United States-Japan policy interests. At the very least, the U.S. government did not conduct its numerous investigations with the required attention and speed.

Between 1962 and 1981, Prestowitz (1989) reported that no less than 20 unfair television-related trade complaints were filed with the U.S. government. The most notorious case of alleged dumping in the history of U.S. consumer electronics pitted Zenith against seven Japanese manufacturers (Hitachi, Matsushita, Mitsubishi, Sanyo, Sony, Sharp, Toshiba) and two U.S. companies (Sears, Motorola). In 1981, Zenith and National Union Electric filed a private antitrust suit against the above-mentioned companies, accusing them of plotting a worldwide, unitary conspiracy to take over and destroy the American consumer electronics industry (*Zenith Radio Corp. v. Matsushita Elec. Indus. Co.*, 1981). The district court ruled that the plaintiffs failed to show any significant evidence that the defendants had entered into a conspiratory agreement or a concerted action to export electronics products to the United States in a manner that could have injured the plaintiffs. Judge Becker added that Congress' intent when enacting these antitrust laws was to protect competition, not competitors. The district court advised the plaintiffs to address their grievances regarding the status of the U.S. consumer electronics industry to Congress, the International Trade Commission, or the President's Trade Negotiator, but not to an antitrust court. But in December 1983, the Court of Appeals for the Third Circuit reversed in part this judgment on the grounds that there was sufficient admissible and factual information (i.e., admissible material fact) to establish the participation of the defendants in the alleged conspiracy (*In re Japanese Electronic Products*, 1983). In March 1986, the U.S. Supreme Court overturned the court of appeals' decision, concluding that Zenith had not presented persuasive evidence supporting petitioners' engagement into an illegal conspiracy that caused respondents to suffer a cognizable injury (*Matsushita Elec. Indus. Co. v. Zenith Radio*, 1986). On remand, the Court of Appeals for the Third Circuit found no additional evidence that would lead it to conclude that the Japanese manufacturers had engaged in a predatory pricing conspiracy and, therefore, dismissed the charges (*In re Japanese Electronic Products Antitrust Lit.*, 1986). In April 1987, the Supreme Court refused to rehear the case.

Why were such apparent unfair trade practices tolerated for so long by the Japanese and the U.S. governments? On the Japanese side, it has been well established that collusive pricing was actively encouraged by MITI and was carried out as an industrial policy instrument to boost the growth of the electronics industry. Industrial policy "is a summary term for the activities of governments that are intended to develop or retrench various industries in a national economy in order to maintain global competitiveness" (Johnson, 1984, p. 7). Traditional Japanese industrial policy tools have included: infant industry protection, controls over foreign direct investment, R&D subsidies, low-interest loans, promotion of business-government cooperation, buy-Japanese programs, and unfair trade practices, such as dumping (Okimoto, 1989; Organisation for Economic Co-operation and Development [OECD], 1972).

A series of laws passed between 1956 and 1971 allowed MITI, among other things, to coordinate television-related R&D activities and approve price-fixing schemes of Japanese television manufacturers. For instance, as early as 1955, the Ministry established a Computer Research Committee to promote transistor R&D (Zampetti, 1994; see also Burton and Saelens, 1987b), and in 1971, it sponsored a research project to explore the use of integrated circuits in color television receivers (LaFrance, 1985). In 1956, the largest Japanese manufacturers formed the Electric Appliance Market Stabilization Council, a production cartel, to control prices on the domestic market and then established the Television Export Council to replicate these tactics abroad (Choate, 1990). Once the cartel set a price, it was submitted to MITI, which approved it as the official "check price." The cartel was expected to abide by these minimum export prices to avoid a potentially damaging price war between its members on the U.S. market and prevent litigation under the U.S. Antidumping Act of 1921 (Yamamura and Vandenberg, 1986). But apparently unknown to MITI, Japanese manufacturers were practicing "double-pricing" schemes—prices that were lower than the approved "check prices" by offering secrete rebates to U.S. importers, thereby violating the intent of the export cartel, Japanese tax laws, and American customs laws all together. The Fair Trade Commission of Japan, which often clashed with MITI, conducted a series of investigations in the 1950s and 1960s and found cartel members guilty of "agreeing ... on the bottom prices, margin rates, and distributors' prices for both color and black-and-white televisions ... [and of] substantially restricting competition" (Yamamura and Vandenberg, 1986, p. 254). The Japanese manufacturers were never prosecuted, though.

On the American side, the U.S. government would probably argue that it simply adhered to the traditional principles of free trade and that it was not its role to protect domestic companies against their foreign competitors. For instance, the International Trade Commission attributed the declining status of U.S. consumer electronics to "severe and sustained price competition" (Yamamura and Vandenberg, 1986, p. 259), not to Japanese collusive and predatory pricing as U.S. manufacturers had contended. Under the terms of the Trade Expansion Act of 1962, U.S. policymakers did not view favorably the petitions of certain manufacturers for government assistance to thwart the competitive advantage of Japanese firms (Millstein, 1983). The Act was devised to stimulate the expansion of trade in a free and nondiscriminatory environment (Trade Expansion Act, 1962). Prestowitz (1989) concluded that the "The policies of the U.S. government, which encouraged imports and transfer of U.S. technology and capital abroad, were not conceived for the benefit of the long-term strategic position of U.S. industry" (pp. 349-350).

History of HDTV Industrial Policy: The Bills

It is against this backdrop that nine bills were introduced in Congress between 1989 and 1990 to facilitate the development of a domestic HDTV manufacturing industry. These HDTV bills, primarily authored by Democrats, called for a relaxation of antitrust laws, administrative and financial assistance in cooperative R&D projects, tax relief for R&D investment, and the formulation of a national HDTV agenda. But by August 1990, five bills were still pending in Subcommittees, and they were unlikely to be addressed any time soon for the reasons spelled out in the next section. A brief description of each bill follows.

On March 2, 1989, Representatives Don Ritter (R-Pennsylvania) and Mel Levine (D-California) introduced H.R. 1267 to stimulate U.S. manufacturing of HDTV products. This bill comprised six main provisions:

1. Tax incentives for R&D of HDTV technology

2. Relaxation of antitrust laws to facilitate cooperative HDTV development and manufacture

3. Federal funding for cooperative HDTV projects

4. Coordination of federal procurement of HDTV technology

5. Submission of a report on ways to ensure the development and viability of domestic HDTV in international trade

6. Additional funding to the Federal Communications Commission (FCC) to adopt an HDTV broadcast standard (H.R. 1267, 1989).

On March 21, 1989, Rep. George Brown (D-California) and his colleagues sponsored H.R. 1516 to improve the competitiveness of the U.S. consumer electronics industry and coordinate cooperative R&D ventures in HDTV. The Secretary of Commerce through the Director of the National Institute of Standards and Technology (NIST) was instructed to provide guidance and funding for joint R&D projects. This federal assistance aimed to solve technology and manufacturing problems, speed up the commercialization process of HDTV, and establish transfer and sharing strategies among participants without endangering intellectual properties. The bill appropriated $100 million for each fiscal year (FY) from 1990 to 1994 (H.R. 1516, 1989).

On May 9, 1989, Reps. Mel Levine and Don Ritter introduced the third HDTV bill, H.R. 2287. This proposed legislation established a SEMATECH-like consortium, called TV Tech, to stimulate R&D and manufacturing of HDTV technology (for a recent analysis of SEMA-TECH, see Grindley, Mowery, and Silverman, 1994). An Advisory Council on Federal Participation was created to advise the Directorate of TV Tech on appropriate activities to be taken for boosting the growth of a domestic HDTV industry. Funded by both the private and public sector, TV Tech was barred from transferring technology to foreign-owned companies without the express permission of the Directorate and the Advisory Council (H.R. 2287, 1989).

On May 9, 1989, Senator John Kerry (D-Massachusetts) sponsored S. 952 to encourage the formation of joint ventures for the research, development, and manufacture of domestic HDTV technology. Title I of the bill amended antitrust law restrictions in the Communications Act of 1934 and the National Cooperative Research Act of 1984. Title II required the Secretary of Commerce to submit to the President and Congress a report outlining actions to be taken to promote the development and viability of a U.S. HDTV manufacture industry (S. 952, 1989).

On May 16, 1989, Sen. Albert Gore (D-Tennessee) introduced S. 1001, which was incorporated in S. 1191 in August 1989. Like S. 952, this proposed legislation required the Secretary of Commerce to present to the President and Congress a report identifying the conditions for establishing HDTV components and equipment manufacturing industries (S. 1001, 1989).

On June 15, 1989, Sen. Ernest Hollings (D-South Carolina) and several co-sponsors introduced S. 1191, the Technology Administration Authorization Act of 1989, to increase federal funding of the Department of Commerce's Technology Administration. This bill, which passed the Senate in October 1989 but failed to become law, was not really HDTV-specific. Instead it was designed to be an advanced technology package promoting the development and application of many new products and processes. Specifically, S. 1191 authorized appropriations for the NIST and fund joint ventures in advanced electronics, including advanced television. Of the $100 million to be appropriated to the NIST Advanced Technology Program for FY1990, at least $30 million were earmarked for HDTV technology. Foreign-owned companies were ineligible for federal assistance (U.S. Senate, 1989c).

On March 21, 1990, Rep. Robert Roe (D-New Jersey) and his colleagues sponsored H.R. 4329, the House companion of S. 1191. Cited as the American Technology Preeminence Act of 1990, this bill contained various provisions to enhance the competitiveness of U.S. advanced technology industries (U.S. House of Representatives, 1990):

1. To fund programs and activities of the Department of Commerce's Technology Administration and the National Institute of Standards and Technology (NIST) for FY1990, FY1991, and FY1992.

2. To enable the Secretary of Commerce to support joint ventures in emerging technology fields, including high resolution information systems. Authorizations were set at $50 million for FY1990, $100 million for FY1991, and $250 million for FY1992.

3. To establish a High Resolution Information Systems Board (a) to monitor the development of domestic HDTV and related advanced electronics industries; (b) to stimulate cooperation between the government and the industry for such a development; and (c) to develop policies for coordinating U.S. procurement of high resolution information systems technology.

4. To extend the National Cooperative Research Act of 1984 (renamed the National Cooperative Research and Production Act of 1990) to cooperative manufacturing agreements.

While being an overall advanced technology package, the American Technology Preeminence Act included most provisions of HDTV-specific bills to promote the development and manufacture of domestic HDTV products. H.R. 4329 passed the House in July 1990 but did not move further in the legislative process.

On May 9, 1990, Rep. Sam Gejdenson (D-Connecticut) introduced H.R. 4764, which mandated a minimum of domestic content for any HDTV receiver sold in the United States. This ratio would have reached 50 percent after four years of market penetration (H.R. 4764, 1990).

Finally, on May 24, 1989, Rep. Edward Markey (D-Massachusetts) and his colleagues sponsored H.R. 4933, the ninth HDTV bill. It appropriated $2.2 million for FY1990 and $2.45 million for FY1991 to the FCC for establishing an advanced television task force, purchasing testing equipment, and hiring additional personnel (H.R. 4933, 1990). The bill passed the Subcommittee on Telecommunications and Finance in July 1990 but then died.

Other bills also proposed general antitrust laws modifications that would have allowed HDTV cooperative research and manufacture: H.R. 1024 (to authorize the establishment of cooperative innovation arrangements without infringing antitrust laws); H.R. 1025 (to amend the National Cooperative Research Act of 1984 to permit joint ventures for producing and marketing products and services); H.R. 423 (to relax antitrust laws to promote joint manufacture and distribution of products); H.R. 2264 (to extend the National Cooperative Research Act of

1984 to joint research, development, or production ventures); and H.R.
4715 (to establish the Technology Corporation of America as the
American equivalent of Japan's MITI and Europe's EUREKA).

The Controversy

The incident that triggered off the chain reaction leading the Bush
Administration to back away from any federal financial support for
HDTV and causing HDTV legislation to lose support in Congress was
the American Electronics Association (AEA)'s $1.35-billion proposal to
bolster the revival of the U.S. consumer electronics industry, unveiled
at the May 9, 1989 hearing of the Senate Committee on Commerce,
Science, and Transportation. As Alfred Sikes, then Head of the
National Telecommunications and Information Administration, put it,
"That was the watershed event that caused HDTV to decline" (Davis,
1990, p. 228). Based on a report prepared by the Boston Consulting
Group (*Development of a U.S.-Based ATV Industry*), the AEA proposed
the creation of a five-year, three-prong government-industry partnership for the U.S. reentry into consumer electronics. The three steps
were: (1) industry defines problems and recommends solutions; (2)
government offers incentives and assistance; and (3) companies invest
in R&D and begin manufacturing (AEA, 1989). Should U.S. companies
choose not to participate in such a revival and build a successful
HDTV manufacturing base, the AEA warned, their foreign competitors will take advantage of HDTV technology to erode U.S. market
share in related electronics products, such as semiconductors and computers. The AEA advocated an extensive government-led HDTV
industrial policy (1) to fund technology development grants ($100 mil-
lion a year for three years through the Defense Department's Defense
Advanced Research Projects Agency (DARPA); (2) to appropriate at
least $50 million to the Department of Commerce's NIST; and (3) to
provide $500 million for low-cost loans and $500 million for loan guarantees (AEA, 1989). So the proposed aid package totaled $1.35 billion.
Adding insult to injury—from the Republican Bush Administration's
perspective—the AEA had not even informed Secretary of Commerce
Robert Mosbacher about the specifics of the plan before its
Congressional testimony on May 9 (Davis, 1990). Mosbacher himself
was a witness at the same Congressional hearing and preceded Pat
Hubbard, the AEA Vice President for Science and Technology Policy, in
order of appearance.

A further irony is that Mosbacher had made it clear at this Senate Committee hearing that while he was considering changes in antitrust policies and capital gains tax laws for the development of HDTV and other high-tech products, he opposed "a government-led industrial policy" (U.S. Senate, 1989a, p. 25), which was precisely what the AEA was demanding. This position was consistent with the statements he made at his confirmation hearing in January 1989 and later at a hearing before the House Subcommittee on Telecommunications and Finance in March 1989 (Farnsworth, 1989; U.S. House of Representatives, 1989a). "The reason I am against industrial policy," he argued, "is because to me it does mean that the government takes the lead, picks winners and losers, gets into the marketplace, and is a player all the way from research and development down through manufacturing" (U.S. Senate, 1989a, p. 40). Mosbacher added that he had "serious doubts as to whether throwing money at the problem is the answer" (p. 41). Several members of the Committee including Chairman Ernest Hollings, Sen. Albert Gore, and Sen. John Kerry, all Democrats, took turns to interrogate him about his perceived inaction in developing a U.S. HDTV strategy. As the following exchange between Kerry and Mosbacher illustrates, this hearing was highly political because it dealt as much with the commercialization of HDTV as with the proper role of government in promoting high-tech industries. Opinions were divided along party lines.

SENATOR KERRY: So the private sector is dying to move forward, is it not?

SECRETARY MOSBACHER: Well, I hope so.

SENATOR KERRY: It is dying while trying to move forward.

SECRETARY MOSBACHER: If they were really ready to move forward, they would have. I think, frankly, the problem is that they are hoping, what I use to call in the private sector, Uncle Sugar will fund it. I do not think they should depend on that. I think it has to be a combination of working with the government, but it has to be that the private sector will still take the lead. I think they also have to know that they are not going to be cut off at the knee by not being backed as far as the policy and the continuity, which we have not always done historically.

SENATOR KERRY: The bill that I am submitting with respect to the antitrust laws does not cost any money. All it does is change the antitrust format and the Clayton restrictions so they can move forward. Why are you not committed to that at this point?

SECRETARY MOSBACHER: Because we have an interagency ongoing review now

that this Administration has been in a little over 100 days, and we have not
completed our work in this area. I am not saying we will be or will not be,
but that is why I am not going to stand up here and say we [will] when we
still have a lot of work to do on it. (U.S. Senate, 1989a, p. 44)

While Kerry and other Democrats urged Mosbacher to move more
quickly on coordinating a national policy for HDTV and adopting specific measures, the Secretary resisted any of these "industrial policy"
demands at this and other hearings (U.S. House of Representatives,
1989a; U.S. Senate, 1989a). Above all, he avoided proposing what
many members of Congress, especially Democrats, sought from the
Administration: federal money. He only pledged to "come back with
specific recommendations" (U.S. Senate, 1989a, p. 45) on areas such as
tax credit, capital gains, and antitrust restrictions. Mosbacher was
scheduled to present a report to the House Subcommittee on
Telecommunications and Finance on July 1, 1989.

The response of the Bush Administration to the AEA proposal was
swift and unequivocal. On May 16, 1989, Mosbacher reiterated that
"We cannot approach HDTV by asking 'Uncle Sugar' ... to put up the
money, tempting as that may be" ("EIA Opposes," 1989, p. 1). He
added that while the government is willing to "remove obstacles to
give encouragement, ... there is no way the American public would
support such a proposal" (p. 1) in view of the huge budget deficit.
Instead he urged the industry to commit more of their own resources
to the development of HDTV technology. *The Wall Street Journal's* Bob
Davis (1990) described the mood at a meeting in the White House
Chief of Staff John Sununu's office soon after the "Uncle Sugar"
remarks. Present at this meeting to discuss HDTV were Robert
Mosbacher, Wayne Berman (Counselor to the Secretary of Commerce),
John Sununu, Dan Quayle, Nicholas Brady (Secretary of Treasury),
Michael Boskin (Chairman of the Council of Economic Advisers),
Richard Darman (Director of the Office of Management and Budget),
and other top aides.

> According to several participants, Mr. Boskin, Mr. Darman and others
> took turns chewing out Mr. Berman for singling out HDTV for special
> treatment, and they cajoled Mr. Mosbacher, a longtime friend of the
> president, to take a broader approach to technology. The message was
> clear: HDTV was out, and so was anyone who pushed it too hard.
> (Davis, 1990, pp. 228-229)

Undoubtedly, the so-called "Iron Triangle" (Darman, Boskin, and Sununu) expended considerable energy to derail HDTV initiatives because they considered them akin to industrial policy (Carey and Harbrecht, 1990; Markoff, 1989; Pollack, 1989). Sununu went as far as to declare that "There are no stronger free-market supporters in Washington than Boskin, Darman, and me" (Carey and Harbrecht, 1990, p. 60).

The presentation of the AEA proposal and the subsequent hard line taken by the White House produced several "casualties." First, the Defense and Commerce Departments abandoned their efforts to produce reports or recommendations for the promotion of HDTV R&D and manufacturing. For instance, a May 1990 draft document from the Defense Science Board that recommended an increase in DARPA funding to at least $100 million per year for five years to develop, among other things, "a domestic base for high resolution technologies" was buried and never made it out of the Pentagon. As to Mosbacher, he never delivered his recommendations to the House Telecommunications and Finance Subcommittee as scheduled on July 1, 1989. In fact, no hearing took place on that date.

Second, HDTV legislation stalled. By January 1990, none of the HDTV bills had come to a final vote, and most of them were bogged down in Committees and Subcommittees. Rep. Matthew Rinaldo (R-New Jersey), a member of the House Telecommunications and Finance Subcommittee, attributed that stagnation to Congress' reassessment of its involvement in HDTV. He added that at this point "there [was] no consensus in the subcommittee on whether legislation [was] necessary" (Gatski, 1990, p. 1). Clearly, the Administration's withdrawal from any HDTV policy plans forced Congress to reconfigure HDTV legislation into a broad advanced technologies package (e.g., H.R. 4329), but this was not the only reason for the loss of momentum. In July 1989, the Congressional Budget Office released a report entitled *The Scope of the High-Definition Television Market and Its Implications for Competitiveness*, which questioned the anticipated effect of HDTV on the U.S. electronics sector and the size of the HDTV market. Reviewing three market forecasts (see Chapter 8), CBO concluded that "either the markets were unlikely to be big enough to have the hoped for effects, or the sequences of events asserted by the studies were not sufficiently developed to warrant the conclusions drawn" (Congressional Budget Office [CBO], 1989, p. v). It also expressed some skepticism about the timing and the size of the HDTV consumer market by contesting the assumption that HDTV will become an instant success. The pessimistic results of this inquiry dampened Congressional action on HDTV legislation.

In May 1990, Craig Fields, the Director of the Defense Department's Defense Advanced Research Projects Agency (DARPA), became the third casualty. While Mosbacher and Berman heard the White House's message loud and clear—the executive branch would not cheerlead for specific technologies like HDTV—Fields did not get the message as clearly and continued to pressure the Department of Defense for support of HDTV research (Seel, 1992). Rep. Les AuCoin (D-Oregon) compared Mosbacher's disappearing act from the HDTV scene to a vaudeville show in which "someone yanked him with a hook" (Davis, 1990, p. 229), and this despite letters mailed by 30 members of Congress in October 1989 urging the Bush Administration to take a more proactive role in developing an HDTV domestic industry (Burgess, 1989). As to Berman, he now believed that "HDTV is just one piece of a puzzle" (Burgess and Richards, 1989, p. C1) and that "there are a lot of other technologies that are equally important" (Pollack, 1989, p. 31).

Fields, on the other hand, continued to campaign aggressively for a *dual-use approach* for HDTV technology (and other technologies) until his demotion in May 1990 ("Defense Department," 1988). He argued that DARPA's investment in HDTV display technology would benefit the Defense Department by reducing the cost of high-definition screens for military use and at the same time would contribute to stimulating the development of a U.S. commercial HDTV manufacturing base. Fields felt very strongly that the defense industrial base could not be disassociated from the consumer industrial base. In June 1989, when the High Definition Systems (HDS) program formally began, DARPA awarded federal grants for the development of HDTV displays to five U.S. companies ("Five U.S.," 1989). In August 1989, Fields complained to Thomas Murrin, the Commerce Department's Deputy Secretary, about what he perceived to be a personal vendetta orchestrated by the Administration to discredit him. He stated in a letter partially published in *The Wall Street Journal*:

> Anyone who proposed any ideas for solving the competitiveness problem can be silenced by accusing him of supporting "industrial policy." … And anyone who suggests an approach tailored to the unique cir-

cumstances of trade with particular countries ... is a "Japan basher."
It's hard to succeed against such a quiver of verbal arrows. (Davis,
1990, p. 230)

In November 1989, the White House began pressuring the
Department of Defense (DoD) to cut back DARPA funding for research
on HDTV and other advanced technologies. John Markoff (1989) of *The
New York Times* reported that "Several Congressional staff members
and officials at government agencies ... had been told that Deputy
Defense Secretary Donald Atwood had in recent days sent an internal
document, called a Program Budget Decision, to Dr. Fields that calls
for placing a number of recently released HDTV contracts on hold" (p.
D9). A high-ranking Pentagon official later explained that this "memorandum had been sent 'in error'" and that HDTV "is not going to get
out" (Richards and Burgess, 1989, p. E1). But in April 1990, the Defense
Department quietly announced that it would divert $20 million of the
$30 million earmarked by DARPA for HDTV research for other uses,
such as foreign aid (Markoff, 1990). Yet Fields was not deterred, and at
about the same time he committed $4 million of DARPA money to
Gazelle Manufacturing, a small Californian company that produces
gallium arsenide chips, which might enhance the performance of computers ("The Government's Guiding," 1991). According to Daniel
Greenberg, editor of the newsletter *Science & Government Report,*
"Sununu went bananas" (Bylinsky, 1991, p. 65) with this latest instance
of industrial policy investment.

This was the coup de grace for Fields, and in May 1990 Atwood
transferred him from DARPA to a "new, important job within the
Defense Department," Deputy to the Director of Research and
Engineering for Defense Management Report Implementation
("DARPA Director," 1990, p. 70). According to Congressional and
Pentagon staffers, Fields was given three choices: accept this dead-end
position, resign, or be fired for insubordination (Davis, 1990). In July
1990, he left the DoD to head Microelectronics and Computer
Technology Corp., an industry consortium. By forcefully removing
Fields from his position, the White House had successfully eradicated
the last vestiges of HDTV industrial policy support within the
Executive Branch of the U.S. government.

The Outcome

None of the nine HDTV bills ever became law. But by early 1991, to the surprise of many, the Bush Administration changed course and unveiled a "technology policy" plan to safeguard U.S. competitiveness in critical industries (e.g., R&D support for new technologies). This policy reversal can be attributed to a series of factors, such as continuous Congressional pressure, lobbying from industry groups, the swelling trade deficit with Asia, the U.S. looming recession, and even perhaps the desire to erase the perception held by many that the White House mishandled the HDTV issue in the late 1980s. Because of the sliding U.S. economy, federal support for key electronics industries became a more attractive policy option for the Administration than it was in the past. The architect of this new Bush strategy, Allan Bromley, Assistant to the President for Science and Technology, convinced "the White House that it can plunge deeper into supporting key technologies without intervening in the market's selection of winners and losers" (Faltermayer, 1991, p. 49). This new attitude in the White House contributed to the passage of the American Technology Preeminence Act of 1991.

On April 23, 1991, Rep. Tim Valentine (D-North Carolina) introduced H.R. 1989, cited as the American Technology Preeminence Act of 1991 and modeled after H.R. 4329. On May 9, 1991, Ernest Hollings sponsored the Senate version, S. 1034. H.R. 1989 passed the House on July 16. S. 1034 passed the Senate on November 27. President Bush signed the bill into law on February 14, 1992. The American Technology Preeminence Act was designed to support a *variety* of advanced technologies and did not specifically target the development of HDTV. In fact, "high-definition television" is never mentioned in the legislation. Instead lawmakers substituted "high-resolution information systems," a term far more neutral than HDTV. It refers to "equipment and techniques required to create, store, recover, and play back high-resolution images and accompanying sound" (American Technology Preeminence Act, 1991, p. 7). The Act (1) authorized the appropriation of funding to the Department of Commerce's Technology Administration ($10 million for FY1992 and $10 million for FY1993) and the NIST ($210 million for FY1992 and $221 million for FY1993); (2) amended the Advanced Technology Program section of the NIST Act and set eligibility requirements for receiving federal assistance under the program; and (3) required the completion of several studies and reports. More important for this chapter, Title V of the Act established a 13-member High-Resolution Information Systems Advisory Board within the Executive Office of the President's Office of Science and Technology "to monitor and, as appropriate, foster the development of United States-based high-resolution information systems industries" (p. 22). The Board was required to submit an annual report to the President and Congressional Committees. But it never met because given the hotly debated nature of the subject matter the Bush Administration never appointed the members. By the time the Democrat Clinton Administration moved into office in January 1993, it had decided to work closely with the industry and therefore there was little need to implement the language of the law (P. Windham, U.S. Senate Committee on Commerce, Science, and Transportation, personal communication, June 18, 1996).

The story of HDTV industrial policy does not quite end with the passage of the American Technology Preeminence Act of 1991. After President Clinton took office in January 1993, he endorsed the principle of a stronger government-business relationship and embarked on an overt industrial policy to support America's high-tech industries and make the nation more competitive ("Do Not Adjust," 1993; Rowen, 1992). Not only has Clinton advocated a more active role of government in promoting R&D activities, but he has been willing to offer federal assistance for the commercialization of new technologies. For instance, in February 1993, he outlined a plan to spend $17 billion on advanced technologies over four years ("Do Not Adjust," 1993).

Since 1993, there has been a DARPA funding redux for flat panel display (FPD) technology. The Clinton Administration (through Secretary of Defense William Perry) has capitalized on the dual-use technology approach to support the development of a U.S.-based FPD industry, a strategy that Fields had unsuccessfully attempted to implement during his tenure as DARPA Director. "Any initiative under the dual use strategy, rather than maintaining defense-unique producers, seeks to foster the creation of a viable domestic industry that is competitive in global markets and able to meet defense requirements drawing on the commercial technology base" (The Flat Panel Display Task Force [FPDTF], 1994, p. I-2). In its October 1994 report, the Defense Department's Flat Panel Display Task Force (1994) observed that Japanese firms dominated the world FPD industry with an estimated market share of over 90 percent, while U.S. firms' share did not exceed 3 percent. In 1993, the size of the world FPD market was estimated to be $6.5 billion, with forecasts of $20-$40 billion by the year 2000 (FPDTF, 1994). Not only may a weak domestic FPD production base increasingly impair the competitive position of the U.S. electronics sector, the report argued, but it may adversely affect the Department of Defense (DoD), which depends heavily on that sector for its military applications. Therefore, in pure MITI style, it concluded that a robust FPD domestic industry was vital for the national and economic security of the United States.

This rationale led the DoD to intensify its involvement in the High Definition Systems (HDS) program, established by DARPA in 1989. From 1989 to 1993, the program heavily emphasized research and development of display technology. But then when President Clinton took office, the program also began providing assistance for the development of a *domestic* display manufacturing infrastructure. For instance, in October 1993, the DoD helped establish the U.S. Display Consortium (USDC), an industry-led group of display manufacturers, users, and components suppliers, and provided it with seed money (Markoff, 1996). In April 1994, it created the National Flat Display Initiative. This program was designed to (1) support U.S. FPD R&D activities; (2) build high-volume process and manufacturing infrastructure; (3) encourage capital investment for FPD production through R&D incentives; and (4) assist in market development (FPDTF, 1994; see also http://eto.sysplan.com/ETO). In March 1996, the Xerox Corporation, a USDC member, announced the formation of Dpix, a company that will manufacture high-resolution flat panel displays (Markoff, 1996).

Through the Defense Department, Commerce Department, Energy Department, NASA, and the National Science Foundation, the U.S. government spent $445 million on FPD R&D programs from FY1991 to FY1994, with DARPA accounting for about 85 percent of the support. DARPA funding on flat panel displays and related technologies totaled $542 million from FY1989 to FY1996: $5 million for FY1989; $30 million for FY1990; $74.5 million for FY1991; $75 million for FY1992; $151.2 million for FY1993; $79 million for FY1994; $79.4 million for FY1995; and $48.3 million for FY1996 (FPDTF, 1994; M. Hartney, DARPA, personal communication, June 17, 1996). The President's budget request for FY1997 was $45 million. Clearly, the funding situation of the HDS program has been shaped by three phases of political engagement. In 1989 and 1990, amidst the heated Congress-White House-Fields-Mosbacher squabbles about HDTV industrial policy, the program received a meager $35 million. But in 1991 and 1992, the Bush Administration adopted a more conciliatory position and allowed funding of R&D programs for emerging technologies. This change of attitude resulted into a two-year $149.5 million package for DARPA's support of FPD technology. Finally, since 1993, the Clinton Administration has embraced the dual-use technology rationale and earmarked significant subsidies ($358 million until FY1996) for commercial and R&D FPD programs, although funding dwindled to $48 million in 1996 under the pressure of budget cutbacks. Consistent with the more interventionist Democrat philosophy, the Clinton Administration supported greater cash infusions into the HDS program to achieve certain economic goals than did the Bush Administration, which privileged a marketplace solution.

Discussion and Conclusions

It would be tempting to lay all the blame for the failure of HDTV industrial policy on the Bush Administration. Not only was this not the case, but the Administration had some valid reasons for opposing such efforts. While it is true that White House officials took every possible step to halt HDTV initiatives and bury the dossier, their industrial policy-averse attitudes did not necessarily conflict with the views of many Americans. Industrial policy has always been a controversial topic in American politics because it calls into question the basic free trade and market economy principles that have established the United States as an economic power since 1945 (Johnson, 1984). There is a real concern about government arbiters as substitutes for market forces. "The United States does not view industry as a matter of national security as Japan does" (Prestowitz, 1989, p. 102). For the Bush Administration, the HDTV controversy boiled down to an assault on free trade principles and Congressional efforts to promulgate government-led industrial policy. In light of the growing trade imbalances between the United States and Japan, some economists advocated that "managed trade," which involves government intervention through unilateral, bilateral, or multilateral agreements, was a more realistic approach than free trade. Others, however, contended that "freer trade," not protectionism, was necessary to promote employment, increase national wealth, and expand consumer choice. The Bush Administration generally adhered to the principles of free trade, although it did not always do so consistently as demonstrated in the auto parts dispute between U.S. and Japanese companies in 1992 (see Rowen, 1992). The White House's position on the commercialization of HDTV was to do little or nothing to alter the market forces, thereby antagonizing the Democrat-dominated Congress which demanded federal action and funding. Administration officials did not want to single out commercial technologies for federal financial support and pick "winners and losers" in the process (as George Bush had insisted in his 1988 campaign) (Cloud, 1989). They were adamantly opposed to creating any type of industrial policy that would directly aid U.S. commercial firms at the expense of their foreign competitors. While Mosbacher supported a revision of antitrust laws and capital gains tax to encourage HDTV development and manufacturing, he *never* favored involving the federal government in any direct financial assistance programs other than seed money.

The AEA committed political suicide by seeking an unrealistically large aid package ($1.35 billion), and therefore contributed to the demise of any further attempts to promote HDTV industrial policy by giving political ammunition to the White House against it. This move may have alienated even the most fervent supporters of HDTV industrial policy. The AEA asked for too much too quickly at a time when such a request was politically untenable, despite the backing of some in Congress. Had they been more reasonable and politically savvy in their demands, some federal assistance could have been approved.

Even Craig Fields bears some responsibility for the debacle. As DARPA Director, Fields embarked on a solitary mission, to promote the commercial development of an HDTV industry through Defense federal grants, although he never considered DARPA activities as "technology policy" or "industrial policy" (see "The Government's Guiding," 1991). His mistake was to use public channels (e.g., Congressional hearings) to convey his ideas instead of operating behind the scenes as traditionally done in the past (U.S. House of Representatives, 1989a, 1989b; U.S. Senate, 1989b). DARPA activities are rarely publicized beyond the nomenclature of the Pentagon (B. Davis, *The Wall Street Journal*, personal communication, June 14, 1996). To secure a larger funding base, Fields decided to take his case directly to lawmakers. But his unconventional approach—going public with a traditionally private agenda—backfired by exposing him to criticism from his peers within the Defense Department and the Bush Administration. Had Fields acted less overtly, he would have probably kept his job but would not been able to lobby Congress for HDTV-related DARPA funding or disseminate his views beyond the Defense Department. In retrospect, this was a no-win situation for him. In addition, Fields was so obsessed with HDTV that he became "impatient and abrasive in pursuing his goals" (Carey and Bartimo, 1990, p. 31). By all press accounts, his tenure as Head of the Microelectronics and Computer Technology Corp., which he left in June 1994, was equally controversial (see Burrows, 1994; Port and Burrrows, 1994).

In retrospect, the implementation of a full-blown industrial policy for the commercial development of HDTV was doable but failed because of the inability of *all* parties to pull in the same direction and set common, realistic goals. The successful reentry of the United States into the highly competitive consumer electronics market necessitated more than just a government aid package. It demanded a multifaceted reform, including educational, economic, and political restructuring (see Dupagne, 1990). There is no indication that either the industry or the government was prepared to make such a long-term commitment.

Despite the definite normative advantages of a healthy consumer electronics sector (e.g., employment), it is uncertain that HDTV industrial policy instruments as proposed in the bills would have been successful and would have developed a substantial manufacturing base in the United States. The U.S. television industry ceased to exist in July 1995 when Zenith Electronics Corporation, the last U.S. TV set maker, agreed to sell a controlling interest (57.7 percent) to the Korean company LG Electronics (Goldstar) (Feder, 1995). This latest acquisition confirms the misgivings that some policymakers had about HDTV industrial policy in the late 1980s: Why should public dollars subsidize HDTV technology when there is no guarantee that would-be American HDTV manufacturers would not sell their facilities to foreign competitors in the long term or would not continue to move offshore to benefit from lower labor costs? The trend of buyouts in U.S. consumer electronics (e.g., Thomson's acquisition of RCA-GE's consumer electronics division in 1987) has done little to reassure them. It is equally unclear whether the problems that have plagued the U.S. television manufacturing industry would have disappeared with the implementation of an HDTV industrial policy.

Unlike the United States, Japan and Europe have adopted industrial policies to promote the development of HDTV, consistent with a long tradition of government intervention in the consumer electronics industry. Since the mid-1980s, MITI and the Ministry of Posts and Telecommunications (MPT) have actively promoted uses and applications of Hi-Vision/MUSE technologies domestically and abroad (see Chapter 3). Likewise, under the aegis of the European Community, EUREKA countries established the EU 95 project in 1986 to create and champion a European HDTV system (see Chapter 4). Contrary to popular wisdom, though, most of the funding invested in Japanese and European HDTV projects—an estimated $1.3 billion in Japan (Johnstone, 1993) and $850 million in Europe (HDTV Directorate, 1994)—has emanated from the private sector. By 1986, Nippon Hoso Kyokai, Japan's public broadcasting organization, provided less than 15 percent of the total $700 million investment in HDTV technology (Choy, 1989).

But in the early 1990s, when the United States edged out its European and Japanese rivals by developing a digital HDTV transmission standard, there was suddenly no need for, or interest in, HDTV industrial policy. The United States was able to leapfrog the Japanese in digital transmission technology by building on Japanese initial research and coming up with a superior product—a strategy that Japan has excelled at in the consumer electronics area (Seel, 1992). Without resorting to public funding, the members of the Grand Alliance (GA) (AT&T, the David Sarnoff Research Center, General Instrument, the Massachusetts Institute of Technology, North American Philips, Thomson Consumer Electronics, and Zenith Electronics) successfully produced the most sophisticated television system available at the time (see Chapter 1). GA participants spent an estimated $200-$300 million on HDTV technology. For many, this technological lead proved that competition works and that market forces, rather than government, are to be trusted with the development of high-tech industries (e.g., Beltz, 1994). But while it is true that competition brought about digital television prototypes in America, the final U.S. digital transmission standard was created through a private joint venture, not through competition. The Grand Alliance is a consortium of seven European and American entities, and collaboration between these former competitors was indispensable for designing successfully the digital HDTV system (see Chapter 7). The National Cooperative Research Act of 1984 has gone a long way to ensure the creation of this grand alliance by exempting "reasonable" research joint ventures from antitrust violations (Braun, 1995). One can even argue that the HDTV cooperative research model in the United States was not unlike the one used in Europe (EUREKA 95) minus the public subsidies.

In addition, because of global trade, industrial policy in consumer electronics may not be as workable today as it was in the past. The real winner of the Grand Alliance technology is not the United States or U.S. companies—which will only collect royalty fees, but Thomson of France and Philips of the Netherlands—which will manufacture the product for consumer use. In an era of growing dominance by these multinational consumer electronics giants, it becomes increasingly problematic to define the economic "national interest" of any single country and formulate an industrial policy consistent with that interest (see Chapter 4). The G-8 nations are bound together in an economic web that extends to all nations of the globe.

HDTV Policymaking in
the United States

When this process began the Commission hoped for a new broadcasting standard that would mark the next evolution in broadcast television. The standard we have before us, however, amounts to nothing less than a revolution.

—*Reed Hundt, Federal Communications Commission Chairman, 1996*

As noted in the preceding chapter on industrial policy, nations with substantial high-technology industries are wrestling with the dilemma of developing technology policies that help, rather than hinder, their respective national economies. Christine Ogan (1992) has identified three approaches that nations can pursue:

1. Develop policy in reaction to technology. Ogan cites the in-home use of VCR recording as a case where the U.S. Supreme Court ultimately had to decide on the legality of recording over-the-air programs. Since the development of new technologies usually outpaces the ability of governments to anticipate their ultimate applications, most nations are playing a catch-up game in trying to create policy after the fact.

2. Develop policy that anticipates technological change. For the reasons outlined above, this is far easier said than done, especially by government officials far removed from the innovation process. As the Japanese have learned with their (primarily) analog HDTV standard, it is possible for government-dictated industrial policy to foreclose innovation and lead a nation to adopt technology that is soon obsolete.

3. *Develop policy that evolves with technology.* Ogan cites as an example the policy changes that have evolved with the transformation of the U.S. cable industry from a community antenna service, designed to relay over-the-air broadcast signals, into multichannel program origination service.

This chapter will illustrate how U.S. high-definition television policy has reflected all three aspects. It has been *reactive* in response to initial Japanese and European activities in HDTV R&D. U.S. policy has been *proactive* in the creation of the Advisory Committee on Advanced Television Service and the test center it formed to determine the best HDTV transmission format for this country. Finally, national policy has *evolved* since ACATS was created in 1987 to ultimately embrace a digital transmission scheme that was thought to be technically impossible at the outset.

The Broadcast Policymaking Model in the United States

Communications policymaking is a very complex subject to study. There are literally thousands of policy "actors" in the United States alone—FCC regulators, elected officials, industry executives, consumer activists, judges, to name just a few categories. Significant parts of the literature in the communications law and policy paradigm are concerned with making sense of the interrelated activities and motivations of this mass of individuals and organizations.

The classic work in the subcategory of broadcast media policy regulation is Krasnow, Longley, and Terry's *The Politics of Broadcast Regulation* (1982). Haeryon Kim (1992) noted that the model "has been the dominant conceptual framework for research in broadcast policymaking for almost two decades" (p. 154).

The authors have devised a model (referred to as the KLT model hereafter) with three key actors: The Federal Communications Commission (FCC), Congress, and the "regulated industries." They interact with an additional three actors of varying influence: the courts, the White House, and citizen groups (see Figure 6.1).

The model was first developed by Longley in his 1969 political science doctoral dissertation and included only the first three primary actors (Kim, 1992). The first edition of the book added the latter three

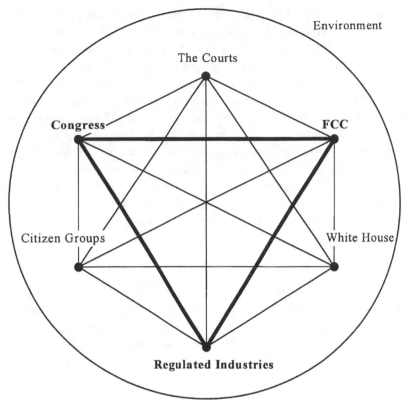

3 primary actors in bold

FIGURE 6.1. The Krasnow, Longley, and Terry (KLT) broadcast policymaking model[1] (*Adapted from:* Krasnow et al., 1982).

entities to the model: the courts, citizen groups, and the White House.

The model is based upon the concept of using general systems theory to define complex interrelationships by describing a bounded environment (e.g., the U.S. political system) and the communication linkages between six categories of actors. The delineation of a boundary or "bubble" around the system is useful in isolating the elements under study, but such arbitrary national limits are problematic in a world where traditional state boundaries are being superseded by transnational organizations and communications. The boundaries of the traditional model are anachronistic and need to be revised in light of two decades of changes in transnational economics, politics, and information flow. These revisions will be noted in Chapter 9.

The systems theory approach to political activity can be traced back to the work of David Easton at the University of Chicago in the 1950s and 1960s. Easton's model of a political system identified *inputs* (e.g., what various institutional actors want from the system) and *outputs* indicated by governmental regulations and allocative decisions (Easton, 1968; Mahan and Schement, 1984). However, rather than being a linear design, Easton designated his model of political life as a "system of behavior" that could respond to environmental stress (in the nonecological sense) via informational feedback loops to various actors and decision makers (Easton, 1965). For Easton, "politics is a system of human interaction that exists in an environment, is open to that environment, and responds to it through a feedback mechanism by varying its structures and processes" (Kim, 1992, p. 156).

As adapted by Krasnow et al. from Easton's systems theory perspective, the broadcast policymaking model has the following characteristics:

1. *It is a dynamic rather than static system.* Rather than seeing regulation as just a series of static outputs (e.g., published FCC regulations), Krasnow et al. (1982) stated that their model reflected a "dynamic process involving many participants, often with different goals, all influencing each other and producing specific policies and standards for broadcasters. Our focus is on regulation as a process rather than as an outcome" (p. 1).

2. *Turbulence is induced in the process by rapidly changing technology.* Krasnow et al. view technological change as a key factor in stirring the policymaking pot. These changing conditions create stress in the status quo and motivate actors to ask for regulatory relief—or conversely, relief from regulation—from governmental agencies such as the FCC. The relief may take the form of increased regulation sought by the National Association of Broadcasters (NAB) in response to the threat of direct broadcast satellite (DBS) services, or actors such as the regional Bell operating companies (RBOCs) might seek reduced regulation in the form of relaxed cable-telco cross-ownership rules so that they might offer cable television services.

At first glance, this viewpoint seems technologically deterministic—unless one is describing an Eastonian system with feedback loops such as the KLT model. The term "technological determinism" has acquired epithet status among communication scholars since the 1980s due to a perceived simplistic, linear, cause-and-effect misapprehension about the role of technology in modern society.

In fact, the international global conflicts over the standardization of high-definition television demonstrate how technological innovation is embedded in the politico-economic cultures of East Asia, North America, and Western Europe (see Chapter 1). The contemporary observer of telecommunications policymaking must not only have an international focus but must be willing to utilize traditional policy-making models to comprehend the complex relationships *within* nations. The KLT model is still quite useful in comprehending the *intra*national policymaking interplay between the FCC, Congress, and the Executive branch of government in the United States.

HDTV Standardization and the Advanced Television Systems Committee (ATSC)

Corey Carbonara (1992) noted that contemporary American inter-est in HDTV began after the Japanese Hi-Vision demonstration at the 1981 Society of Motion Picture and Television Engineers (SMPTE) con-ference in San Francisco.[2] In 1983, the Advanced Television Systems Committee (ATSC) was formed by the National Association of Broadcasters, the Electronic Industries Association, the Institute of Electrical and Electronics Engineers, the National Cable Television Association, and the Society of Motion Picture and Television Engineers (Carbonara, 1992). The ATSC (similar to its predecessor, the NTSC) was formed by industry interests to "promote the standardiza-tion of high-definition television" in the United States (James C. McKinney, then-Chair, ATSC, personal communication, November 4, 1993). The ATSC also represents the United States at international meetings on HDTV matters, and had a significant role in documenting the technology created by the HDTV Grand Alliance.

The ATSC's annual budget is raised through dues levied from its membership which consists of over 50 consumer electronics firms, broadcasters, universities, film studios, telephone companies, and pro-fessional societies. The ATSC is an association of these constituent groups and has no legal status in connection with the FCC. However, former ATSC Chair James McKinney was one of the 25 members of the FCC's Advisory Committee on Advanced Television Service. In 1983, the ATSC represented the first American political faction showing an interest in HDTV.

Also in 1983, the Society of Motion Picture and Television Engineers (SMPTE) organized a subsidiary Working Group on High-Definition Electronic Production (WGHDEP) led by Universal Studios executive Richard Stumpf (Carbonara, 1992). SMPTE has taken a leadership role in the standardization of film formats, and WGHDEP did the same thing for HDTV, establishing the SMPTE 240M standard based upon the Japanese 1125/60 system for U.S. television and film production.

At the 1986 Dubrovnik meeting of the CCIR, the U.S. State Department and the ATSC supported the adoption of the Japanese 1125/60 Hi-Vision standard, but the Europeans blocked the proposal as noted in Chapter 1. The broadcasting community became frustrated by the ATSC's lack of progress in creating an American response concerning EDTV or HDTV. Some of this frustration can be gleaned from trade articles of the era that noted NAB and MSTV efforts to "help ATSC bring proposed systems out of the woodwork," and to "aid the stalled work of a second ATSC technology group on improved NTSC-compatible systems" ("ATSC: Looking at," 1985, pp. 68-69). In fairness to the ATSC, it should be noted that the group functioned with limited staff and funding. The research panels it created built a foundation for comparable groups (often with similar membership) that made dramatic progress under the FCC's Advisory Committee on Advanced Television Service.

In the wake of the 1986 Dubrovnik rejection (a "postponement" at the time) of the NHK 1125/60 proposal, the issue of HDTV standardization became increasingly politicized. By early 1987, the ATSC had spun its wheels for four years and political pressure increased in the United States for the creation of a uniquely American response to HDTV standardization.

John Abel, former-Vice President for the National Association of Broadcasters, noted, "I think the feeling was that the Advanced Television Systems Committee did not have enough government influence—it really had zero government influence at that time" (J. Abel, personal communication, March 19, 1994). Broadcasting industry leaders felt that a larger organization with more political clout was necessary to define a U.S. HDTV standard.

The Advisory Committee on Advanced Television Service (ACATS)

ACATS was created by the FCC in response to a February 1987 petition from 58 broadcast organizations including the National Association of Broadcasters (NAB) and the Association of Maximum

Service Telecasters (now MSTV) (Federal Communications Commission [FCC], 1987a).[3] The NAB is the larger of the two and represents all broadcasters (including radio), while MSTV's membership consists of 270 local television stations that are affiliates of the national networks (M. White, MSTV, personal communication, November 3, 1993).

One month prior to the filing of the FCC petition, MSTV and the NAB brought engineers from NHK in Japan to Washington to conduct a demonstration of the first U.S. *terrestrial* broadcast of HDTV signals. Held in the Commissioner's meeting room at the FCC, the demonstration drew a standing-room only crowd anxious to see the broadcast ("HDTV: Efforts," 1987). This was a significant event since, until this demonstration, only satellite broadcasting of HDTV had been considered.

MSTV President Margita White articulated two rationales for the broadcasters' petition to the FCC:

> [First] we were very aware that video competitors such as cable and potentially DBS did not have the spectrum constraints that broadcasters had. At that time, the MUSE system, which was the only one vetted, would have required four, 6 MHz channels to transmit a single HDTV signal. [Second] our move, and the industries move, was, in effect, defensive—just to make sure that there would be a window of opportunity if and when [HDTV] was the broadcasters' salvation (M. White, MSTV, personal communication, November 4, 1993).

Howard Miller, PBS senior vice president of broadcast operations, said that, from his perspective, the key factor in the creation of ACATS was the threat to American broadcasters from a "bypass technology" such as direct broadcast satellite (DBS) (H. Miller, PBS, personal communication, March 14, 1994). Both the Japanese and European HDTV systems were based on satellite delivery, and U.S. broadcasters were concerned that an American DBS company transmitting in HDTV might put them at a serious competitive disadvantage. By the late 1980s, a number of companies had proposed U.S. DBS broadcasting, but none of the projects had come to fruition.[4]

However, in September of 1987, in the midst of the FCC's inquiry process on advanced television, Technical Information Services Manager E.B. Crutchfield of the NAB raised the prospect that the Japanese would transmit "high-definition television to the United

States via satellite by the year 1990, about the same time consumer receivers will be available here" and this potentiality was why the search for a U.S. terrestrial transmission standard "was so important" ("Engineers Search," 1987, pp. 75-76). Aside from the allusion to an "invasion" of U.S. broadcast airspace by the Japanese (linked to a plan to sell HDTV sets), the comment reveals the concern about potential Japanese dominance of this new technology. Not only did the Japanese not attempt to transmit HDTV to the U.S. via satellite (which would have needed FCC approval), but they had their hands full just getting their own high-definition DBS service underway by 1990.

Broadcast groups were also concerned over requests to the FCC from groups involved in land-mobile radio (and cellular telephone services) for additional spectrum in the television ultra-high-frequency (UHF) band. As former NAB executive John Abel noted, "The FCC had a proceeding [to study] spectrum sharing in the top 10 [U.S. broadcast] markets. We and MSTV both were very opposed to this" (J. Abel, NAB, personal communication, March 19, 1994). The FCC had previously awarded portions of the UHF spectrum for nonbroadcast radio and the two organizations were concerned that further allocations might preclude the implementation of HDTV broadcasting.

In summary, the primary reasons for the broadcasters' 1987 petition to the FCC for the creation of the Advisory Committee were:

1. Concern over "bypass technologies" such as DBS and cable.

2. A related concern that new technologies like HDTV might require such large amounts of spectrum that they would preclude terrestrial broadcast transmission.

3. Concern over allocations of the broadcast spectrum to other users such as land-mobile radio and cellular telephone.

4. Concern that an industry-based standard-setting organization such as the existing ATSC would not have enough political clout to forge the creation of a U.S. terrestrial HDTV broadcast standard.

The FCC's Response

The chart shown in Figure 6.2 illustrates the ATV policymaking relationships between the FCC and ACATS from 1987 to July 1997. We have already discussed the FCC's ATV policies in Chapter 1—this chart shows the influence of the Advisory Committee's deliberations on related FCC policy output. In some cases there was a direct cause-

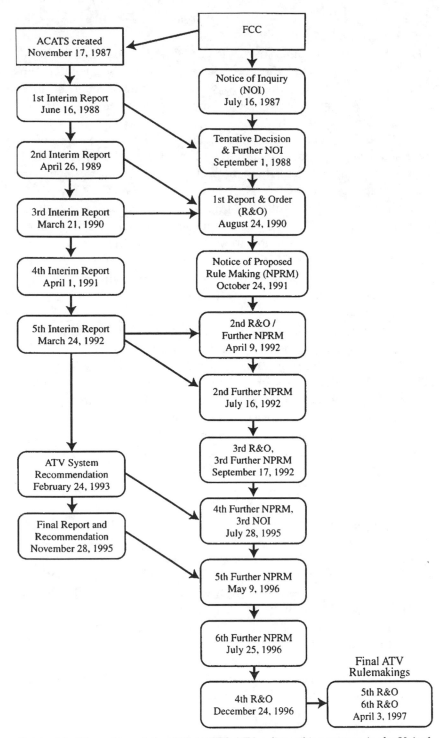

FIGURE 6.2. Chronology of the FCC-ACATS ATV policymaking process in the United States (*Source:* FCC docket MM 87-268).

and-effect linkage, in other areas such as FCC Chairman Alfred Sikes' 1990 decision to pursue a simulcast strategy for the ATV transition, the key loci for policy planning rested with the Commission. The influences of ACATS in the FCC's policymaking process are described in the following chapter.

At the outset the FCC responded to the broadcasters' petition with the issuance of a *Notice of Inquiry (NOI)*, "In the Matter of Advanced Television Systems and Their Impact on the Existing Television Service," adopted on July 16, 1987 (FCC, 1987a). The FCC's use of the term "advanced television" (ATV) instead of "high-definition television" was intentional. Advanced television is a generic description of any system that is superior to NTSC. ATV could also include improved-definition (IDTV) and enhanced-definition (EDTV) technologies in the range of possible solutions along with HDTV. The Commission wanted to keep its options open at the start of the standard-setting process.

The *Notice* set a deadline of November 18, 1987 for the submission of public comments on the issues listed above. The first meeting of the Advisory Committee occurred on November 17, 1987 ("Blue Ribboners," 1987). It was originally cited in the trade press as the "blue ribbon ATV advisory committee," due to the presence of the big-three network presidents and other CEOs from film and cable companies. The 25-member committee (see Appendix to Chapter 7 for the 1995 membership) was appointed by FCC Chairman Dennis Patrick with an eye for the combined political clout of the group.

The trade press heralded the formation of the Advisory Committee with a short article in *Broadcasting* headlined "Advanced TV Brain Trust" that identified the 25 committee members, *ex officio* members, and the chairs of the three working subcommittees ("Advanced TV," 1987).[5] Executives of the cable television industry were not happy about their limited representation (CEO Trygve Myhren of American Television & Communications, president Joseph Collins of Home Box Office, and CEO Frank Biondi of Viacom) on the Advisory Committee ("HDTV: Progress Report," 1987). The industry, led by their lobbying group, the National Cable Television Association (NCTA), felt that the Advisory Committee had a preponderance of broadcast industry representatives. Their displeasure emerged in trade press articles, but the level of representation did not change.

Critical Views of the FCC's Use of Industry Advisory Committees

If the game plan of the television broadcast community was to seize the high ground on ATV policy, they succeeded (with the assistance of Chairman Patrick) in dominating the Advisory Committee. Eleven of the 25 members were CEOs of broadcast companies and five others (including Wiley) represented the industry in various ways.[6] Minority factions included manufacturers (four representatives), cable (three), film studios (one), and telephone companies (one). The subcommittees of ACATS had a more diverse membership, but the ultimate vote on recommending an HDTV system to the FCC was made by the 25 members of the Advisory Committee.

The 16 members of the broadcast management or representative community constituted a majority of the Committee, overshadowing the minority groups. Equally important in terms of political balance was who was *not* on the Committee. There were no consumer or viewer representatives, and in light of the fact that digital technology was overtaking television in the late 1980s, it was surprising that there was no computer industry representation. In 1995, FCC Chairman Reed Hundt added Samuel Fuller of Digital Equipment Corporation and Craig Mundie of Microsoft five weeks prior to the Committee's final report. Richard Wiley indicated in 1993 that the Committee might have had a different composition if it were created at that time (1993):

> I would say in fairness to Dennis Patrick if he were doing it today he wouldn't have picked exactly the same people that he's got. [It is] probably too heavily weighted [with] too many broadcasters and not enough computer representation. (Seel, 1995a, p. 108)

The debate has continued for 60 years over the degree to which the FCC is a "captive" of the broadcast industry that it is supposed to regulate. This criticism has subsided somewhat with the growth of alternative delivery services such as cable and DBS. However, Kennedy-era FCC Chair Newton Minnow stated:

> Those who make policy and regulate must necessarily have frequent contact with the industry in order to be well informed. Under the pres-

ent system, the possibility of improper influence, or at least the charges of such influence is always present. (Moore, 1973, p. 16)

Vincent Mosco (1979) noted that this conflict of interest "is rooted in the Commission's dependence on the industry for technical information" (p. 16). The Hoover Commission in 1947 was critical of the FCC's technical staff and its reliance on the regulated industries for engineering data and advice (Lessing, 1949). A 1958 Senate report criticized the FCC's use of industry advisory committees on technology issues in lieu of adequate long-range planning by the Commission itself (U.S. Senate, 1958). Le Duc (1973) argued that the FCC has to go outside the agency because "there is no effective information-gathering process within the FCC capable of providing the material necessary to evaluate the potential for public service of new communication techniques" (p. 28).

Since the 1980s, the FCC's Office of Plans and Policy—directed by Dr. Robert Pepper—has produced a series of insightful policy reports on the future of broadcasting in a world of new communication options, but its focus is on broad policy issues rather than technology assessment. The Commission's Office of Engineering and Technology—headed by Tom Stanley, followed by Richard Smith in 1994—worked closely with the Advisory Committee throughout the proponent testing process. However, the FCC has neither the funding nor a Congressional mandate to run an advanced engineering facility such as the Advanced Television Test Center. Unless this situation is reversed in the future, the FCC will have to continue to rely on the industry for engineering advice and testing services in the evaluation of new communication technologies.

Executive Branch Involvement in HDTV Policy

The National Telecommunications and Information Administration (NTIA), formerly the White House Office for Telecommunications Policy, was active in studying HDTV in the Fall of 1987. Then-NTIA head Alfred Sikes, a former radio broadcaster from Missouri, acted as the Reagan Administration's point-man in directing high-technology telecommunications policy. Prior to the establishment of the FCC's Advisory Committee in November 1987, the NTIA had developed a more coherent policy for HDTV than either the Commission or the ATSC.

The NTIA had three primary goals concerning high-definition television ("Government Support," 1987):

1. That the chosen HDTV standard would be "spectrum efficient." The necessary channel spacing with NTSC (to prevent cross-channel interference) is very wasteful of the electromagnetic spectrum allotted to television.

2. That the adopted system be as "compatible as possible" with the existing NTSC system. There is an inherent contradiction with this goal—to invent an HDTV system that was fully compatible with NTSC meant that the new technology would inherit the faults of the old. This is why the adopted digital HDTV standard for the United States is incompatible with analog NTSC, although it can be down-converted to it.

3. That the implementation of the new television system "benefit American manufacturers." As NTIA spokesperson R.T. Gregg stated, "We would like U.S. industry to participate in this new technology. If all the HDTV equipment is manufactured offshore, it will have a negative impact on a lot of things, not the least of which is the balance of trade" ("Government Support," 1987, p. 53). This statement is one of the first official indications that the Executive Branch was examining the implications of Asian or European dominance of the U.S. market for HDTV hardware.

An Executive Branch Champion for HDTV: Al Sikes

Alfred Sikes was to play a pivotal role in the promotion of a U.S. HDTV standard at both the NTIA and later as Chairman of the FCC (he served in the latter post from 1989 to 1993). At the NTIA in November 1987, Sikes was urging the FCC to support the broadcasters petition for standardization and a freeze on UHF spectrum allocations to mobile radio ("Government Support," 1987). He also called for the FCC to reach a decision on HDTV standardization by December 1991 and the NTIA nudged American companies to "band together" to conduct research on HDTV technology.

The NTIA is a branch of the Commerce Department, an agency which sponsored the National Cooperative Research Act (NCRA) of 1984. The Act exempted U.S. companies from antitrust laws to promote coordinated high-technology research (Braun, 1995). The U.S. television manufacturing industry had been hammered by foreign competition, primarily from Asian companies, and the Department of Commerce and NTIA thought that HDTV might further exacerbate the balance of payments problem unless U.S. companies played an active part in its development. A secondary concern was that American high-technology semiconductor and computer industries might be frozen out of a dynamic new growth area by virtue of entering the game too late.

As Sikes said at the time, "We see this [the creation of a distinctive U.S. HDTV standard] as an opportunity. Whether it comes about remains to be seen" ("Government Support," 1987, p. 54). Bob Davis of *The Wall Street Journal* has documented the roles of Alfred Sikes at NTIA and Craig Fields at the Defense Advanced Research Projects Agency (DARPA) as two key advocates for government support of HDTV research (Davis, 1990). As noted above, Sikes saw HDTV as an opportunity for the United States to take a proactive stance on a key high-technology issue. Davis (1990) noted that Sikes "recruited allies in Congress, commissioned studies showing a huge potential market, courted Secretary of Commerce Robert Mosbacher and pushed industry to devise a plan to pursue the HDTV market" (p. 227).

HDTV and the Defense Department

Craig Fields at the Department of Defense (DoD) was concerned about HDTV for different reasons. He viewed the Japanese "as an adversary second only to the Soviet Union" (Davis, 1990, p. 227) and saw their potential domination of the world market for integrated circuits (ICs) as a threat to the American supply of chips needed for essential defense hardware. While this worldview seems dated given recent U.S. resurgence in chip-making, at the time the Japanese were dominating the world market for dynamic random access memory chips (U.S. Congress, 1990). Things looked bleak for the long-term survival of U.S. electronics giants such as the Intel Corporation and Texas Instruments, and Fields was concerned about U.S. dependence on foreign sources (e.g., Japan) for the basic IC building blocks of all advanced electronic defense technology.

Fields' intention in 1988-89 was to utilize a small portion ($30 million) of the Defense Department's enormous R&D budget to fund the development of American prototype HDTV display systems. DARPA was interested in high-definition flat panel screens that could be used to display battlefield images in aircraft, tanks, and ships. From the perspective of the Defense Department, HDTV had implications not only for the health of the U.S. semiconductor industry, but also for the ability of American manufacturers to make essential military display systems.

Congress, the White House, and the Debate Over U.S. Competitiveness

The activities of Al Sikes at NTIA and Craig Fields at DARPA to promote the development of a distinctive U.S. HDTV standard attracted the attention of Congress. Beginning in 1989, the U.S. Congress held a series of hearings on HDTV and related technology, trade, and competitiveness issues. Between the House and Senate, nine bills in 1989-90 were introduced that were designed to promote the development of HDTV technology in the United States (see Chapter 5).

For instance, H.R. 1267 contained funding for HDTV research, relaxed antitrust laws for cooperative research, and provided tax incentives for R&D (H.R. 1267, 1989). H.R. 1516 provided $100 million per year for four years of research and had the Department of Commerce act as a catalyst for cooperative research (H.R. 1516, 1989). H.R. 2287 proposed to set up a "Sematech"-like consortium called "TV Tech" to conduct HDTV research (H.R. 2287, 1989). None of the nine bills were passed into law.

In the Senate, John Kerry of Massachusetts authored S. 952 (1989) to support joint ventures in cooperative HDTV research, but the bill died in committee. Despite the number of hearings in Congress on government support for HDTV and other advanced technologies, neither house seemed able to turn concern over U.S. competitiveness into legislation in 1989 or 1990. If any of the pro-HDTV bills had passed the Senate and House, getting them signed into law by the President would have been very problematic during the Bush Administration. Like Ronald Reagan before him, George Bush was philosophically opposed to federal financial support of commercial technologies. Both Republican administrations supported federal backing of R&D in the areas of theoretical science, military technology, or space exploration. These multibillion-dollar projects could be supported philosophically since they were seen by the executive branch as "noncommercial" technologies that could only be funded with the deep pockets of the federal government.

As noted in Chapter 5, the issue of federal R&D support for targeted technologies such as HDTV came to a head in May 1989 with the appeal of the American Electronics Association (AEA) to Congress for $1.35 billion in federal subsidies and loans for HDTV research. Shortly after this proposal, Secretary of Commerce Robert Mosbacher was "strongly" encouraged by the White House to cease any further Commerce Department promotion of HDTV. Craig Fields at DARPA continued to take a strong advocacy role in promoting Defense Department involvement in HDTV R&D and lost his job as a result. The message from the White House came through loud and clear— there would be no Executive Branch or Department of Defense promotion of research and development of narrowly targeted technologies such as high-definition television.

Alfred Sikes took a different tack—he began to successfully lobby for the chairmanship of the FCC with the assistance of his mentor, Missouri Senator John Danforth, Chairman of the Senate Commerce Committee. Confirmed in 1989, Sikes was instrumental in pushing HDTV standardization as a front-burner issue at the FCC. It is difficult to assess why Sikes was nominated by the Bush Administration despite his active role in promoting a pro-HDTV industrial policy at the NTIA. It is possible that his appointment was made before the industrial policy debate came to a head in the spring of 1990. It is equally likely that the Administration saw the FCC standardization process as a way to covertly support a distinctive U.S. HDTV standard without having to create an overt industrial policy that would fund specific electronics industries. In other words, the FCC process could allow the government to pick technological "winners and losers" (ostensibly U.S. companies as winners, foreign companies as losers), but without any overt federal subsidy that would violate the Reagan doctrine on industrial policy.

Digital HDTV

While Congress and the White House waffled on taking action on the issue of technology policy, the General Instrument Corporation developed a radical new method of television transmission that shifted the international balance of power in HDTV technology. The development of a U.S.-invented digital transmission scheme for HDTV silenced critics who stated that the United States was hopelessly out of the race to develop high-definition technology. Even such ardent HDTV advocates as Joe Flaherty of CBS had stated in 1987 that "we'll have digital television the same day we have an antigravity machine" (Brinkley, 1997a, p. 94). It also bolstered the case *against* massive federal subsidies for HDTV research and development. General Instrument developed this technology without any federal R&D aid or industrial policies targeted toward HDTV. However, the development of a U.S. digital transmission scheme did not guarantee long-term success in HDTV manufacturing for American companies—it only meant that they will receive royalties from those who successfully market hardware based on the new standard.

Tables 6.1 and 6.2 list significant HDTV policymaking events in the United States between 1983 and 1997. The early period from 1983 to 1990 reflects Asian and European influences on American industrial policy culminating with intense congressional interest in advanced television technology. 1990 was a "watershed" year for HDTV policy—it marked overt White House efforts to muzzle advanced television proponents in the Executive Branch, and also marked the breakthrough in digital transmission technology by the General Instrument Corporation. During the period from 1990 to 1997, the focus shifted from Congress and the White House to the FCC, their Advisory Committee and the competition to determine the "ideal" U.S. standard.

FCC Policy Development on Advanced Television Standardization

With this chronology in mind, it is important to analyze the FCC's policy output over this period concerning advanced television matters. It is also relevant to decipher how the Commissioners—the Chairman, in particular—put their own stamp on policy output. An overview of the process was provided in Chapter 1; the section below will analyze significant orders from 1988 to 1997 to trace influences on FCC ATV decisionmaking in a longitudinal manner.

Terry and Krasnow in their rebuttal of Kim's criticism (1992) of their model indicated that she had used but one case in her critique and stated:

> Kim's article is not, however, a real test of our model because it analyzes the output of a single agency, the Federal Communications Commission, as reflected in a single policy document, its 1982 DBS *Report and Order.* Our model was never designed for the analysis of a

single output from one regulatory participant. Rather, it is intended to describe and explain the *longitudinal development* of policy in a given area by *multiple participants. To use the model to understand DBS policy requires an expansion of Kim's analysis in both the scope of time examined and the role of other participants* [emphases added] in the formulation and implementation of DBS policy. (Terry and Krasnow, 1992, p. 479)

It is a valid criticism of using only a single case, and only taking the FCC's perspective in analyzing policy output that, of necessity, changes over time. Terry and Krasnow argued that FCC policy on DBS was not a *radical change* as Kim argued on the basis of the single case, but rather demonstrated an *incremental approach* when examined from a longitudinal perspective. There are parallel radical/incremental

TABLE 6.1 CHRONOLOGY OF EARLY U.S. HDTV POLICYMAKING (1983-1990)

DATE	ACTIVITY	KEY PARTICIPANTS
1983	ATSC formed	NAB, EIA, IEEE, NCTA, SMPTE
1983	WGHDEP formed	SMPTE, U.S. film and TV industry
MAY 1986	SMPTE/ATSC support Japanese Hi-Vision standard	SMPTE, ATSC
MAY 1986	"Dubrovnik rejection" of Hi-Vision at CCIR	European Community, CCIR/ITU, NHK, U.S. State Department, ATSC
FEB. 1987	ATV standardization petition sent to FCC	NAB, MSTV, FCC
JULY 1987	ATV *Notice of Inquiry*	FCC
LATE 1987	Sikes at NTIA promotes HDTV	NTIA, U.S. Commerce Department
NOV. 1987	Advisory Committee (ACATS) created	FCC, NAB, MSTV
JUNE 1988	ACATS first interim report	ACATS
SEPT. 1988	ATV *Tentative Decision and Further Notice of Inquiry*	FCC, ACATS
AUG. 1989	Sikes appointed FCC Chairman	White House, FCC
1989-1990	9 bills introduced in Congress to support HDTV research	Congress, industry
MAY 1989	AEA asks Congress for $1.35B.	Congress, industry
MAY 1989	White House orders halt to pro-HDTV industrial policy	White House, Commerce Dept., NTIA, DoD
APR. 1990	Craig Fields fired at DARPA	White House, DoD

NOTE: See list of abbreviations.

TABLE 6.2 CHRONOLOGY OF RECENT U.S. HDTV POLICYMAKING (1990-1996)

DATE	ACTIVITY	KEY PARTICIPANTS
JUNE 1990	First ATV digital transmission system	General Instrument Corporation
AUG. 1990	FCC *1st Report and Order (R&O)*— simulcast strategy outlined	FCC, ACATS
JULY 1991	ATV proponent system testing begins at ATTC	proponents, ATTC, ACATS, FCC
OCT. 1991	ATV *NPRM*—broadcasters get first option on ATV spectrum	FCC, ACATS, broadcast industry
APR. 1992	ATV *2nd R&O/Further NPRM*— 15-year transition to ATV defined	FCC, ACATS, (broadcasters opposed)
FEB. 1993	No clear winner in ATTC testing— plans made for second round	ATTC, ACATS Special Panel, FCC, proponents (R. Wiley intervenes)
MAY 1993	Grand Alliance consortium formed	proponents, ACATS
APR.-AUG. 1995	Grand Alliance ATV system tested at ATTC	Grand Alliance, ATTC, ACATS, FCC
NOV. 1995	ACATS endorses GA system as ATSC Digital Television Standard	ACATS, FCC, GA, ATSC— (ACATS ceases operations)
FEB. 1996	*Telecom. Act of 1996* passed — awards ATV spectrum to broadcasters	Congress, White House, FCC, telecommunications industry
MAY 1996	*5th NPRM*—FCC to adopt ATSC standard	FCC, ATSC
JULY 1996	*6th NPRM* — channels 7-51 for ATV	FCC
NOV. 1996	Accord reached between GA partners, broadcasters and CICATS	FCC, GA partners, Broadcasters Caucus, CEMA, CICATS
DEC. 1996	*4th R&O*	FCC adopts DTV standard
APR. 1997	*5th R&O*	General service rules adopted
APR. 1997	*6th R&O*	Table of Allotments for DTV channel assignments adopted

change issues within the HDTV policy case when examined over time—another rationale for taking a longitudinal view of policymaking.

To make fair use of the model as the authors intended, we will examine FCC policy on advanced television over a 10-year period, from November 1987 to April 1997. We will also analyze the policy perspectives not only of the Commissioners, but that of the varied factions within ACATS as well. Despite the efforts of Richard Wiley to build consensus within the advisory panel, there have been a number of policy issues that created conflict. Some of these points remain unresolved to date such as progressive-scan versus interlace-scan.[7] Thus

the analysis will be longitudinal and will encompass the diverse perspectives of the policymaking participants. The goals of each group of
players will be defined from a game theory perspective to see which
were achieved in each phase of the policymaking process.

The analysis will also integrate an "ecology of games" approach
(Dutton, 1992) to study the influences of other games being played *in
parallel* with the HDTV standardization process by key participants in
the latter effort. It is important to reaffirm that the use of the term
"game" in this context is not to trivialize the process or to inject a cyn-
ical tone. To the contrary, the high stakes in this undertaking indicate
that there has been significant personal, corporate, and governmental
investment in the eventual outcome of the standard-setting process.
Game theory is a useful tool in comparing the goals and strategies of
participants in the ATV policymaking process with the actual outcomes.

Key Policy Decisions—Phase 1

The FCC's key policy decisions can be broadly classified in eight
phases. The first key policy decision on ATV made by the FCC was the
creation of the Advisory Committee itself. The decision to go to a blue
ribbon panel for advice was significant for several reasons:

1. The involvement of industry CEOs gave the panel credibility
with the Executive Branch and Congress. This political clout would be
a future asset in building consensus on the path to take for HDTV standardization.

2. Allowing the U.S. broadcast, cable, and telephone industries to
run the comparative testing drew them into the process and created
interest in HDTV technology. It may also have stacked the deck against
the Japanese in favor of Euro-American alliances such as the
Advanced Television Research Consortium (Philips, Thomson, and
Sarnoff).

3. The development of the Advisory Committee created a structure
by which the FCC could enlist the volunteer aid of over 1,000 technical experts to define a national competition to create an ideal ATV
transmission system. This voluntary approach saved the government
millions of dollars in design and testing costs, but allowed industry to
define the rules of the game.

4. Allowing industry to dominate the process may have marginal-
ized the input of other important players such as consumer and view-

er groups. In fairness to ACATS, consumer groups were not exactly breaking down the door to participate in the early stages of the process.

Key Players

The key players in this phase of the policymaking game were FCC Chairman Dennis Patrick, ACATS Chairman Richard Wiley, Alex Felker of the Commission's Mass Media Bureau (the latter two selected the subcommittee vice-chairs and working parties), and the leadership of the National Association of Broadcasters and the Association of Maximum Service Telecasters (MSTV) who petitioned the FCC for the advanced television *Notice of Inquiry*.

Selected Goals

The NAB and MSTV achieved their goal of having the FCC investigate a national ATV terrestrial standard to allow U.S. broadcasters to compete against the potential threat of HDTV transmission from cable or DBS. A corollary goal was achieved by blocking the potential reallocation of broadcast spectrum to land-mobile radio.

The FCC won also in this phase by responding to the entreaties of television broadcasters, a favored segment of the regulated industries in the Krasnow et al. model. The Commission also won by taking a leadership role in the promotion of new convergent communications technology. The FCC had been criticized in the past for waffling in technology policymaking by failing to mandate a sole standard for new technologies such as AM stereo radio. From an ecology of games viewpoint, the FCC also won by appearing to assist the standing of U.S. broadcasters and manufacturers in the separate, but related, international HDTV standardization game.

FCC Chairman Dennis Patrick won by virtue of delegating this key policymaking role to industry leaders and enlisting their aid in promoting broadcast ATV. Cultivating these relationships may have also led to a post-FCC executive position in industry for Patrick—he became CEO of Time Warner Telecommunications in Washington after stepping down as Chairman in 1989.

On an individual level, Richard Wiley and Alex Felker won the responsibility of assembling the Advisory Committee below the subcommittee chair level. They were able to select working-group leaders on the basis of their prior experience in working with the FCC and their personal knowledge of individual competence.

The losers in this phase of the game were land-mobile radio and cellular telephone services. Their desire to claim additional broadcast television spectrum was frustrated by the FCC's freeze on reallocation until the resolution of the ATV transmission issue. Cable industry executives claimed that they were under-represented on the ACATS panel with only three seats out of 23 total. However, Wiley and Felker sought to rectify this by appointing a number of cable industry representatives to the subcommittees.

Consumer groups were also losers in this phase by not choosing to participate in the work of the ACATS subcommittees. It may have been too much to expect the FCC to appoint consumer representation at the blue-ribbon panel level since an all-industry group was apparently their intention, but there was a distinct lack of interest in the ATV standardization process on the part of consumer groups despite the multibillion-dollar consequences of the potential planned obsolescence of the NTSC standard by the FCC.

Rules of the Game
The rules for this phase of the game were dictated by the Federal Advisory Committee Act, which authorized the creation of panels such as ACATS for a two-year period (R. Wiley, personal communication, November 3, 1993). The FCC re-authorized the Advisory Committee in 1989, 1991, and 1993 for additional two-year intervals.

The Communications Act of 1934 provided the overarching rationale for FCC involvement under the "new uses for radio" (and television) and the "experimental uses of frequencies" provisions of section 303(g).

Key Policy Decisions—Phase 2

With the advisory structure in place in 1988, the FCC began to make a series of key policy decisions. In September, the Commission issued a *Tentative Decision and Further Notice of Inquiry* (FCC, 1988) that among other issues:

1. Specified that existing broadcasters be permitted to implement ATV in order that the benefits of this new technology could be realized by "the public most quickly."

2. Proposed limiting ATV transmission to the existing broadcast spectrum—no additional spectrum would be allocated.

3. Concluded that NTSC transmission should continue after the introduction of ATV service, "at least during a transition period." The

possibility of simulcasting both signals was raised here for the first time.

4. Proposed to conserve spectrum by limiting additional ATV channel allocations to 6 MHz per station for proponent systems that were *"noncompatible"* with NTSC, eliminating the Japanese 9 MHz MUSE system from consideration.

5. Concluded that it was in the public interest not to "retard the introduction of ATV in other services or on non-broadcast media (p. 6521)," but remained "sensitive" to the need to seek compatibility between existing and potential ATV formats (FCC, 1988).

Key Players

These tentative findings initiated the "narrowing" of options process that Howard Miller alluded to above, and simultaneously raised a number of key issues with which policymakers would have to grapple. Here again, game theory provides a useful structure for analyzing the impact of the tentative decisions on the key players.

Broadcasters were the big winners in this phase of the standardization game. The Commissioners indicated that the FCC would follow through with the NAB/MSTV petition to create a terrestrial standard for some type of ATV, meeting their concerns about being bypassed by other delivery systems. "Existing broadcasters" were identified as those likely to implement ATV "most quickly." Finally, the Commission proposed a continuation of NTSC broadcasting during a transition period to ATV—broadcasters would be given a defined period to convert to the new transmission format.

In the past two years it has become apparent how valuable this broadcast spectrum is in the communications marketplace. A 1996 spectrum auction for DBS services generated a $682.5 million winning bid from News Corporation and MCI (McConville, 1996). The award of an extra 6 MHz to existing television broadcasters is a significant windfall, especially if they can avoid having to participate in an auction for it.

On the downside, the Commission declared that it did not want to allocate additional spectrum for ATV outside that already assigned to broadcasting. At the time engineers did not know if ATV signals could be effectively transmitted in a 6 MHz channel—the Japanese MUSE system required 9 MHz for transmission and it was being tested in 1988. Broadcasters were concerned that, if ATV required more than 6 MHz per channel, some stations in urban areas such as Los Angeles where spectrum was very scarce would not be able to receive additional spectrum allocations.

NHK, the Japanese developer of the MUSE transmission system, was the apparent loser in this phase of policymaking. However, it is important to remember that there are three basic elements in any television system: *production, transmission, and reception.* The decision by the FCC to limit ATV spectrum allocations to 6 MHz affected only the MUSE *transmission* component of the Japanese system. In fact, the ambiguous nature of point 5 above, indicating that the Commission would not retard the introduction of other ATV formats in the United States (e.g., NHK's 1125/60 Hi-Vision *production* format) while seeking "equipment compatibility," is a reflection of the debate within ACATS and the Commission over whether to work with the Japanese or compete against them. For economic reasons, the European Community had already decided to compete against initial Japanese dominance of HDTV technology by inventing their own production and transmission standards.

It is interesting to compare the ACATS spectrum recommendations in its first *Interim Report* (Advisory Committee on Advanced Television Service [ACATS], 1988), dated June 16, 1988, with those in the Commission's *Tentative Decision* issued the following September (FCC, 1988). The ACATS report stated that proponent systems would require 12 MHz (two channels), 9 MHz (one channel and a half, or 6 MHz (one channel) and that:

> Based upon current bandwidth compression techniques, it appears that full HDTV *will require greater spectrum than 6 MHz* [emphasis added]. The Advisory Committee believes that efforts should be focused on establishing, at least ultimately, an HDTV standard for terrestrial broadcasting [as opposed to an EDTV or IDTV standard]. (ACATS, 1988, p. 6)

What is unusual is that while the Advisory Committee said in June that full HDTV would probably require more than 6 MHz to transmit, three months later the FCC stated flatly that "noncompatible" systems that required more than 6 MHz would "not be authorized for terrestrial broadcast service." (FCC, 1988, p. 6521)

There are two possible rationales for these contradictory decisions:

1. Time Inc., on behalf of its cable television operations, stated in

the *Tentative Decision* that all national cable TV systems are designed on a 6 MHz channelization scheme (FCC, 1988). Nine or twelve MHz ATV allotments would require a massive redesign of cable system headends and consumer converter boxes. Cable MSOs were opposed to any ATV proposal that deviated from the 6 MHz channel norm.

2. Or it is equally possible that the FCC ignored the advice of the Advisory Committee and imposed the 6 MHz specification to eliminate the Japanese MUSE system from consideration. It is true that the Commission was very concerned about potential spectrum availability in major markets, but at the time of the *Tentative Decision* no other proponent system had demonstrated that it could transmit true HDTV in a single 6 MHz channel.

However, in the *Interim Report*, both Working Party 5 (Economic Factors and Market Penetration) and Advisory Group 2 (Consumer/Trade Issues) raised the issue of the consumer electronics trade imbalance with Japan. The report noted that the 1987 trade deficit with Japan was over $6 billion for televisions and VCRs alone (ACATS, 1988). Both working parties saw the potential introduction of HDTV in the United States as a "second chance" to "re-establish domestic manufacturing of receivers for ATV service" (ACATS, 1988, pp. 24, 29). It is possible that the FCC sandbagged the MUSE system by limiting U.S. transmission bandwidth to 6 MHz under the guise of spectrum efficiency.

These international influences on both the Advisory Committee and the FCC reconfirm the need for incorporating a global perspective in studying these effects. It also strengthens the case for utilizing an ecology of games approach which encourages the consideration of several games in play at the same time, often with common players, but operating with different rules. Thus, the ATV standardization game is being played in parallel with the Japan-United States trade game. U.S. electronics companies and the Federal government are involved in both games, but the rules are very different and the government is acting in a different role in each contest.

Testing Assistance for the FCC

The Advanced Television Test Center (ATTC) became an important player in the standardization process in its role as an impartial referee in assessing the relative merits of the various systems. However, its importance extended beyond just the test phases of 1990-1993 and 1995—the center staff were used as a sounding board concerning a number of technical issues that faced the Advisory Committee.

James McKinney, former Chairman of the Advanced Television Systems Committee, noted that earlier developments in television technology, such as the development of color transmission, had been accomplished by research laboratories maintained by each of the big-three networks or by manufacturers such as RCA (personal communication, May 11, 1994). By the 1980s, these proprietary R&D labs had either been closed or reduced to a fraction of their previous size. Even RCA's Sarnoff Research Center had been buffeted by the 1986 sale of the parent company to GE, and ultimately to Thomson of France the following year. Most U.S. television manufacturers had been driven out of business by the 1980s—only Zenith remained as a source of new video research.[8]

AT&T's legendary Bell Laboratories, the site of the development of the transistor, had been divided in the breakup of the parent company with Bell Communications Research (Bellcore) emerging as a smaller laboratory for the Regional Bell Operating Companies. Only the Japanese and Europeans continued to maintain extensive in-house R&D facilities, which gave their electronics companies a substantial competitive advantage in the race to perfect new hardware for HDTV production and transmission.

In this R&D vacuum, the ATTC played a valuable role as a clearinghouse for ATV technical information and as an objective referee in the Advisory Committee's competition to select the best proponent HDTV system (see Brinkley, 1997a).

New Players at the FCC: A Chairman and Two Commissioners

In 1989, Alfred Sikes, who had done much to encourage the development of HDTV at the National Telecommunications and Information Administration (NTIA), was nominated by President George Bush to succeed Dennis Patrick as Chairman of the FCC. Sikes had served as an Assistant Attorney General in Missouri under John Danforth—who later proved to be an influential friend as a U.S. Senator. Senator Danforth had recommended Sikes for the NTIA position and, as ranking minority member of the Senate Commerce Committee, was influential in gaining confirmation for Sikes as FCC Chairman ("FCC Gains," 1989).

Not coincidentally, given their mutual interest in HDTV, Sikes and Richard Wiley were friends and tennis partners. Wiley noted at the time of Sikes' confirmation that "He [was] very accessible. He [listened] to your views" ("FCC Gains," 1989, p. 29). Sikes was seen at the time as a more moderate Republican, and less of an ideologue, than previous Chairmen Mark Fowler and Dennis Patrick.

In press interviews, Sikes indicated a strong interest in promoting new communication technologies, singling out HDTV in particular. Prior to being sworn in, Sikes had expressed an interest in the Commission's promotion of American "natural competitiveness" overseas by developing policies that facilitated the entry of U.S. companies in domestic markets. An article in *Broadcasting* stated that "He noted that communication companies are often dependent on FCC action—allocation of spectrum, for instance—in order to introduce new products" ("Closing in," 1989, p. 28).

This is a thinly veiled reference to HDTV and indicated that Sikes planned to carry over his HDTV promotion efforts from the NTIA to the FCC. It also reflected growing international influences on the work of the FCC, and that international competitiveness issues would be considered as a factor in making policy decisions on new technologies that came before the Commission. This stance also enhanced Sikes' confirmation as Chairman by a Congress that was in the midst of the debate over what it could do to promote American competitiveness in international markets. Sikes had been an effective spokesperson at the NTIA for an American role in the international contest over HDTV standardization, and his supporters in Congress (on both sides of the aisle) confirmed him as FCC Chairman at a crucial stage in the ATV proceedings.

At the same time, Congress also confirmed new FCC Commissioners Sherrie Marshall and Andrew Barrett to fill Republican seats that had been vacant for all of Dennis Patrick's term as Chairman. The seats had been vacant in dispute between the White House, the FCC, and Senate Commerce Committee Chairman Ernest Hollings (D-S.C.) over the Commission's repeal of the fairness doctrine in August 1987 ("Closing in," 1989). Sherrie Marshall was a partner in Richard Wiley's law firm and directed the FCC's Office of Legislative Affairs during the 1987-1989 "commissioner gap." Barrett came to the Commission with nine years of regulatory experience as a member of the Illinois Commerce Commission ("Closing in," 1989).

Sikes, Marshall, and Barrett joined Democratic Commissioners

Ervin Duggan (who joined in June 1989) and long-term member James Quello to create the first full-strength Commission in two years. The political make-up of the Commission is an important factor in any policy analysis, but it was especially true during the ATV standardization process since parallel industrial policy and international trade games were under way at the same time.

Key Policy Decisions—Phase 3

On August 24, 1990, the Commission adopted its *First Report and Order* which indicated that the FCC wanted to pursue a *simulcast strategy* for ATV in the United States (FCC, 1990). This was a major policy decision since it abandoned previous attempts to design an ATV system that was compatible with the existing NTSC format. An incompatible ATV system would require that both the old and new television standards be *simul*taneously broad*cast*, or *simulcast*, for a number of years as the old technology was phased out.

This narrowing of options eliminated from competition proponents who wanted to augment the NTSC signal by transmitting an additional 3 or 6 Mhz in an adjacent channel to improve on the picture and sound quality, or to add a widescreen image.

The FCC also deferred any consideration of enhanced-definition television (EDTV) systems such as the ACTV system proposed by Philips/Thomson/Sarnoff until after a decision was reached on an HDTV standard. This deferral, in effect, killed any further progress on the Advanced Compatible Television format and the Philips consortium moved on to develop a high-resolution HDTV system.

Key Players

The key player in this major policy decision to adopt a simulcast strategy was FCC Chairman Al Sikes. Planning Subcommittee Chairman Joseph Flaherty stated that the FCC made a "courageous decision" not to try to enhance or augment the existing NTSC format, but to pursue an incompatible HDTV standard (personal communication, May 23, 1994). He credited Al Sikes with being the "driver" and a "real leader" on this issue, leading to a "gutsy decision" that forced the Philips/Thomson/Sarnoff team back to the drawing board to replace their EDTV submission. Brinkley (1997a) credits ACATS chairman Richard Wiley as the impetus behind Sikes' stand against enhancement or augmentation, but the fact remains that Sikes was willing to buck entrenched interests to codify a simulcast implementation strategy.

In an earlier ACATS meeting, Advisory Committee member Robert Wright, the President of NBC, had stated that an *incompatible* HDTV format would be selected "over his dead body" (J. Barnathan, retired ABC executive, personal communication, February 17, 1994). Sikes' desire to select a high-quality, incompatible HDTV transmission scheme over a lower-quality compatible EDTV format championed by the two largest television manufacturers in the United States demonstrates his willingness to use the leadership authority of the Chairman's office. The FCC Chairman is first among equals at the Commission and the simulcast decision illustrates the leadership leverage that accrues to the Chairman's position.

In varying degrees, broadcasters and consumers were the losers in this decision. The creation of an incompatible HDTV standard meant that the installed base of 220 million consumer NTSC television sets would ultimately be made obsolete. Broadcast television stations would ultimately have to convert their studios and transmitters to an incompatible HDTV standard. Television manufacturers such as Thomson and Philips would have preferred an EDTV standard for sets that would be less complex and thus less expensive to make than true HDTV models.

Winners in this decision were cable television operators in the United States. The abandonment of augmentation concepts and deferral of EDTV technologies meant that whatever the FCC ultimately developed, it would fit within a 6 MHz channel scheme. As noted above, cable operators were concerned that the FCC might propose an ATV solution that required split channels to accommodate 9 or 12 MHz NTSC augmentation technologies. The downside for cable operators was the prospect that simulcast meant that every channel in an existing 70-channel system might have to be duplicated for both NTSC and HDTV transmission—doubling the needed channel-carrying capacity of each system.

Key Policy Decisions—Phase 4

The next key decisionmaking phase in the ATV standardization process was the adoption of a *Notice of Proposed Rule Making* (*NPRM*) on October 24, 1991 (FCC, 1991). The *NPRM* did not deal with technical aspects of ATV transmission as did the *First Report and Order,* but rather it proposed a "tentative plan for ATV broadcast implementation." The *NPRM*:

1. Reaffirmed the proposed policy in the *Tentative Decision* that existing broadcasters would get first crack at implementing ATV transmission, but stated that once these initial spectrum allotments were made the process would be opened for applications from "any qualified party." Broadcasters would have to use the allotted ATV spectrum or lose it to other applicants—a "use it or lose it" policy.

2. Defined a three-year application window (from the selection of an ATV standard) for an ATV construction period and a two-year interval for actual construction.[9]

3. Proposed to create a Table of Allotments for the assignment of ATV channels to existing broadcasters. ATV channels would be assigned on a first-come, first-served basis and applicants would be able to negotiate within local markets to exchange allocated channels.

4. Defined for the first time that once the transition to an ATV broadcast system was accomplished, "We expect that broadcasters will surrender one of these two [NTSC and ATV] channels" (FCC, 1991, p. 7026).

Key Players

NTSC broadcasters were the clear winners in this decision. They would get the first option on additional ATV spectrum, with the condition that they apply for the needed permits and build the system within the specified five years. The FCC held open the option that additional parties could apply for ATV spectrum, but only after existing broadcasters had staked their claims. The Commission also made provisions to protect ATV spectrum for noncommercial applicants (e.g., public broadcasters).

The FCC also achieved several important goals with this decision. First, the return of the NTSC spectrum would bolster the Commission's assertion that the transition to ATV was, in fact, an *improvement* in the existing television broadcast service, and not a *new service*. Under the Supreme Court's *Ashbacker Radio Corp. v. FCC* (1945) decision, the FCC would have been required to hold comparative hearings if the ATV spectrum assignment had been considered to be a new service. The Commission, as a result, has always taken pains to characterize any national conversion to advanced television as an "improvement" to existing broadcasting services. The other FCC goal concerns the eventual reversion of NTSC spectrum to the nation for reallocation for other uses. Television broadcasting is a notorious hog of electromagnetic spectrum—every AM radio signal in the United States could be transmitted in the spectrum used by one television station. Not only is television a large consumer of spectrum, but much of the spectrum between channels is wasted to prevent co-channel interference.

If the FCC could regain control of the VHF spectrum by moving all ATV services into the ultrahigh frequencies (UHF), it would free a tremendous amount of spectrum for alternative communication uses. Land-mobile radio and cellular telephone services are two primary users of that spectrum, but new technologies are emerging (e.g., personal data communication devices) that could also use the returned spectrum. This goal of the ATV standardization game is a key issue in understanding the FCC's statements on spectrum issues and was a primary factor in congressional and White House interest in spectrum auctions.

The potential losers in this decision were low-power television (LPTV) operators in major markets. They could be bumped from their spectrum assignments if their channels interfered with ATV assignments to VHF and UHF broadcasters. LPTV stations have very small audiences and they are treated as second-class citizens by the FCC and "high-power" broadcasters.

A Final Note from Al Sikes

The 1991 *NPRM* closed with a statement from FCC Chairman Alfred Sikes. He acknowledged:

> All of us have been encouraged by the extraordinary developments of the last year and a half which seem to put the U.S. in the position of offering the world the first digital broadcast television system. However, recent statements concerning large screen NTSC as an alternative to HDTV have raised concerns that at least some in the broadcast industry regard the economics as unattractive. ... If the record indicates, however, that broadcasters, guided by their view of future economics, are losing interest in HDTV, then valuable UHF spectrum could be used for new land mobile services. (FCC, 1991, p. 7036)

This statement reflected a carrot-and-stick approach to HDTV pol-

icy on the part of the Chairman. The carrot recognized the digital tri-
umph of the proponent systems, but the stick was an overt threat to
broadcasters. If they declined to pursue full HDTV, then the Commission might reopen the proceedings for transfer of UHF television
spectrum to nonbroadcast users. The need for this statement also illustrates that there was dissension in the broadcast community over the
economic viability of a national conversion to HDTV. While the major
television networks were active supporters of the Advisory
Committee, and network affiliate stations contributed $1.3 million
toward the operation of the Advanced Television Test Center, there
were still many broadcasters who were convinced that HDTV conversion would bankrupt them (Clifford, 1992).

Finally, the Sikes statement revealed the power that the FCC
Chairman holds as one of the key players in the policymaking game.
Sikes was telling the broadcast community that he was prepared to
play hardball with them if they were recalcitrant in using the offered
spectrum. The broadcasters' strategic advantage was the knowledge
that FCC Commissioners and Chairmen change frequently (except for
James Quello), and that the Commissioners in a 1996 FCC might see
things very differently compared with the 1991 Commission.

Key Policy Decisions—Phase 5

On April 9, 1992, the Commission adopted a *Second Report and
Order/Further Notice of Proposed Rule Making* that contained a number
of significant policy issues (FCC, 1992a). Among them were these key
decisions:

1. In a change from the 1991 *NPRM,* broadcasters would have two
years to apply for an ATV channel, followed by a three-year construction period.

2. Broadcasters were again put on notice that "when ATV becomes
… the prevalent medium" (p. 3341), they would convert to the new
transmission system by surrendering one of their two simulcast channels and they would cease broadcasting in NTSC.

3. A requirement for 100 percent simulcasting of NTSC and ATV
would be adopted at "the earliest appropriate point."

The *Second Report and Order* also solicited comment on several
additional scheduling issues:

1. The Commission proposed to set a "firm date" for national conversion to ATV transmission of "15 years from either selection of an ATV system or the date a Table of ATV Allotments is effective, whichever is later" (p. 3342). This 15-year timetable would be reviewed by the FCC in 1998 to determine if it were appropriate.

2. The FCC also proposed to establish a 100 percent simulcast requirement "no later than four years after the ATV application/construction period has passed" (p. 3342).

3. The Commission proposed to afford broadcasters "some initial flexibility" during the simulcast phase to full ATV transmission.

Key Players

The Commissioners played a key role in shaping the significant content of this *Report and Order.* For the first time, a timetable of 15 years for conversion was established, and the Commission clearly specified that the additional spectrum allocation was transitional—6 MHz from each applicant would revert back to the FCC at the end of the simulcast period.

Commissioner Ervin Duggan (now president of PBS) appended an eloquent statement to the *Report* with his forthright interpretation of the fundamental policy issues involved:

> These decisions are not just significant, they are fateful. We are, in essence, decreeing the creation of a whole new television broadcasting industry and the shutting down of the old one. We do not do so lightly. I believe that all of us realize that the move from conventional to advanced television will be expensive, difficult and time-consuming; that full conversion to advanced television broadcasting is likely to take more than a decade. (FCC, 1992a, p. 3371)

On the subject of the 6 MHz spectrum reversion plan, Duggan reinforced that the ATV allocation was not a "giveaway," and added that "The spectrum we will eventually assign for advanced television development, in short, has a strong tether attached to it; squatters' rights on the second channel are not what this Commission had in mind" (p. 3371).

Duggan acknowledged that the Commission had not arrived at these decisions independently and credited the Advisory Committee on Advanced Television Service, its volunteer members, and Chairman Richard Wiley for their recommendations. FCC Chairman Al Sikes was singled out for his "tireless efforts to bring advanced television from the drawing board to the laboratory, and eventually, into American homes and the global marketplace [and he] deserves to be remembered and honored years from now (p. 3371)."

Nested Game Theory and the FCC—Industry Dynamic

George Tsebelis (1990) has described what he terms a "nested" two-player game where the play of the principal parties is influenced by other parallel games simultaneously underway. This type of game is diagrammed below with the Commissioners of the FCC as player 1 and broadcast industry representatives (NAB, MSTV) as player 2. The two-person game is being influenced by nested games—the FCC with other potential spectrum users such as personal communication services (PCS) as player X-1, and the network and local affiliate constituents of broadcast representatives as player X-2.

The goals and strategies of the two primary players—the Commissioners of the FCC and the broadcast industry—cannot be understood without an examination of the goals and strategies of the

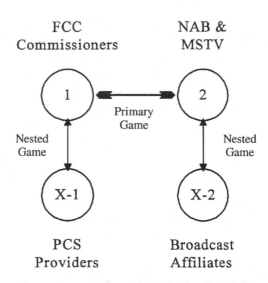

FIGURE 6.3. Primary and nested games (*Source:* Tsebelis 1990).

players in the nested games as well. Thus when the Commissioners take an apparent hard line with broadcast representatives on the issues of spectrum allocation, possible spectrum auctions, flexible spectrum use, and the eventual return of the NTSC spectrum allocation, they are also playing to the PCS and land-mobile community. The Commissioners seem to be aware that any actual or perceived giveaway of spectrum to television broadcasters could lead to a legal challenge from alternate users who are willing to bid for spectrum in an open auction. The Commissioners seemed to bend over backwards in the *NPRM* and *Second Report and Order* to emphasize, as Commissioner Duggan did, that the ATV allocation was not a "giveaway" of spectrum. In a telecommunications environment where wireless services are growing rapidly with a corresponding increase in spectrum value, the Commissioners appeared to be taking great pains to avoid a potential *Ashbacker* challenge to the ATV allocation.

Broadcast representatives such as the NAB and MSTV had to design their FCC game strategy keeping in mind the divergent interests of their network and local affiliate clients (who pay their salaries). From a longitudinal perspective, the NAB/MSTV game strategy changed from concern over potential HDTV competition from DBS and cable in 1987 to affiliate concern in 1992 that a forced HDTV conversion might put many of them out of business. Many local affiliates could not calculate a way to amortize the multimillion dollar cost of conversion, and they apparently communicated this to their NAB/MSTV representatives in Washington. This attitude changed somewhat with the belated realization that standard-definition digital transmission (SDTV) might permit additional revenue streams with multichannel broadcasting, but this did not sink in until after 1995.

In short, an analysis of the primary FCC-broadcast industry game requires the inclusion of the nested games influencing the primary players. Only then do the NAB/MSTV policy flip-flops from positive to negative and back again to positive on HDTV conversion make sense.

Broadcaster Reactions to the *Second Report and Order*

If Commissioner Duggan wanted to stimulate debate among broadcasters over the FCC's advanced television policies, he certainly got his wish. The FCC *Report and Order*, not coincidentally, was adopted just prior to the 1992 National Association of Broadcasters convention in Las Vegas and some of the proposals created quite a stir among the attendees.

The key issue was the proposed 15-year timetable for conversion. The potential conversion to HDTV was no longer an abstract issue—the FCC had defined a conversion window from a planned 1995 adoption of a standard to the scheduled shutdown of NTSC television broadcasting in the year 2010.

Trade journal coverage of progress toward ATV standardization did a 180-degree turnabout in two weeks from the end of March until mid-April when the NAB show began. On March 23, 1992, General Instrument conducted a private demonstration of their DigiCipher HDTV technology on Capitol Hill for invited members of Congress, government officials, and industry executives. Their DigiCipher system was used to transmit a digital, 1050-line, true-HDTV signal from WETA-TV in the Maryland suburbs to a $50 conventional TV antenna on the U.S. Capitol building (Lambert, 1992a). Midway through the demonstration, the antenna feed was switched to a cable signal from the Capitol's own antenna/cable system. By all accounts the image and sound quality were excellent.

Chairman Al Sikes and Commissioners Barrett, Duggan, and Quello were present for the demonstration. Afterwards Sikes stated, "It's exciting and encouraging to see the first digital broadcast. A year and a half ago, you could have got any odds you wanted that digital wouldn't work" (Lambert, 1992a, p. 10).

Richard Wiley paraphrased Astronaut Neil Armstrong in saying, "It's one giant step for General Instrument and one giant leap for the future of television" (Lambert, 1992a, p. 10). The demonstration *was* quite an achievement, coming less than 21 months after General Instrument stunned its competitors with the 11th-hour application proposing a digital solution. Finally, after Congressional gloom and doom in the late 1980s over U.S. competitiveness, American industry appeared to have won a major battle in the international competition to develop high-definition television. As National Cable Television Association President James Mooney said, "It now seems likely that the standard will be home grown" (Lambert, 1992a, p. 11).

Even broadcasters appeared to be enthusiastic. MSTV President Margita White, one of the original petitioners for HDTV standardization, stated, "What has been accomplished in just a few years is amazing. It's very important—one more step toward HDTV over the air, a goal we've worked very hard to achieve for many years" (Lambert, 1992a, p. 11).

However, on the eve of the NAB convention three weeks later, broadcasters had pondered the proposed 15-year conversion window in the FCC's April 9th *Second Report and Order* and many of them were not very happy about it. In a trade journal article headlined, "HDTV: Too Close for Comfort?" broadcasters complained about the five-year application proposal and the 15-year simulcast window. Julian Shepard, General Counsel of MSTV, stated that:

> This is not a perfect situation for broadcasters ... I can appreciate their skepticism and concerns; we certainly have not advocated that type of timetable. Our position all along has been: don't make rules prematurely, wait and see what the consumer response is and postpone decisions that depend on that factor until the information is available. (Flint, 1992, p. 4)

Organizations like MSTV and the NAB were being forced to shift their ATV game strategy from a defense against potential competition from DBS and cable to a defense against an early and expensive conversion to HDTV. By early 1992 they were giving mixed messages that varied from vestiges of initial enthusiasm (Ms. White's statement in March) to cautious concern over the FCC's proposed timetable for implementation (Mr. Shepard's April comment). Thus as the danger of being bypassed by HDTV services delivered from satellite or cable receded, some broadcast affiliates became more concerned about being forced into an early conversion to HDTV that might bankrupt some of them.

We say "some" affiliates because there were wide disparities in opinion on the proposed timeline—there were some major-market affiliates who were making plans to begin HDTV conversion as soon as the FCC set the standard. In the face of these mixed signals from their membership, organizations such as the NAB and MSTV redefined their desired payoffs in the FCC's standardization game.

Their strategy at the time was to:

1. Continue to pursue the second 6 MHz allocation for "advanced television service." They sought maximum flexibility in defining what this phrase meant.

2. Try to delay the actual setting of the national HDTV standard by asking ACATS to investigate alternative technologies such as COFDM transmission.[10]

3. Solicit Congress and the FCC to allow maximum flexibility in the interim use of the additional 6 MHz channel during the simulcast period. This usage may include the transmission of 4-6 compressed SDTV channels on the new spectrum until HDTV broadcasting becomes the prevalent mode of transmission.

Thus, for some broadcasters the payoffs for participating in the FCC's standardization process have shifted since this game began in 1987. As Ken Elkins of Pulitzer Broadcasting noted in April 1992, "In my opinion, right now it is extremely difficult for broadcasters to come to any reasonable conclusion on a satisfactory return on investment for HDTV" (Flint, 1992, p. 14).

While Elkins' remark must be seen in the context of the effects of the 1990-1993 recession on broadcasters, the fact remains that network affiliates were not at all sure that they would be able to recoup a potential multimillion-dollar investment in HDTV hardware, let alone paying for the new digital spectrum as some members of Congress wished.

Key Policy Decisions—Phase 6

Congress re-entered HDTV policymaking in a significant way with the passage of the Telecommunications Act of 1996, signed into law by President Clinton on February 8, 1996. Congress had struggled with the passage of an omnibus reform of the now-antiquated Communications Act of 1934 throughout much of the 1995 session. The impasse yielded to election-year pressure to demonstrate to the American electorate that Congress could pass proactive legislation to effectively deregulate telecommunications. A central figure was majority leader and presidential candidate Robert Dole, U.S. Senator from Kansas.

Senator Dole, among others in Congress, was adamantly opposed to a "giveaway" of additional spectrum to television broadcasters for the conversion to HDTV. He stated, "I question whether telecommunications reform is worth the television broadcaster's price. Let's for the sake of the taxpayers and for the sake of the American consumers, fix this one corporate welfare provision before we have to vote on it" (Andrews, 1996, p. C4). Dole's opposition to the awarding of the ATV conversion spectrum held up the passage of the Telecommunications Act until he received written assurances from all the Commissioners of the FCC that they would not award any of the "second-channel" spectrum until Congress had a chance to hold hearings on the issue (Stern, 1996a). Dole and fellow Republican Senator John McCain of Arizona were actively pushing for spectrum auctions, citing FCC valuations of the digital spectrum that ranged from $11 billion to $70 billion (Stern, 1996b,c).

The White House had also entered the fray with a proposal to auction the existing NTSC spectrum in the year 2002 with a reversion of the spectrum to the federal government by 2005, raising $17 billion for budget deficit reduction in the process (Stern, 1996d). This was alarming to the broadcast industry as it meant that the planned 15-year transition to ATV might shrink to nine years, at the end of which interval all existing NTSC transmitters would have to be turned off. The broadcast industry fought back with a series of anti-auction public service announcements (produced and distributed by the National Association of Broadcasters for member stations), and through print advertisements and op-ed pieces. The print ads were headlined, "Congress has a strange idea to balance the budget: Turn free TV into pay TV," and asked readers to "say no to the television tax" ("Congress Has a Strange Idea," 1996, p. C2).

Consumer groups were also active in the auction debate. The nonprofit Benton Foundation responded to the broadcaster's media campaign with a full-page "Open Message to the Nation's Broadcasters" that appeared in the trade journal *Broadcasting & Cable* ("An Open Message," 1996). They excoriated the broadcast industry stating:

> For an industry using the public airwaves—and therefore obligated by law to serve the public interest—the ads you've been running about a TV tax are shameful. There's no pending TV tax. There's no real threat to free TV. You know that and so do we. (p. 53)

The Benton Foundation also is a financial supporter of the Media Access Project (MAP), a group that represents consumer and educational interests in Washington. MAP had issued *Pretty Pictures or Pretty Profits*, a 12-page white paper arguing against the free award of digital spectrum to broadcasters (Sohn and Schwartzman, 1995), and they sought to dedicate the proceeds from any auction to funding children's television, public television, and linking schools to the Internet.

The Telecommunications Act of 1996 did, in fact, award the digital spectrum without charge to all existing broadcasters. Senators Dole and McCain made blustering noises throughout the spring of 1996 about reviving the auction issue, but even they gave up in the face of the NAB's ad campaign. Dole blasted the NAB just before their annual convention, accusing them of "bullying Congress and running a multi-million-dollar ad campaign to mislead the public" (Stern, 1996e, p. 6). By that time however, even Dole's supporters on the issue such as Senator McCain had conceded defeat on the issue of using digital auctions as a means of raising funds to balance the budget. When Dole announced in mid-May that he was leaving the Senate to focus on his Presidential campaign, there was jubilation in the broadcast community—the auction threat faded with his resignation (Stern, 1996f).

On May 9, 1996, the Commissioners of the FCC approved a *Fifth Notice of Proposed Rulemaking* that proposed a mandatory conversion to the ATSC advanced television standard (FCC, 1996a). Amidst all the uproar over the auction issue this news item appeared as a small sidebar in the trade press, but represented a significant policy step for the Commission (McConnell, 1996b). During the 1996 NAB convention, Chairman Reed Hundt was speaking of "authorizing" the ATV standard, rather than mandating its use by all U.S. broadcasters. Industry leaders such as Robert Wright of NBC (who had been at best a lukewarm supporter of ATV conversion) were now of one mind concerning this point—they wanted a mandatory standard that every U.S. broadcaster and television manufacturer would have to follow (McConnell, 1996b). Commissioners Chong, Ness, and Quello all agreed with the creation of a mandatory standard, forcing the hand of Hundt in support of this key policy matter. With the digital spectrum auction a dead issue and the FCC leaning toward a national mandatory conversion from analog NTSC to a digital broadcast system the prospects for HDTV finally seemed very bright in the United States.

Key Players

It is rare in telecommunications policymaking to have all six of the KLT model entities involved in a particular case. The ATV auction controversy involved five (Congress, the FCC, regulated industries, the White House, and citizen/consumer groups such as MAP), and perhaps the courts will get involved before ATV signals are universally broadcast. The Krasnow et al. model is useful in describing broad categories of primary participants and examining their complex interactions, but game theory is also useful in studying the rules of the policymaking game.

The clear winner in this game of brinksmanship between television broadcasters and Congress was the industry. Despite the strange alliance of the conservative pro-auction Republican right and the liberal consumer groups such as MAP, the industry was able to utilize the power of advertising to mobilize public opinion through a media campaign based upon half-truths. Even such a politically powerful Senator and Presidential candidate as Robert Dole was helpless in the face of this ad blitz. One issue that is rarely mentioned is the unwillingness of most politicians to alienate the broadcast industry. Dole was willing to take on the industry in an effort to remold his image as a populist in the Presidential campaign, but most other members of Congress were not. The 1996 Presidential campaign must be seen as a parallel game that influenced both Dole's and the Clinton White House's positions on using auctions as a deficit reduction tool. Despite much rhetoric about the national sacrifices needed to reduce the federal deficit, when push-came-to-shove neither group was willing to face down the powerful broadcast lobby over the issue. As Republican House Speaker Newt Gingrich noted, "The practical fact is, nobody's going to take on the broadcasters" on the spectrum auction issue (McAvoy and West, 1995, p. 38).

The passage of the Telecommunications Act of 1996 did reflect Congressional power in defining the fundamental rules of competition for players in the field of electronic communication. The Commissioners of the FCC had to acknowledge this Congressional power by acquiescing to Senator Dole's letter asking them to delay any action on spectrum allocation until after Congress had debated the issue. In the final analysis, the broadcast industry flexed its political muscle by demonstrating the power of single-issue advertising in mobilizing public opinion against spectrum auctions.

Key Policy Decisions—Phase 7

On December 24, 1996, the FCC adopted its ATV *Fourth Report and Order* (FCC, 1996c) accepting a modified "Advanced Television Systems Committee (ATSC) Digital Television Standard" as the future television transmission system for the United States. It was the culmination of a nine-year struggle to establish an advanced television standard for the United States. What was extraordinary about the digital

standard as adopted was the intentional deletion of 18 carefully crafted image scanning parameters (Table 3 of the ATSC documentation,
see Table 1.1 in Chapter 1) and any specification of narrow or
widescreen aspect ratios (see FCC, 1996c, Appendix A). It was as if
someone had developed a unique high-speed aircraft and then decided to delete any mention of how its wings were to be designed. This
decision reflected a fundamental difference of opinion between television manufacturers and certain elements of the U.S. computer industry, and the FCC's final decision was based on an eleventh-hour compromise refereed by Commissioner Susan Ness. It makes an ideal case
for game theory analysis.

Key Players

After eight years of debate and system testing, the FCC's Advisory
Committee on Advanced Television Systems (ACATS) had recommended the Grand Alliance's digital television system to the
Commission in November 1995. The issue of interlaced versus progressive scanning had dogged the Grand Alliance (GA) from its inception—television manufacturing and broadcaster members of the
alliance favored interlaced scanning and claimed that it would be less
expensive to manufacture. Proponents of progressive scanning included Dr. Jae Lim of MIT, a key member of the GA design team, and many
U.S. computer hardware and software companies such as Apple, Intel,
Compaq, and Microsoft (Brinkley, 1997a). They claimed that progressive scanning (used on most computer screens sold worldwide) would
provide higher picture resolution and reduce image flicker, which
causes eye strain with interlaced scanning. They also asserted that
allowing interlaced scanning to be one of the variants offered with the
ATSC standard would inhibit the planned "migration" to an all-pro-
gressive-scan system in the next century. What was unsaid is that these
members of the computer industry wanted a standard adopted that
would promote the future development of the PC-TV, a computer-tele-
vision hybrid that could access the Internet as well as receive local
broadcast signals.

Microsoft, Intel, Apple, and Compaq, amongst others, formed a
group named the Computer Industry Coalition on Advanced
Television Service (CICATS) and began to lobby the Commission and
Congress in early 1996. As Microsoft spokesperson Mark Murray
noted, "We believe this fight is far from over. These groups [CICATS
and the others] are lining up against a restrictive standard and in favor

of a much broader technology standard that would provide greater benefits to consumers at lower cost" (Beacham, 1996a, pp. 1, 8).

In a surprise twist they were joined in July 1996 by the National Cable Television Association (NCTA) (McConnell, 1996c). Up to that point, the U.S. cable industry had been an active partner in the development of an advanced television standard through the participation of CableLabs in the work of the FCC's Advisory Committee. One can speculate that there are several possible rationales for this shift in the position of the cable industry: (1) the realization that under proposed FCC simulcast rules, they would have to provide an additional ATV channel for every present NTSC channel carried, requiring a doubling in system capacity without increasing revenue; and (2) newly developed standard-definition (SDTV) technology will permit terrestrial broadcasters to transmit four to six SDTV channels simultaneously over the air, in effect becoming mini-multichannel providers in competition with cable news and sports channels.

As noted in Chapter 2, the Directors Guild of America (DGA) and the American Society of Cinematographers (ASC)—joined together as the Coalition of Film Makers—also opposed the proposed ATSC standard, criticizing the 1.78:1 aspect ratio as too narrow. They wanted a ratio of 2:1 mandated instead and enlisted the aid of industry professionals, such as film director Steven Spielberg, to lobby FCC Commissioners concerning this fundamentally aesthetic issue (McConnell, 1996e).

Thus arrayed on one side of the dispute over the proposed standard were CICATS, the U.S. cable television industry, and the Coalition of Film Makers who all wanted aspects of the standard changed. Allied against them were those with a vested interest in the adoption of the standard as specified by ACATS and the ATSC. This group included the politically influential National Association of Broadcasters (NAB) and the Consumer Electronics Manufacturers Association (CEMA).

The fundamental progressive-interlace scanning issue came to a head in October 1996 when FCC Commissioner Susan Ness set a November 25th deadline for broadcasters and television manufacturers to resolve the standardization conflict with computer industry representatives (McConnell, 1996e). Ness pushed the issue because the FCC Commissioners wanted an agreement on the ATV standard by the end of 1996. The two groups reached an accord on the date of the deadline that resolved the interlace-progressive scanning issue by deleting *any* specification of scanning format in the proposed standard. The 18 specified variations of HDTV and SDTV video formats were deleted, as were any aspect ratio requirements (Landler, 1996a). This compromise facilitated the FCC's adoption of a digital advanced television standard on December 24, 1996, and smoothed the way for the American computer industry to compete for the replacement of 220 million television sets in the United States alone (Brinkley, 1996b).

Commissioner Ness was aware of the importance of creating a place for the computer industry in this potentially huge market. In a statement released with the *Fourth Report and Order*, she asserted, "It hastens convergence—transporting us into a competitive world of computer-friendly television sets and broadcast-friendly computers. *Our decision also provides a springboard for global leadership in high definition digital equipment and programming* [emphasis added] (FCC, 1996c, p. 17808). Representatives of the computer industry made it clear that the international TV industry was a target for digitally-based hardware and software. Andrew Grove, CEO of Intel Corporation, stated that the computer industry was engaged in a "war for eyeballs" with television programming, and that it "must look outside our own backyard for new users" (Brinkley, 1996b, p. C11). This may explain why Intel and Microsoft were willing to go to the mat with the coalition of television broadcasters and manufacturers over the scanning issue in the proposed ATV standard.

The primary winners in this phase of the policymaking process were the computer industry backers of the CICATS coalition. While they were unable to mandate a progressive-scan ATV standard as they desired, they were able to force the television manufacturers consortium (CEMA) and their broadcaster allies to withdraw any mandated scanning parameters. The withdrawal of a mandated 16:9 aspect ratio meant that computer companies could manufacture standard 4:3 displays in whatever high or low definition mode they wished. As part of the compromise, CICATS also agreed to not impede the ATV standardization process "before the FCC, by judicial review, legislatively or otherwise" ("Compromise Letter", 1996, p. 1). In a pocketbook issue dear to broadcasters, they also agreed not to support any congressional action to auction the spectrum allocated for digital transmissions. Intel and Microsoft are two of the largest and most politically-influential high-technology corporations in America and these pledges removed several significant barriers to eventual adoption of the ATV standard.

Television manufacturers and their allies can also be considered winners in this game of political compromise as the settlement allowed the FCC to move forward in December 1996 with the *Fourth Report and Order* establishing the U.S. digital standard. The manufacturers planned to include the ability to decode all 18 transmission variants (from the withdrawn Table 3) in their forthcoming digital television sets and the new open standard would not preclude this. They were more concerned about any further delays in standardization that might postpone getting the new digital sets on the market. Broadcast organizations such as the NAB and their television network constituents were now solid supporters of DTV terrestrial transmission standardization as they warily gauged the growing success of alternative all-digital services such as direct broadcast satellite (DBS). They now viewed the transition to digital transmission as a basic issue of future industry survival (McClellan, 1996, see also Harris Corporation, 1996).

Primary losers in the compromise negotiations were the Coalition of Film Makers. Because the compromise between the television and computer industries eliminated *any* required aspect ratio, the ability to mandate a wider 2:1 ratio was removed with it. At a time when the FCC was looking for an ATV standard that was more open and accommodating toward a wider range of transmission modes, the filmmakers' futile appeal for mandated aspect ratios (tied to those on the film negative) was a matter of swimming upstream against a prevalent open-standards current. Noted director Spielberg and other Hollywood luminaries had personally lobbied Commissioners Hundt and Chong on this issue, but in the *Fourth Report and Order* they both supported the FCC's open standard philosophy.

This phase in the ATV policymaking process demonstrated that oppositional forces such as CICATS had enough political clout to force changes in the digital standard adopted by the FCC. It is worth noting that the dominant voices in the opposition camp were not citizen groups such as the Media Access Project or even professional organizations such as the American Society of Cinematographers, but rather economically powerful multinational companies such as Microsoft and Intel. In an era when global audiovisual electronics manufacturing is dominated by non-American companies, the international success of U.S. computer hardware and software companies could not be ignored by FCC policymakers concerned about American competitiveness in advanced television technology.

Key Policy Issues—Phase 8

The FCC concluded its deliberations on advanced television with the adoption of two rulemakings on April 3, 1997. As noted in Chapter 1, the *Fifth Report and Order*:

1. Required all broadcasters to provide a "free digital programming service the resolution of which is comparable to or better than that of today's service" (FCC, 1997a, p. 13). After all the effort that had gone into two decade's worth of research and development on creating higher-quality images, it is remarkable that the Commission was not prepared to mandate any requirements concerning true high-definition broadcasting.

2. The rulemaking also required the affiliate stations of the top four networks (ABC, CBS, Fox and NBC) in the top 10 U.S. markets to be on the air with a DTV signal by May 1, 1999 (FCC, 1997a). Their affiliates in the top 30 markets had until November 1, 1999 to be up and running, and all remaining commercial broadcasters had to be simulcasting by May 2002. The Commission proposed the year 2006 as a planned ending date for NTSC broadcasting.

On the one hand, the FCC was very generous to broadcasters in creating an open standard that allowed them great flexibility in their choice of DTV broadcast options, data transmission services, and initial simulcast requirements. On the other hand, the Commission had greatly accelerated the transition period from the original 15 years to 10. The six years originally planned for station licensing and construction had shrunk to less than two years for network affiliates in the top 10 markets.

Key Players

The key player in the decision to not mandate high-definition broadcasting and make the DTV Standard as open as possible was Chairman Reed Hundt. As Brinkley (1997a) has documented, in 1996 Hundt floated the idea of the FCC not mandating *any* digital television standard, but was rebuffed by the other commissioners. To not issue a DTV standard might invite a repeat of the AM stereo debacle where the lack of FCC action on a national standard inhibited the development of radio technology. However, Hundt's position on mandated broadcast standards was that the government's role was to be as least intrusive as possible and that the marketplace should dictate the mix and quality of digital services desired. The language in the *Fifth Report and Order* eliminating the requirement for high-definition programming reflects Hundt's point of view. The other FCC Commissioners (even the Republican nominees) were not as doctrinaire on this "let the marketplace decide" position and this stance in the rulemaking is evidence of Hundt's clout as chairman.

The other obvious winners in this decision were broadcasters. One of the key rationales for awarding them the new spectrum was that the new HDTV images would be dramatically better than conventional NTSC television. At the eleventh hour in the rulemaking process they were given the spectrum they desired *and* the freedom to do what they wished with it in terms of image quality and mix of services.

Broadcasters seemed to be the losers in the FCC's planned timetable for the transition to DTV, but it must be remembered that the NAB and the Association for Maximum Service Television (MSTV) submitted an 18-month implementation plan similar to that adopted by the Commission (McConnell, 1997). They had been alarmed by Chairman Hundt's call for a 12-month rollout in the top 10 markets and countered with their voluntary proposal. This type of accelerated implementation would have been unheard of prior to 1995 when the FCC estimated the licensing and construction period at a standard six years. The accelerated schedule was palatable to broadcasters in 1997 because of the Commission's "market-staggered" approach that ordered wealthy network owned and operated (O&O) affiliates in the top 30 markets to convert first. It meant that over 50 percent of the nation's television households could receive at least one digital signal by late 1999.

The FCC offered four reasons why the "aggressive construction schedule should be implemented" (FCC, 1997a, p. 35):

1. Broadcast DTV might fail if not "rolled out quickly." Other digital service providers such as cable or DBS might introduce incompatible systems that would inhibit the adoption of the over-the-air standard.

2. The accelerated schedule would mitigate any "disincentives" (p. 36) that early adopters would incur in offering the digital service. In other words, if most of the large O&O stations in the top markets are going to be transmitting DTV signals by 1999, there is less incentive for late adopters in these markets to wait until transmission hardware

prices drop at the end of the adoption curve (see Rogers, 1995). On the contrary, with such a short adoption time frame there would be competitive advantages to being first with DTV service in each market.

3. The "rapid build-out works to assure the recovery of broadcast spectrum occurs as rapidly as possible" (p. 36). This reflects the not-so-hidden prime agenda of the FCC in regard to the reversion of the NTSC spectrum for auction to reduce the federal deficit. Thus the accelerated conversion to DTV dovetailed with plans by the Commission, Congress, and the White House for a complete U.S. conversion to DTV broadcasting by the year 2006.

4. "A rapid construction period will promote DTV's competitive strength internationally, as well as domestically." This is an interesting rationale for the national conversion of television stations in the United States to the DTV standard. It indicates that the Commissioners of the FCC considered the international trade implications of the success of the U.S. DTV Standard in the world marketplace in their deliberations on implementation.

The *Report and Order* noted that emerging European and Japanese digital television variants were incompatible with the U.S. DTV standard. The National Telecommunications and Information Administration (NTIA) had "argued, that absent quick action, America might relinquish its technological lead to international competitors" in digital television (p. 36). The FCC stated that "rapid introduction of digital television in the U.S. will facilitate its adoption abroad" (p. 36). This sentiment is shared by Joseph Flaherty of CBS who predicted that the U.S. DTV Standard has about a 75 percent chance of emerging into a Western Hemisphere television standard, and perhaps a 50 percent chance throughout the Asia Pacific area (McConnell, 1996h). The planet may be in for another round of technopolitical competition over television standards as the DTV Standard battles the European DVB system and whatever future digital production/transmission standard emerges from Japan.

Discussion and Conclusions

Using Ogan's technology policy typology outlined at the beginning of this chapter, it is clear that the initial U.S. response to the development of advanced HDTV technology by Japan was *reactive*. While some high-definition research had been accomplished by CBS and

RCA, it was insignificant compared to the billion-dollar R&D investment by NHK and Japanese consumer electronics firms such as Sony and Toshiba.

It is ironic that the United States was quite willing to adopt the Japanese Hi-Vision/MUSE standard until the Europeans blocked the proposal at Dubrovnik. Once the prospect of one global standard for HDTV was lost, the United States (via the FCC's competition) went on to devise a digital transmission standard that had the same effect as the strategy proposed in the 1989 Commerce Department study— blocking Asian and European analog HDTV technologies in the U.S. market (National Telecommunications and Information Administration, 1989).

One may even argue that the FCC competition to select a U.S. HDTV transmission system represented a de facto American industrial policy—one designed to negate the huge Japanese investment in analog HDTV, while creating a potential multimillion dollar windfall for those Grand Alliance companies that will share in potential patent royalties. The European Union (EU) will also share in these royalties through the participation of Philips and Thomson.

Thus the FCC's competition represented a *proactive* policy initiative by providing a federal framework for a significant research and development effort by American corporations and U.S. subsidiaries of European electronics companies. The Japanese lead in HDTV technology was blunted by their decision to stick with the analog Hi-Vision system despite the development of digital alternatives by U.S./European competitors.

The FCC, through its blue-ribbon industry Advisory Committee, was able to broadly define its high-definition television goals and leave it up to the international competitors to discover the best means of implementing them. It was a policy that provided for a uniquely American solution of holding a high-technology "bake off" to define the best transmission method, while simultaneously avoiding charges that it was creating a U.S. industrial policy subsidized with taxpayer dollars.

Lastly, U.S. HDTV technology policy has been *evolutionary* in the sense that it has shifted over time with changes in the political and technological environment. The creation of the influential Advisory Committee reflected concern that the existent ATSC did not have enough political clout to oversee the creation of an HDTV television standard. When the Advisory Committee began work in 1987, it was uncertain that HDTV could be transmitted over the air within the confines of U.S. broadcasting spectrum requirements. The process of how the Advisory Committee winnowed its technology options is a fascinating study of the juncture of politics, economics, and high-technology electronics.

Outside the Bubble: Limitations of the Model

The case of high-definition television illustrates the limitations of the Krasnow et al. (KLT) broadcast policymaking model. Key influences on policy issues are outside the "environmental" bubble that encompasses the traditional six policymaking entities in the model. A study of HDTV policymaking events within the bubble simply will not make sense without their international context.

A few examples from the historical analysis above:

1. The formation of the Advanced Television Systems Committee (ATSC) in 1983 does not make sense without incorporating the effect on American broadcasters and manufacturers of the initial U.S. demonstration of Japanese Hi-Vision technology at the 1981 SMPTE conference in San Francisco.

2. An analysis of American support for the Hi-Vision system at the 1986 Dubrovnik CCIR conference would be incomplete without accounting for the desire of U.S. film and television program producers for a single worldwide television standard that would facilitate the international electronic distribution of American-made media programs. These exports have also been a source of contention with efforts by EU nations—France in particular—to limit what they call American cultural encroachment.

3. Understanding the American perception that the nation had lost its competitive edge in international high-technology must be seen in the context of the 1986 European rejection of Hi-Vision and the EC decision to create its own unique HDTV technology to protect its internal electronics community. The creation of the Advisory Committee on Advanced Television Service in 1987 was a direct result of U.S. concern over being left out of the international race to develop a commercially viable HDTV standard.

These influences are not a one-way flow from outside the U.S. The activities of ACATS have been monitored by foreign governments and electronics companies. The development of an American digital transmission standard in 1990 had widespread repercussions around the world. The Europeans canceled their analog MAC research program after it became apparent that digital HDTV transmission was feasible. Even the Japanese, until early 1997, gave conflicting signals about whether they would adapt their 1125-line system to conform to the emerging U.S. (and potential world) standard of 1080 active lines.

In short, the KLT model does not incorporate international policy influences in the case of the standardization of HDTV in the United States. The increasing globalization of, and concentration of ownership in, national media industries (both in hardware and software) is a factor which must be included in any policymaking model.

Summary of Significant FCC Decisions on HDTV

In a series of rulemakings from 1987 to 1997, the FCC and its Advisory Committee gradually narrowed the options for advanced television in the United States. In consecutive order, the most significant of these decisions were:

1. A 6 MHz limit on the second channel allocated for ATV service.

2. The selection of a simulcast conversion strategy instead of pursuing an augmentation or EDTV solution.

3. The adoption of an "existing broadcasters first" approach to awarding supplemental ATV channels.

4. The creation of an independent test facility, funded by industry, that would evaluate proponent systems in a process designed by the Advisory Committee.

5. The initial selection of a three-year period for ATV licensing and a two-year period for the construction of a transmitter and other facilities.

6. The proposal for the creation of a 15-year window for national conversion to HDTV broadcasting, and the eventual reversion of the existing NTSC spectrum to the FCC at the end of the simulcast period.

7. The selection of a digital standard that would permit multiple programming streams in varying levels of image definition.

8. The decision in the *Fifth Further NPRM* to mandate a national conversion to the ATSC Digital Television Standard was significant because it committed the United States to a wholesale change-over in terrestrial broadcast technology with multibillion-dollar consequences.

9. The role of Commissioner Susan Ness (in spurring an agree-

ment by the Grand Alliance partners and the consumer electronics industry with the oppositional forces of the Computer Industry Coalition on Advanced Television Service) was instrumental in removing the last serious roadblock in the path of the creation of a national ATV standard. By dropping mandated video parameters in the proposed DTV standard, this compromise represented the final hurdle to be surmounted in the standardization process.

10. The eleventh-hour FCC decision *not* to mandate a minimum amount of HDTV 16:9 transmission by broadcasters was surprising. Chairman Reed Hundt was able to prevail in his view that the television marketplace should dictate the ultimate success or failure of HDTV versus its standard-definition digital offspring. U.S. broadcasters will have significant flexibility in their choice of digital services and transmission modes.

While the Advisory Committee and its subcommittees played a key role in providing technical advice on issues such as the simulcast versus EDTV controversy during this period, the FCC Commissioners were the key decision makers in each of these cases. Chairman Sikes made the "gutsy call," as Joseph Flaherty put it, to pursue the simulcast route at a time when the proponent systems were still in the prototype stage. Commissioner Ness assumed a leadership role in forging the last-minute November 1996 compromise needed to establish a digital television standard.

The role of the Commissioners during this period reinforces the hierarchical decisionmaking model illustrated in Figure 7.1 in the next chapter. While the army of engineers and other specialists at the subcommittee-level would fill hundreds of binders with their recommendations on ATV implementation, it ultimately was the purview of the Commissioners to make hard choices about spectrum allocation, selecting an equitable testing process, and defining a schedule for how the nation would implement an advanced television system.

The digital spectrum auction controversy is a classic case that illustrates the utility of the KLT model in classifying the six broad categories of primary players in the policymaking game. There are a number of different players in each category and game theory is useful in parsing out the goals and strategies of these players. What emerges in this case is the political savvy of the broadcast industry in its use of a mass media campaign (through the medium they control) to influence Congress. The nested game in this case is the American political process and the role that mass media play in it (especially broadcast television). Speaker Gingrich was correct in this case—ultimately no elected official wanted to "take on the broadcasters," especially in an election year.

As nested game theory indicates, Congress, the Commissioners and the Washington representatives of the broadcast industry were being influenced by parallel games with other players that affected the policymaking process. Tsebelis (1990) was correct in noting that complex multiparty policy games need to be broken down into two and three party games for analytical purposes. Analysis needs to begin at the nested two-party microlevel and followed upward for ultimate macrolevel study.

There is one additional important section of the U.S. policymaking puzzle to examine and that is the role of the Advisory Committee on Advanced Television Service. It is the subject of the chapter that follows.

7 The Role of ACATS in U.S. HDTV Policymaking

We had to pick a winner—this was a competition and we were walking on eggshells all the time. They [the proponents] were all gaming the process like crazy. I had six years of trying to practice law while refereeing a wrestling match.

—*Richard Wiley, Chairman of the Advisory Committee on Advanced Television Service, 1995.*

Richard Wiley was working in his eleventh-floor office at his Washington law firm on the afternoon of Thursday May 24th, 1993 when he received a telephone call that greatly perturbed him.[1] For the previous five years he had served as Chairman of the Advisory Committee on Advanced Television Service (ACATS), a voluntary group of broadcast, cable, and computer industry executives who were to advise the Federal Communications Commission (FCC) concerning the introduction of advanced television in the United States (see Appendix for a list of committee participants). Wiley served as Chairman of the FCC from 1974 to 1977 and presently is one of the principal partners of Wiley, Rein & Fielding, a large law firm that, among several areas of practice, represents broadcasting clients before the Commission.

The firm occupies eight floors of an office building at 1776 K Street—a prime location that is halfway between the

White House and the FCC. Many communication law firms are located in "communication gulch" in the same K Street neighborhood, but few have the political clout of Wiley, Rein & Fielding, and few Washington lawyers are as well-connected as Dick Wiley. When FCC Chairman Reed Hundt took office in late 1993, he spoke at a dinner attended by Washington communication lawyers and joked that when he used the FCC's phone system to make a conference call he got "Dick Wiley on the phone" and that every time Commissioner Andrew Barrett tried to use call waiting, "he gets Dick Wiley on the phone" (McAvoy, 1993). Hundt was joking, but the humor was based on the viability of Richard Wiley's connections to the FCC and other federal agencies in Washington.

At the time of the distressing call, Wiley was 58 years old, a very active attorney who routinely worked 11-hour days and kept in shape by playing tennis on weekends. His warm and ebullient demeanor, ideal for a negotiator, masks a focused and tenacious personality when pursuing a task at hand. The phone call upset Wiley because it appeared that executives of the companies and institutions that had competed for ACATS (and presumably FCC) approval of their proponent technologies, were unable to find a way to combine their systems in a "grand alliance" as Wiley had urged. Without an agreement to merge, the Advisory Committee would be required to schedule a second round of testing to identify one of the proponent systems as the winner of the Committee's competitive process. This prospect might delay the establishment of a standard by almost another year in a protracted process that had already taken over five years since the establishment of ACATS in November 1987. Wiley was troubled by this turn of events and left his office to travel to the nearby Grand Hotel where the group was meeting. He was a skilled negotiator and resolved to make one final effort to see if he could forge a compromise among the dissenting parties. The results of this intervention and Wiley's larger role in the standard-setting process will be discussed in this chapter.

This chapter examines the critical role of the Advisory Committee on Advanced Television Service in providing guidance to the FCC on potential HDTV transmission standardization in the United States. The "blue-ribbon" ACATS membership voted on the acceptance of a new U.S. advanced television standard on November 28, 1995, but much of the work involved in designing a system to evaluate various proponent HDTV systems was accomplished by three key subcommittees and their various working parties. While all 25 members of

ACATS ostensibly had an equal voice in ATV matters, Chairman Richard Wiley played an instrumental role in guiding the work of the subcommittees and negotiating conflicts between proponent companies. To clarify the role of the Advisory Committee in the standard-setting process, in-depth interviews were conducted with Wiley, other key participants within ACATS, and other broadcast organizations.

The interviews sought to answer three questions:

1. What was the role of the Advisory Committee and its subcommittees in the adoption of an advanced television standard for the United States?

2. How did the Advisory Committee reconcile the often contradictory goals of computer industry interests versus the television manufacturing faction within the Grand Alliance HDTV development consortium?

3. Which U.S. broadcast policymaking model best characterizes the role of industry advisory committees such as ACATS in guiding the FCC on vital issues like communications standardization?

ATV, HDTV, and the Digital Revolution

"High-definition" television is a relative term. It is only "high" in resolution compared to the current standard. When the monochrome NTSC standard was adopted in 1941 its 525 scanning lines were considered "high"-definition contrasted with early TV systems of less than 300 lines (Carbonara, 1992). In this chapter, HDTV and advanced television (ATV) are used interchangeably, although technically HDTV is a specific subset or type of ATV. Advanced television is a generic term used by the FCC to designate any proposed transmission system that "results in improved television audio and video quality" compared to NTSC (FCC, 1988, p. 6544). High-definition television refers to a subset of ATV technology that can display at least 1000 vertically measured lines in interlaced mode and has a 16:9 aspect ratio.

Sharper widescreen pictures and improved audio are the most apparent characteristics of HDTV, but the most radical change will be invisible to the viewer—U.S. HDTV signals will be transmitted digitally. As recently as 1989, digital broadcast transmission of television images and audio was thought to be technically impossible in this century (Paul Misener, Wiley, Rein & Fielding, personal communication, June 4, 1996). In 1990, engineers from the General Instrument

Corporation developed a method of transmitting a television signal using digital modulation (Kupfer, 1991). It was a major technical breakthrough and vaulted the United States back into the international race to perfect HDTV technology.

Television and digital devices such as computers are constructed from similar electronic components, but until 1990 their respective display screens operated very differently. U.S. prototype HDTV sets tested in 1995 more closely resembled computer monitors than conventional television sets. The development of U.S. digital high-definition technology is a textbook case of the *convergence* of computer and television worlds. The new screens will be able to display high-resolution text as well as broadcast or cable programs. This is a neglected issue in the debate over HDTV, but the textual presentation ability of HDTV sets has great potential, especially if a large number of homes are ultimately linked via fiber-optic lines to the Internet in the next century.

Media Convergence

Formerly distinct modes of information and entertainment dissemination such as newspapers, magazines, television, film, and telephony have begun to converge in a new digital multimedia distribution universe in recent years.[2] Early computer terminals could display only text, but since the 1970s digital displays have incorporated color graphics, sound capabilities, and, most recently, the ability to process motion media in the form of animation, television, and film images.

The fundamental incompatibility between computer displays and television screens has been the scanning rate. Televisions scan at a rate of almost 60 screen fields per second—or approximately 30 frames per second.[3] To save on transmission bandwidth, the National Television System Committee (NTSC) specified in 1941 that half of each frame (or one field) would be scanned first, then interlaced with the rest of the frame (field two) scanned over it. While this technique reduced the bandwidth needed to broadcast early TV signals to within 6 MHz, it created a number of unwanted artifacts in the process. Interfield flicker is compensated by the human eye's persistence of vision, especially if the image is viewed from some distance away. The 1953 grafting of a color subcarrier to the black and white image added a number of other troublesome artifacts such as a marquee-effect dot "crawl" around the edges of differently colored picture elements (U.S. Congress, 1990).

On the other hand, computer screens typically operate at a higher frame rate and are scanned progressively (i.e., one complete frame after another), thereby reducing flicker that can cause eye strain when viewing text for long periods. Computer color signals are also processed differently than composite television images by using three separate component red, green, and blue channels (RGB) which are reassembled at the display, yielding much higher resolution color images.

The struggle to reconcile the interlaced/progressive-scan disparity is of great importance in the ultimate convergence of the worlds of television and computers. It is a fundamental barrier to the development of a hybrid telecomputer. This struggle is also central to efforts to design a digital high-definition television system that is both inexpensive for the consumer (e.g., interlaced-scan), and yet "extensible" in terms of incorporating future improvements (e.g., progressive-scan). The larger conflict in the macrocosm is mirrored in the political battles that took place in the formation of the HDTV Grand Alliance.

Conceptual Framework

The Krasnow, Longley, and Terry (KLT) model (1982) outlined in Chapter 6 (see Figure 6.1) will be used for macrolevel analysis of the parties involved in U.S. HDTV standardization, and their patterns of interaction. This model is useful from a "big picture" (no pun intended) viewpoint that clusters policymaker players in broad categories such as "regulated industries" for analytical comparison with other participants. The difficulty with studying *just* the macroview is that, in the act of viewing solely the big picture, the actions and motivations of key individuals get lost in the process. In this chapter we are going to combine the KLT model macroview with a game-theoretic perspective that provides insights on the microlevel of technology policymaking.

Game theory as elaborated by William Dutton (1992) will be used for this microlevel analysis of the actions of key players in the standardization game. Dutton has developed an "ecology of games" variant to traditional game theory that he derived from the work of sociologist Norton Long. Rather than trying to identify one monolithic elite or pluralistic policy game, an ecology of games approach seeks to identify a number of *parallel, interdependent games underway simultaneously.* Dutton (1992) explained that an ecology of games theory constitutes:

a move away from a focus on the macrolevel, particularly the structure of power, as a means to explain policy outcomes, to a *focus on the microlevel, particularly the goals, rules, and strategies guiding the behavior of individual actors* [emphasis added], in order to explain outcomes at the macro- or aggregate level (p. 306).

Space in this book does not permit an analysis of the goals and motivations of the 1,000+ participants in the Advisory Committee's HDTV standardization activities, but it is possible to study a cross section of key individuals. Thus, while the KLT model enhances the macro systems view of the policymaking process, the ecology-of-games perspective focuses on parallel games simultaneously under play and the actions of key figures *within* each game. The actions of the FCC's Advisory Committee can be studied as a macrolevel interaction between the telecommunications industry and the FCC, while the activities of key participants such as Richard Wiley can be studied at the microlevel using Dutton's model.

The Advisory Committee on Advanced Television Service

The FCC's Advisory Committee on Advanced Television Service has played a central role in policymaking concerning ATV service in the United States. Policy decisions recommended by ACATS have had a ripple effect on advanced television plans in Asia and Europe as well (West, 1995b). The decision of the Committee to hold a national competition to select an ideal ATV transmission system ultimately led to a global breakthrough in the digital transmission of video signals.

In its consultative role to the Commissioners and the Chairman of the FCC, the Advisory Committee was an influential factor in the drive to create a national standard for advanced television in the United States. While the Commissioners are the ultimate arbiters of U.S. policy in this area—and FCC Chairmen Dennis Patrick, Alfred Sikes, and Reed Hundt have played critical roles during their terms—the Advisory Committee has been the forum for negotiating a number of key technical and policy issues that are the basic building blocks of the U.S. HDTV standard. The Advisory Committee made its *Final Report and Recommendation* to the FCC on November 28, 1995 and officially ceased operations at that final meeting (ACATS, 1995).

Narrowing the Options

PBS Senior Vice President Howard Miller, an active participant in the work of ACATS, felt that the advisory committee was an ideal example of "government-commercial sector cooperation" (Howard Miller, personal communication, March 14, 1994). As opposed to Japan and the European Community where technological solutions for HDTV were imposed by a relatively small groups of researchers, the process in the United States began with casting a wide net for potential transmission solutions, followed by a winnowing process. Miller stated that the Commission's use of ACATS was a "unique process" that utilized policy recommendations to "gradually narrow the options" available.

The process was significant in this case. One of the difficulties of government formulation of industrial policies is the selection of which technologies to support. The central problem with industrial policy is not the conventional shibboleth of government officials selecting technological winners and losers, but rather avoiding the premature creation of policies built upon nonworkable or obsolete technologies. This is the dilemma that Japan and Europe faced with their policy infrastructure built on now-obsolete analog HDTV standards. Betting on the wrong technological horse is the fatal flaw of industrial policies that attempt to anticipate technological change.

Miller indicated that the fundamental concept undergirding the creation of ACATS was that the Committee would act as an open forum for all ideas. All potential HDTV terrestrial transmission solutions would be considered by the Committee, but eventually the group would have to reach consensus on one transmission solution to recommend to the FCC. This openness to various technological options is one of the key differences between the North American approach to HDTV and that taken in Asia and Europe, although the Digital Video Broadcasting consortium in Europe appears to be taking a broader view.

Organizational Structure of the Advisory Committee

Much of the fundamental policy work of ACATS was accomplished by three subcommittees and their constituent working parties: nine for the Planning Subcommittee, four for the Systems

Subcommittee, and two for the Implementation Subcommittee (see Figure 7.1).

ACATS member James McKinney noted that "most of the crucial (technical) decisions" were made at the subcommittee level by engineers and other technical experts. McKinney wryly added, "someone's not going to call (ACATS member) Rupert Murdoch if they have a question on a technical issue" (James C. McKinney, Advanced Television Systems Committee Chairman, personal communication,

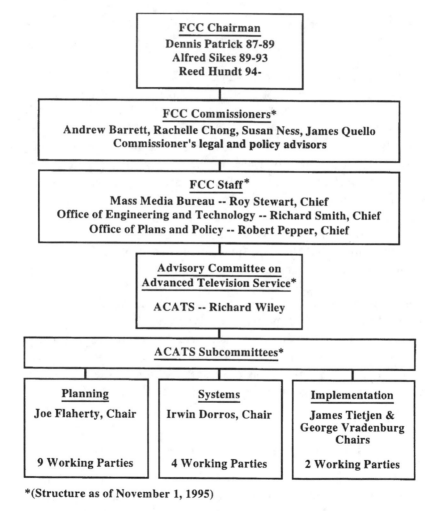

FCC Chairman
Dennis Patrick 87-89
Alfred Sikes 89-93
Reed Hundt 94-

FCC Commissioners*
Andrew Barrett, Rachelle Chong, Susan Ness, James Quello
Commissioner's legal and policy advisors

FCC Staff*
Mass Media Bureau -- Roy Stewart, Chief
Office of Engineering and Technology -- Richard Smith, Chief
Office of Plans and Policy -- Robert Pepper, Chief

**Advisory Committee on
Advanced Television Service***
ACATS -- Richard Wiley

ACATS Subcommittees*

Planning	**Systems**	**Implementation**
Joe Flaherty, Chair	Irwin Dorros, Chair	James Tietjen & George Vradenburg Chairs
9 Working Parties	4 Working Parties	2 Working Parties

*(Structure as of November 1, 1995)

FIGURE 7.1. FCC and ACATS decision-making hierarchy.

May 11, 1994). After reaching a consensus at the subcommittee level, their recommendations were passed along to the blue-ribbon panel chaired by Richard Wiley.

McKinney's comments highlight both the critical role that technical advisors played in creating ATV policy recommendations, and Richard Wiley's utilization of the expertise of the subcommittees. Wiley had a "key and major role," according to McKinney, in drawing upon technical advice from the engineering experts on the various subcommittees and then forging policy options for a vote of the full Advisory Committee. If there were consensus among the subcommittee members concerning a technical issue, Wiley would strongly recommend approval of it by the full ACATS panel.

The three subcommittees within ACATS were:

1. *Planning.* Chaired by CBS Vice President for Technology Joseph Flaherty, the Planning Subcommittee was responsible for "establish[ing] a test parameter plan" (ACATS, 1989, p. 5) by which ACATS could competitively winnow proponent systems to one ideal standard for ATV terrestrial transmission. In essence, this group was charged with designing the process by which ACATS would select a standard to recommend to the FCC.

2. *Systems.* The Systems Subcommittee was responsible for "specifying a test procedures plan and a test management plan" (ACATS, 1989, p. 5) that would define how and when proponent testing would be conducted to select an ideal candidate. This subgroup was chaired by Dr. Irwin Dorros of Bell Communications Research (Bellcore).

3. *Implementation.* Chaired by attorney George Vradenburg, the Implementation Subcommittee was charged with examining the "policy and regulatory issues associated with the introduction of advanced television service and with the [proposed] transition process" (ACATS, 1989, p. 6) from NTSC to ATV broadcasting.

Richard Wiley's Role in Consensus-building

Several influential members of the Advisory Committee have commented on Richard Wiley's ability to build consensus within the subcommittees, within the blue-ribbon panel, and even amongst the proponent companies. As Planning Subcommittee Chair Joseph Flaherty put it,

There aren't many managers who are as successful at it as he is. I think his secret is that he spends a great deal of time at it. He continually talks to people who have divergent views and he keeps them talking to one another. He gives people their head and therefore he elicits a lot of new ideas. He does build a consensus, but it's by spending a lot of difficult time at it. (Joseph Flaherty, CBS Senior Vice President for Technology, personal communication, May 23, 1994)

Wiley spent extensive time on the phone with Flaherty and other members of the Advisory Committee, as well as FCC officials, proponent company executives, and test center managers seeking to forge a group consensus on fundamental issues. Nearly all technical, testing, and scheduling issues were resolved at the subcommittee level and Wiley was usually able to walk into blue-ribbon panel meetings with a consensus in hand for their approval. As Wiley saw his role:

I was a manager, if you will, telling them what I wanted to do. I would facilitate the structure of their meetings—structuring what we were going to do, setting out what the assignments would be, how we were going to do it, what issues needed to be discussed, [and] setting agendas. (Seel, 1995b, p. 8)

Wiley's desire to hammer out agreements on contentious issues in advance of full ACATS membership meetings, in effect, transferred a substantial amount of decision-making authority to the ACATS subcommittees, the FCC staff, and to Wiley himself as the manager of the process. This is not to say that he unilaterally made key decisions himself, but that he had a very influential role in deciding *where* the rules of the standardization game would be created—at the subcommittee, committee, or commission level.

Proponent System Testing

The testing process to select a winning proponent system was a game nested within the larger ATV policy game conducted by the FCC (see Chapter 6). It was originally designed to be a "winner takes all" game with potentially significant financial stakes for the winner. Even if the royalties were only $10 per HDTV set, and assuming that 25 million sets were sold per year in the United States, this would translate into $250 million in annual income for the winner or winners of the FCC competition.

The Advanced Television Test Center (ATTC) was founded in the Washington suburb of Alexandria, Virginia in March of 1988. The budget for the operation of the ATTC was provided solely by the broadcast, cable, and electronics manufacturing industries—no federal funds were allocated. Fifty-nine percent of the budget came from ATTC members, 11 percent from CableLabs, 25 percent from proponent company test fees, and 5 percent from other sources (Joseph Widoff, ATTC Deputy Executive Director, personal communication, November 5, 1994).[4] The original budget for testing was projected at $10 million over a three-year period from 1988 to 1991, but delays in testing and the advent of digital HDTV pushed the total over $25 million by the end of 1995. In the spring of 1992, broadcast organizations conducted a campaign that raised $1.3 million from network affiliate stations to support the operations of the Test Center.[5]

An enormous amount of volunteer work was performed by members of various ACATS working parties to design and construct an equitable test process in cooperation with the staff of the ATTC. Hundreds of hours of labor were donated by participating companies and their employees to prepare test procedures and test materials. The type of innovative lab work performed by CBS Labs and NBC/Sarnoff in the creation of two color television standards in the 1940s was replicated in the early 1990s by engineers and technicians scattered throughout the United States. This standardization process was decentralized on a national and international level with input from Canada and Mexico. For instance, subjective testing was performed by the Advanced Television Evaluation Laboratory (ATEL) in Ottawa, Canada. ATEL used college students and other "non-expert" (in a professional sense only) viewers to evaluate ATV picture and sound quality. These data were then compared with ATTC "expert" viewer evaluations of the same systems (ACATS, 1992a).

The U.S. cable television industry was also an active participant in ATV testing through a branch of CableLabs (based in Boulder, Colorado) that was co-located with the ATTC in Alexandria. The CableLabs ATV test data were integrated with ATTC results forwarded to the Advisory Committee.

On the Hot Seat

The testing program at the ATTC was delayed by the Advisory Committee for four months (from April until July 1991) to allow Zenith/AT&T and the Advanced Television Research Consortium

(ATRC) to change their proposed systems to digital technology (ACATS, 1991). After General Instrument submitted its all-digital system in June 1990, ACATS voted to allow other proponents to submit digital systems as well.

Proponent testing began with the analog Advanced Compatible Television (ACTV) system developed by the ATRC group (Philips/Thomson/Sarnoff). Each proponent system was scheduled for eight weeks of tests with an intervening two- or three-week interval for system set-up and tear-down. The role of the Test Center was to identically evaluate each system through a series of qualitative objective evaluations and subjective expert-viewer tests. The results were provided in numerical form, and the ATTC did *not* evaluate the data— the job of evaluating the data and picking the winner of the competition was delegated to the Special Panel of the Advisory Committee. ACATS would then recommend that the winning system be adopted by the FCC as the standard for ATV broadcasting in the United States (and, most likely, the rest of North America).

The proponent systems were tested in the following order:

1. *Advanced Compatible TV (analog),* July-August 1991 (ATRC—Philips/Thomson/Sarnoff)

2. *Narrow MUSE (analog),* September-November 1991 (Nippon Hoso Kyokai [NHK]—Japan Broadcasting Corporation)

3. *DigiCipher (digital),* December 1991-January 1992 (General Instrument/Massachusetts Institute of Technology)

4. *Digital Spectrum Compatible,* March-April 1992 (Zenith/AT&T)

5. *Advanced Digital HDTV,* June-July 1992 (ATRC—Philips/Thomson/Sarnoff)

6. *Channel Compatible DigiCipher,* September-October 1992 (General Instrument/Massachusetts Institute of Technology)

Three of the proponents had problems during the testing period. The Narrow MUSE system yielded "poor results" in four key test parameters according to NHK's own analysis of the test data (Lambert, 1992a). This poor showing led to the withdrawal of the Narrow MUSE system from the ACATS competition after the Special Panel reviewed testing results in February 1993 (Sukow, 1993b).

Zenith/AT&T discovered they had two misprogrammed chips in the modulator of their system only after the start of testing. Zenith technicians at the ATTC replaced the chips with new ones, but were ordered to take the new ones out "at the insistence" of ATTC officials—testing regulations did not permit any system alterations after testing had begun (Cole, 1992). Zenith/AT&T were allowed to modify their system at the end of the scheduled period and conduct 10 additional days of tests (at $200,000 in extra costs paid by the consortium). Other proponents questioned the fairness of this extra testing time, and this factor may have led to a legal challenge if Zenith had won the initial competition.

The Advanced Television Research Consortium (Philips et al.) had a more serious problem: They were not ready for testing on their scheduled starting date of June 3, 1992. They had problems integrating all of the system components and notified Richard Wiley that they were going to be late. Despite the additional time given to Zenith, ACATS declined to delay the start of their test window, and the Test Center had to create an abbreviated test schedule for the Philips group (Lambert, 1992c). This decision might also have been the basis for a future legal challenge to any FCC decision on a winner in the ATV competition.

Richard Wiley was aware that these types of equity issues were a potential legal minefield. He noted:

> We had to pick a winner—this was a competition and we were walking on eggshells all the time. They [the proponents] were all gaming the process like crazy. I had six years of trying to practice law while refereeing a wrestling match. (Seel, 1995b, p. 11)

New York Times reporter Joel Brinkley has documented the ATV testing process in great detail (1997a) and he noted sardonically that few of the proponent companies would admit to any design flaws in their systems—they indicated they were just "implementation errors" (p. 210). All of the proponent systems encountered problems at the Test Center but testing was finally completed as scheduled and the results were forwarded to the Advisory Committee for analysis in the fall of 1992.

The ACATS Special Panel

Richard Wiley appointed a 25-person Special Panel in May 1992 to analyze the test results and identify a winning system. Chaired by Robert Hopkins, Executive Director of the Advanced Television

Systems Committee, the Special Panel was composed of ACATS sub-committee chairs who had no affiliation with any of the proponent companies (ACATS, 1992b).

The mission of the Special Panel was to analyze the test data from the ATTC, CableLabs, and ATEL and recommend a winner to the Advisory Committee. ACATS would then recommend the winning system or systems to the FCC for adoption as a national ATV standard. The use of a special technical advisory panel suggests that Advisory Committee blue-ribbon panel members, largely corporate executives, lacked the technical expertise to analyze the lab data and delegated the job to an intermediary group.

By May of 1992, speculation appeared in the trade press that the proponents might combine the best elements of their systems in a "hybrid" candidate that would be submitted to the FCC for approval. Richard Wiley had encouraged such a combined approach since the issuance of ACATS *Second Interim Report* in 1989 in which a footnote alluded "to the emergence of a standard based on a combination of attributes found in several candidate systems" (ACATS, 1989, p. 17). Wiley added in a 1993 interview:

> You just can't believe the meetings that we have had here in this con-ference room [in our law firm]. During the testing when one would miss the deadline, or the computer chip failure that Zenith had, or that Sarnoff [ATRC] was going to be several weeks late, and each one would say "Wait, you can't do that!" We would have conference calls and hear everyone out—trying to have an open system, which is the way it's supposed to work. I hoped to keep my eye on the ball and keep everyone in play with the thought that I had in the back of my mind that someday we could put them all together. I have to admit I thought about that for a long time. (Seel, 1995b, p. 11)

Wiley's choice of words is noteworthy, stating that he perceived his role as similar to that of a referee in a sporting event by keeping his "eye on the ball" and keeping "everyone in play." This mirrors his description above of proponent behavior as "gaming the process like crazy" in the ACATS competition. While the referee and game analo-gies are perfectly accurate in a competition such as this, what is unique is the Advisory Committee's ability to structure the rules of the game, as well as Wiley's influence on the ultimate outcome in his role as ACATS Chair.

If Wiley was enthusiastic about merging proponent technologies, the four consortia were far more interested in going for the gold ring of FCC acceptance of *their* system—and only their system. The Special Panel planned to meet in the Washington suburb of Tysons Corner, Virginia, from February 8-13, 1993, to attempt to reach an agreement on the best system (Sukow, 1993a). A meeting of the ACATS blue-ribbon panel was scheduled for February 24—with letters of invitation specifying that members had to attend in person to vote on the recommended U.S. ATV standard, as alternates were ineligible to cast ballots.

The Report of the ACATS Special Panel

In early February 1993, the Special Panel reviewed all the data from the ATTC, CableLabs, and ATEL in Canada and decided that it could not decide on a winner. The Special Panel arrived at six significant findings (ACATS, 1993):

1. Digital ATV service is achievable in the United States;
2. Narrow MUSE is unsuitable for terrestrial broadcasting due to spectrum considerations;
3. All of the digital proponent systems showed superiority over analog systems;
4. None of the digital systems showed overall superiority over other digital systems;
5. None of the digital systems demonstrated overall inferiority compared with other digital systems; and
6. Due to improvements that proponent systems had made since testing, the Special Panel recommended that the digital candidates undergo supplemental testing.

Trade journals noted that the subjective quality of the four proposed digital systems (based upon comparisons by expert and non-expert viewers) was "nearly identical." However, some proponents felt that there were enough differences in coverage areas, modulation techniques, and interference susceptibility for the Panel to select one winner (Sukow, 1993a).

The Special Panel also specified several findings in relation to *interoperability*—a term that signifies the degree of proponent system convergence between television and computer technology. A high level of interoperability meant that ATV digital television signals could be easily transcoded to digital computer images and sound—and vice versa.

The Special Panel recommended four key findings that related specifically to interoperability (ACATS, 1993):

1. An all-digital approach is important in satisfying the selection criteria relating to interoperability.

2. The proponents of four digital systems had proposed a flexible packetized data transport structure with universal headers and descriptors, an important issue in digital network communications. (Packetization is a concept originally developed for digital data transmission over telephone networks. Small groups of data packets have leading bits—headers—that are used to route the packet to a destination and descriptors tell how to decode the data bits at the receiver.)

3. Two of the proponent systems use progressive scan and square pixels, a third system uses a progressive scan transmission format and has a migration path for square pixels in the future, and the fourth system has an option for progressive scan transmission. (Note that only two of the systems had true progressive scan. Square pixels are more readily manipulable than rectangular ones in the digital domain, and are important for digital interoperability.)[6]

4. *A transmission format based on progressive scan and square pixels is beneficial to create synergy between terrestrial television and national public information initiatives, services, and applications* [e.g., the National Information Infrastructure/the Internet][emphasis added] (ACATS, 1993, p. 5).

The last point is the most significant. The Special Panel agreed with the computer industry that progressive scanning and the use of square pixels was critically important. Incorporation of these two key factors in the ATV system specification would mean that digital HDTV sets could function as terminals connected to the information superhighway simply by adding a communications card in the back of the set and attaching a keyboard or other input device.

This is the prototypical "telecomputer" that futurist George Gilder (1994) has been advocating for the past five years. At this point digital interoperability could be more valuable to a potential consumer than either widescreen pictures or improved audio. However, despite the recommendation of the Special Panel, two of the four remaining proponents still advocated interlace scan due to higher line resolution

(1000+ interlaced lines versus progressive scan systems at 720 active lines) and the claim of lower manufacturing costs.

The Advisory Committee adopted the report of the Special Panel at its February 24, 1993 meeting and ordered the additional testing program recommended by the panel (ACATS, 1993). However, Richard Wiley dangled the option of the formation of what he termed a "Grand Alliance" as an alternative to further testing. If the proponents could agree to merge the best elements of their respective systems, a Technical Subgroup of ACATS would evaluate the results and recommend ACATS and FCC approval as a de jure U.S. ATV standard. Wiley set an initial deadline of March 15th for the proponent consortia to reach a partnership agreement or begin the additional test period as individual entities. He noted in an interview:

> They began to worry about whether I would look at the system as it was tested or at the system as it was improved. And I said, "I'll tell you what we'll do. We'll have a second round of testing, based just on the improvements that each of you has put forth by a certain date, and that's going to be very expensive." They all groaned. And then I said, "But I'll give you an alternative, and that's to form a single so-called Grand Alliance system." And at first it didn't seem that it was going to be possible. (McConnell and West, 1995, p. 40)

Enter the Computer Industry

Initially, the U.S. computer industry showed little interest in HDTV policy. Most of the early research in ATV technology was analog-based, and American computer firms appeared to have little interest in ACATS and vice versa. However, in the late 1980s, semiconductor and computer companies noted the rapid development of digital video technology in the United States and overseas, and their interest in HDTV policy grew accordingly. The development of HDTV digital transmission by General Instrument (GI) triggered a massive push by other proponent companies to develop equivalent systems. Prior to that innovation, scientists and engineers at the Massachusetts Institute of Technology (MIT) had been advocating what they termed "open architecture" technology for HDTV that called for television sets to be built more like computers (with circuit boards that could be inserted to add additional functions). Such concepts were dismissed as too expensive by traditional manufacturers such as Zenith. However, the development of GI's digital transmission technology caught the attention of electronics firms around the world. The domination of the ACATS testing program by digital proponent systems cemented that interest, particularly over the key interlace/progressive scan issue.

John Sculley, then CEO of Apple Computer, wrote a letter to the FCC in late May 1993 expressing his concern that the Commission might select a non-progressive scan HDTV format (Sculley, 1993). Not so coincidentally, Representative Edward Markey of the House Finance and Telecommunications Subcommittee scheduled a hearing on this issue for May 27, 1993—three days after ACATS May 24 deadline for the start of system retesting (U.S. House of Representatives, 1993).

Game Theory and a Potential Grand Alliance

After the Special Panel decided that there was no clear winner in the ACATS competition, the three remaining proponents faced a quandary that is interesting from a game theory point of view. Each proponent consortium had two strategic options to pursue between the adoption of the Special Panel report by ACATS on February 24, 1993 and the deadline for the start of retesting on May 24:[7]

1. The three-month interval could be used to bolster one's system by incorporating successful aspects of the competition's technology. The test reports from the ATTC were available to all proponents, and these data could be used to refine system technology within the limits imposed by patent law. All of the proponents had already modified and improved their systems before the Special Panel report. However, each player knew that the others were improving their competitive positions, by copying their successful attributes as well. The potential reward for taking this option was a 1 in 3 chance of winning and taking home all potential royalties. (The partnership of General Instrument and MIT had a 1 in 2 chance since two of the four remaining systems were theirs.)

The downside to pursuing a winner-takes-all strategy was the prospect of litigation over irregularities in the testing process (e.g., Zenith/AT&T's test extension and the denial of a late start to the ATRC group). The winner and/or members of ACATS and the ATTC might all be sued by disgruntled losers. This litigation could drag on for years and delay the implementation of ATV service in the United States until well into the next century.

2. The second option would be for the proponents to go along with Richard Wiley's concept of a "grand alliance" and merge their technologies into one system. Such a strategy would probably avert the threat of litigation delays and might yield a more technologically effective solution than any single system alone. Potential royalties would have to be split three ways, but a third of a loaf is better than none.

The downside in this option is the potential difficulty in deciding which elements of each proponent's system should be incorporated into a "grand alliance" format submitted to ACATS and the FCC. Royalties might be calculated based on respective proportions of technology contributed to the final system. The negotiation of an equitable agreement could be problematic.

Richard Wiley's Role in Negotiation

Wiley mailed each of the proponents a letter after the February ACATS meeting explaining the two choices and the pros and cons of a "grand alliance," which the parties knew he favored. From Wiley's point of view, such an alliance would benefit ACATS as well. Another round of winner-takes-all testing could be avoided and the Committee could wrap up its six-year odyssey with a final year of system integration and testing. While Wiley has described a merger scenario (in the McConnell and West interview cited above) that was simply in the proponents' best interests, he also felt that it was the path of least resistance for ACATS and the FCC.

As Wiley commented on a potential alliance:

I encouraged it. I thought it would save a year or two ultimately in the recommendation we could make. I saw how they were getting closer [in terms of their system specifications]. I know what will happen with the four of them out there—I saw their lobbyists—and I know that any decision we make would be challenged on the "Hill," in the courts, [and] at the FCC. (Seel, 1995b, p. 15)

As the May deadline approached, negotiations intensified with Wiley holding meetings in his office every two weeks with representatives of the proponent companies. Ultimately, the proponents met in a room at the Grand Hotel in Washington, DC, during the week prior to the ACATS deadline of Monday the 24th in an attempt to reach a merger agreement. By late Thursday, May 20th, they were ready to give up and go home without an alliance. The stumbling block was the fundamental schism between interlaced scan favored by ATRC (Philips/Thomson/NBC/Sarnoff) versus progressive scan championed by Zenith/AT&T and MIT. General Instrument had both bases covered with one proposed system in each format. The interlaced camp noted that while a nonprogressive standard inhibited interoperability, this factor was offset by their prediction of lower manufacturing costs with interlaced scanning—although this has been disputed by the computer industry.

Finally, after a fruitless day of negotiations the representatives of the three consortia called Wiley at 4:00 p.m. and invited him to come over to the Grand Hotel and make a last attempt at conciliation.

Wiley described what happened next:

I came over there and met with them around a round table. It was rather awesome because I thought this was my shot, because I was really an advocate of the "grand alliance" in my mind. Although I gave them two choices, I thought this was the way to proceed. The way I proceeded—it was sort of a spur of the moment thought—was let's start with what we agree on. I was listening to them [and] interlaced/progressive was one of the things [that was a major point of disagreement]. What could we agree on?

It was amazing once we started at that end of the periscope. "Do we agree on high-definition television being the goal? Do we agree on high line number progressive scanning being part of the goal? Do we agree on square pixels being part of the goal?" They all agreed that a 1,000 [line], or high line rate, all-progressively-scanned system was the way to go.

Once you have the goals in mind it was easier to do. No doubt the progressive/interlaced thing was difficult—even there they agreed on all sets over 34 inches [in diagonal picture size] being progressive, and all film [transfers] being progressive scan, even now. We kept narrowing the issues of disagreement so they were talking to one another again. It had been sort of angry—I'm not saying that I created an atmosphere of calm, but it did help. (Seel, 1995b, p. 16)

Tempers had flared over the scanning issue. Thomson/Philips/

Sarnoff representatives were adamant that they would not accept an all-progressive system in the early stages of HDTV broadcasting—they wanted a high line number (over 1000) interlaced standard (Joseph Donahue, Senior Vice President, Thomson Consumer Electronics, personal communication, November 29, 1993). The state of TV camera technology in 1993 could not accommodate a high-line-rate progressive system (such cameras have subsequently been developed). Any initial progressive system would have only 720 lines, but could ultimately improve to 1080 lines with new developments in television/computer technology.

Dr. Jae Lim of MIT was equally adamant that adopting an interim interlaced standard might impede the migration to an all-progressive system, perhaps indefinitely. Joseph Donahue, the Thomson representative at the meeting, described Dr. Lim as "very obstinate, very extreme" in his insistence on an all-progressive standard (Joseph Donahue, personal communication, November 29, 1993). Dr. Lim is a Professor in the Department of Electrical Engineering and Computer Science at MIT where he worked on a number of prototype HDTV systems. His position on progressive scan is congruent with that of the U.S. computer industry: They are concerned that the acceptance of interlaced scanning for HDTV broadcasting will inhibit the expected convergence of computers and television.

The proponents agreed to establish a dual standard that incorporated both interlaced and progressive scanning depending on the application. Interlaced would be utilized at first to simplify manufacturing on low-end TV sets (with progressive scan used primarily for larger screen sizes)—with a long-range goal of "migrating" gradually to an all-progressive system. It was a difficult compromise but it made the alliance possible.

Wiley left the hotel that afternoon thinking that the Grand Alliance was a reality, but he was prematurely optimistic:

> I thought we had it pretty well set, but then they started calling my house at 7:00 a.m. the next morning [Friday, May 21st], and there was disagreement. They were going to have two different plans and I said, "No, no, no we don't want that. Let's go with the plan—we can drop a footnote [in the draft press release] that Jae Lim would have preferred all-progressive from the outset. Let's get a footnote in instead of having two different plans."
>
> I wanted them to get together once and for all. And even the fol-

lowing Monday (the 24th), when I thought I had put it to bed Friday, we were on a conference call for seven hours. It was incredible. I kept saying "We've got a press conference, we've got a press conference." I kept moving the press conference back. He [Jae Lim] was pretty responsive, as it turned out, as long as [his position] was represented. He could understand the benefits of going ahead with the dual format. (Seel, 1995b, p. 17)

The press conference went on as planned at the FCC with Wiley making the announcement of the formation of the Grand Alliance with interim FCC chairman James Quello. Dr. Lim's dissent was indicated in a footnote on page two of the press release that stated, "MIT believes that a digital video broadcast standard that exclusively utilizes progressive scan transmission, from the beginning, is in the best interests of the United States" (U.S. House of Representatives, 1993, p. 18). Wiley commented on the rationale for holding the FCC conference:

My reason for having the press conference was I wanted to lock it in and get it [the Grand Alliance] going. What has really pleased me since that time—once the struggles and animosities that were evident in that Grand Hotel room [were over] —they've worked really well together. I think we've gotten a lot further by their cooperation than we could have ever gotten with their working apart. (Seel, 1995b, p. 17)

Congressional Reaction to the Grand Alliance

Representative Markey held a hearing on high-definition television on Thursday May 27th—three days after the Grand Alliance press conference (U.S. House of Representatives, 1993). A key issue in the hearings was the interlaced/progressive schism within the U.S. high-technology community. Michael Liebhold of Apple Computer testified (in place of then-CEO John Sculley) on behalf of the Computer Systems Policy Project (CSPP), an association of 13 computer industry CEOs.

Liebhold had several criticisms of the Grand Alliance system:

1. The inclusion of interlaced scan, while simplifying HDTV conversion for manufacturers and broadcasters, would seriously impede interoperability with computer systems and the National Information Infrastructure (NII).

2. Liebhold stated that the Grand Alliance interlaced specification

would "assure a major role for Japanese and European video equipment vendors" (U.S. House of Representatives, 1993, p. 90), raising the specter that foreign companies might be the primary beneficiaries of the inclusion of interlaced scanning.

3. He noted the "serious protest" to interlaced scanning from MIT in the ACATS press release and suggested that the Advisory Committee was not considering the "NII stakeholder communities"— schools, hospitals, publishing concerns, and research institutions.

4. He implied that ACATS was a captive of the television manufacturing industry and that a thorough review of the interlaced/progressive interoperability issues would be best conducted by another FCC advisory committee, the White House Office on Technology Policy, the National Institute of Standards and Technology, or the Defense Advanced Research Projects Agency (U.S. House of Representatives, 1993).

Game Theory and the Exclusion of Players

One of the most interesting aspects about Liebhold's statement to Congress is not just the questioning of ACATS role as an impartial referee, but the rules of the game itself. By addressing Congress for a change of venue to another FCC committee or other executive branch agencies, Liebhold, Sculley, and the other CSPP computer industry executives were signaling a vote of no confidence in the Advisory Committee process. The relegation of "their" spokesperson, MIT, to a dissenting footnote in a press release was used as evidence of the marginalization of their position on progressive scanning.

The Krasnow et al. model is useful in considering other policy venues for players who feel thwarted by the standard-setting process at the FCC. The options are to take one's case to Congress, as the computer industry did, or to seek redress through the National Institute of Standards and Technology in the Department of Commerce or other executive branch agencies. The last option would be to resort to legal action in the courts, a prospect that Richard Wiley indicated that he was concerned about (in the context of system proponent challenges) since it could delay the setting of an ATV standard for many years.

The computer industry, due to its economic clout in the United States and being one of the bright spots in U.S. electronics manufacturing, was able to make its case through CSPP to Congress. Apple, Microsoft, Compaq, and other computer companies still feel strongly that the inclusion of interlaced scanning in the U.S. digital standard is a serious mistake. They formed a new group called the Computer Industry Coalition on Advanced Television Service (CICATS) and used Congressional hearings and press conferences to articulate their position on this fundamental technological issue (Beacham, 1996a; Karr, 1995).[8]

The frustration of the computer industry at holding a minority position in the Advisory Committee's standard-setting process led them to voice their concerns in another policymaking venue—the U.S. Congress. From this perspective of simultaneous policy activity occurring in distinct venues such as Congress and the FCC, the telecommunications policymaking process is an ecology of separate, but interrelated, games as Dutton defines them. Activities in each arena—the FCC, executive branch agencies, the courts, the regulated (and unregulated) industries—constitute nested games within the macro game of U.S. policymaking.

The Debate over Conversion Costs and Flexible Use

John Abel, then-NAB Vice President, also testified at the Markey hearings. While he praised the achievements of ACATS and the Grand Alliance in creating an innovative digital standard, he was concerned about broadcaster costs in being forced by the FCC to stimulate demand for consumer ATV hardware. Abel reiterated the point that total HDTV conversion costs for each local affiliate might run as high as $14 million and:

> It will cost the same whether the station is in Boston or Billings [Montana], and this cost is simply prohibitive for most small and medium market TV stations, especially when many of these small TV stations are not even worth half the cost of conversion. Advertisers are not going to pay more for their advertising just because our costs have gone up, so this is a dilemma. (U.S. House of Representatives, 1993, p. 93)

Abel had a valid point: It would be ironic for the FCC to adopt a new HDTV standard to promote the public interest, convenience and necessity, when in fact it might do the opposite by driving small-market television stations out of business. Localism is a distinctive characteristic of American broadcasting and any concept that threatens it is examined very carefully by the FCC.

Richard Wiley's rebuttal was that the FCC ought to investigate a phased implementation of HDTV broadcasting:

> I think this is going to be an evolutionary thing. I'm not sure that every station ought to have to do this at the same time—the FCC might have a roll-out that might recognize that high-definition conversion will start with the networks and the bigger stations and get it out later to Peoria, [Illinois], where I'm from. Currently that's not their [FCC] policy, but it will be looked at again.
>
> If I were doing it, I'd say: "Here's six megahertz, free, to you broadcasters. We want you to use it for high-definition television and other advanced services. But you're not going to have a 24-hour a day HDTV programming schedule the first moment you're in there, so if you want to use it for interactive television, or you want to use it for digital services—video or nonvideo, things like that. [You will have] a second revenue stream and you're getting free spectrum." I'm not sure it's all as dire as it's being portrayed by the lobbyists in Washington stirring up scare stories. (Seel, 1995b, p. 19)

The issue of conversion costs for small-market broadcasters is a fundamental policy concern and will be a controversial topic as the FCC revises its timetable for the shutdown of NTSC broadcasting and the reversion of the analog spectrum.

ACATS Recommends the Grand Alliance System to the FCC

The Grand Alliance (GA) submitted its dual interlace/progressive system to the Advanced Television Test Center in April of 1995 for final testing and evaluation. The system had been scheduled for final testing in February and there was some concern on the part of the Advisory Committee when there appeared to be last-minute integration problems with the hybrid GA system (Cole, 1995). Despite the complex process of integrating the technologies developed by the former competitors, the system easily passed its final tests at the ATTC. The ACATS Technical Subgroup evaluated the test results and made a recommendation in October 1995 to the blue-ribbon Advisory Committee for approval of the Grand Alliance system as the national ATV standard (ACATS, 1995).

On November 28, 1995, the Advisory Committee convened in the eighth-floor hearing room of the FCC's 1919 M Street headquarters. The Committee voted to accept the findings of its Technical Subgroup who evaluated the test results from the ATTC. The *Final Report* stated:

Specifically, the Grand Alliance system meets the Committee's performance objectives and is better than any of the four original digital ATV systems; the Grand Alliance system is superior to any known alternative system; and the ATSC (Advanced Television Systems Committee) Digital Television Standard, based on the Advisory Committee design specifications and Grand Alliance system, fulfills the requirements for the U.S. ATV broadcasting standard.

Accordingly, the Advisory Committee on Advanced Television Service recommends that the Federal Communications Commission adopt the ATSC Digital Television Standard as the U.S. standard for ATV broadcasting. (ACATS, 1995, p. 19)[9]

The eight-year "long march" of the Advisory Committee to create an American standard for HDTV was over—it officially ceased to exist at the conclusion of the meeting. Few of the participants were aware at the outset that the task would take as long as it did, and consume hundreds of thousands of collective hours in meetings, testing, and negotiations. As Richard Wiley pointed out, "It's been eight years. When then-FCC chairman Dennis Patrick offered me the job, he said: 'This and will be about a two-year project'" (McConnell and West, 1995, p. 32).

Brinkley (1997a) credits John Abel of the NAB as the "father of HDTV in America" (p. 64) and there is some merit to this assessment for his role (and that of the NAB) in initiating the Advisory Committee process. However, Abel spent much of his effort after 1990 stonewalling the work of Wiley and the committee to serve the ends of the NAB. If true paternity is to be assigned to someone for the establishment of an advanced television service in the United States the honor (or blame from some corners) should go to Richard Wiley. He held a voluntary group together for almost a decade through many crises that might have led others to simply throw up their hands and concede defeat. The task was a political and managerial minefield and Wiley, his assistant Paul Misener, and the ACATS volunteers negotiated it successfully.

Discussion and Conclusions

ACATS and the Role of the FCC

The Advisory Committee on Advanced Television Service and its 1,000 volunteer participants did a remarkable job during the eight-year process of designing and conducting the competition that created a

U.S. standard for digital high-definition terrestrial broadcasting. The Committee played a key role in defining the key policymaking questions that the FCC needed to resolve, and it also created the organizational structure needed to establish an ATV standard. The ACATS process of casting a wide net for feasible technical solutions, then gradually winnowing out the less desirable systems, avoided the technological dead-ends (e.g., analog solutions) perpetuated by faulty industrial policy in Europe and Japan. It was a unique case of industry-government cooperation that many thought impossible in an era of intense competition in the international electronics marketplace.[10]

The parent "blue-ribbon" Advisory Committee can be criticized as being dominated by the U.S. broadcast industry. While consumer representation was noticeably absent on the panel, the Committee did attempt to include representatives of the computer industry after it became apparent in 1990 that a digital standard was feasible. However, given the original mandate for the Advisory Committee from the FCC, the makeup of the group named by the Commission in 1987 reflected the terrestrial broadcast industry that would be most affected by the creation of a new transmission standard. The Committee did make an earnest effort to create a digital standard that would be interoperable with computer technology by specifying a square-pixel standard that incorporated progressive scanning on larger screen displays. Several influential members of the computer industry are still displeased by the inclusion of interlaced scanning in the proposed ATV standard, but the Grand Alliance system is designed to transmit both scanning formats (Karr, 1995; McConnell, 1995c). Despite the protests of representatives of Compaq, Microsoft, and Apple, the interlaced scan option remained in the standard.

From a "microscopic" policy viewpoint, Richard Wiley played a key role in managing the Advisory Committee from its inception in November of 1987 through the testing phase at the ATTC, the creation of the Grand Alliance research consortium, and the delivery of the final report in November 1995. As noted above, he played a central role in the negotiations that forged the GA coalition. It is also interesting to note whom Wiley credited with influential roles within the Advisory Committee:

> Over the years, the dominant player has been Joe Flaherty of CBS, but there have been other main players. I would cite Jim McKinney and Bob Hopkins of the Advanced Television Systems Committee, who did such great work on the documentation, and the futuristic laborato-

ries—the Advanced Television Test Center, headed by Peter Fannon—
and the great work that Mark Richer of the Public Broadcasting Service
did as my representative. People like [MSTV consulting engineer] Jules
Cohen, who has been so valuable. (McConnell and West, 1995, p. 32)

Wiley's description of the participants as "players" is also interesting from a game theory perspective. Of all the individuals cited above
by Wiley, only Flaherty and McKinney were members of the blue-ribbon panel. This reinforces the assertion that most of the heavy lifting
in the standard-setting process was accomplished at the subcommittee
and testing lab level and that the blue-ribbon panel was largely a fig-
ure-head group appointed for their collective corporate clout. The
blue-ribbon panel routinely approved the consensus created at the
subcommittee level by Wiley, a pattern that continued through their
unanimous (Craig Mundie, the Microsoft representative on the blue-
ribbon panel, abstained) acceptance of the Technical Subgroup's proposal to select the Grand Alliance system (McConnell, 1995b).[11]

It should also be noted that while the FCC Commissioners have
delegated most of the testing and other technical work to the Advisory
Committee, they have played important roles at key junctures in the
ATV policymaking process. Alfred Sikes, FCC Chairman from 1989-
1993, also was very active in promoting HDTV as head of the National
Telecommunications and Information Administration (NTIA) prior to
moving to the Commission ("Government Support," 1987).[12]

Once appointed to the FCC, Sikes was an activist Chairman in
regard to ATV. Sikes was instrumental in forging the Commission's
simulcast decision that essentially mandated that the U.S. ATV system
be incompatible with NTSC and would eventually supplant it after the
simulcast period (FCC, 1990). Planning Subcommittee Chair Joseph
Flaherty characterized this as a "courageous decision" in the face of
broadcaster opposition to incompatible formats and credited Sikes
with being the "driver" and "real leader" on this fundamental policy
issue (Joseph Flaherty, personal communication, May 23, 1994). This
decision also deferred any U.S. consideration of compatible enhanced-
definition (EDTV) systems such as that proposed by the
Philips/Thomson/Sarnoff consortium ("FCC to Take Simulcast
Route," 1990). Richard Wiley later noted that this decision avoided
what he characterized as a policy error on the part of the European
Union of adopting EDTV as an intermediary standard because it could
inhibit the eventual diffusion of true HDTV. Wiley said, "We threw out

that idea early on. FCC Chairman Al Sikes made that decision, of going for the gold" (McConnell and West, 1995, p. 40). Sikes' decision-making role underscores that the FCC chairman is the first among equals on the Commission. Of the five commissioners, the chairman has disproportionate power to shape policy, often along ideological/political lines.

Dutton's ecology of games approach to the study of the policy-making process is helpful in analyzing the "goals, rules, and strategies guiding the behavior of individual actors in order to explain outcomes at the macro- or aggregate level" (Dutton, 1992, p. 306). Careful analysis of the goals and motivations of key players such as Richard Wiley and Alfred Sikes are important in understanding policymaking events that occur at the macrolevel.

The Macro View of ATV Policymaking

Simultaneously analyzing the U.S. standard-setting process from the macro perspective proved by the Krasnow et al. model is also instructive. The key axis is defined by the standard-setting game played between the FCC and what Krasnow et al. defined as "regulated industries." The Advisory Committee can be plotted along the middle of the axis as an industry-dominated group created and sanctioned by the FCC. Other related ATV organizations such as the Advanced Television Test Center and the Advanced Television Systems Committee are both funded and operated by U.S. broadcast/cable corporations and could be placed near the industry end of the axis.

Congress is the third part of the basic policymaking triad, but until 1995 it played mainly an observational role with many hearings held on ATV but little legislation enacted. This changed in 1996 with the passage of the Telecommunications Act (see Chapter 6). The FCC waited until it saw which way the political wind was blowing in Congress prior to issuing a definitive rulemaking on advanced television, especially concerning any auction of either the ATV or NTSC spectrum. Krasnow et al. were correct in defining this basic triad as the key telecommunication policymakers. The White House and citizen groups have had an influence on the process but their input has come late and to little effect.

The Role of Industry in Policymaking

The creation of the U.S. ATV system can be seen as a textbook example of government-industry cooperation to define a national standard for HDTV technology—despite the long time interval exacerbated by the development of digital transmission.

This is not to say that this should be the model for creating all new media standards in the United States. The process took too long given the pace of current communication technology development, especially that involving digital systems. In many contemporary cases, the involved industries have created a voluntary standard to prevent a repeat of the incompatible VHS-Betamax videocassette debacle. The 1995 agreement between Toshiba and Sony/Philips over a common international standard for the high-capacity Digital Video Disk (DVD) is a case in point (Dickson, 1995a). The agreement was reached within a year after the two groups created incompatible technologies for video compact disks that will probably supplant existing CD-ROMs. Strong pressure was exerted by global computer manufacturers to reach an agreement for this fundamental digital distribution medium. However, leaving standard-setting up to a regulated industry always raises the specter of the AM stereo controversy in the United States where the lack of a national standard inhibited the successful diffusion of the technology (Braun, 1994).

While the Krasnow et al. model is useful in studying the policy-making process in the United States, it is inadequate to explain international policy influences in a telecommunications environment that is increasingly global in scope. The system boundary labeled "environment" is needed to define their area of study, but we will argue that such a boundary is very porous in an era of expanding multinational corporations and the fluctuating influence of transnational standard-setting bodies such as the International Telecommunication Union.

The Advisory Committee on Advanced Television Service (ACATS)

The role of the 25-member "blue-ribbon" Committee was to advise the FCC on advanced television issues and recommend an ATV standard for FCC consideration. The list of members has changed over time—the list below includes the members of record as of the endorsement of the *Final Report and Recommendation* on November 28, 1995.

Richard Wiley, Wiley, Rein & Fielding, (ACATS Chairman)*

Frank Biondi, Viacom International*

Joel Chaseman, Chaseman Enterprises International*

Bruce Christensen, Public Broadcasting Service*

Joseph Collins, American Television Communications Corp.*

William Connolly, Sony Corporation of America*

Martin Davis, Wellspring Associates, Inc.*

Irwin Dorros, Bell Communications Research*

James Dowdle, Tribune Broadcasting Company*

Ervin S. Duggan, Public Broadcasting Service

Joseph Flaherty, CBS Inc.*

Samuel Fuller, Digital Equipment Corporation

Stanley S. Hubbard, Hubbard Broadcasting*

James Kennedy, Cox Enterprises Inc.*

James C. McKinney, Advanced Television Systems Committee*

Craig Mundie, Microsoft Corporation

Rupert Murdoch, Fox, Inc.

Thomas Murphy, Capital Cities/ABC Inc.*

Jerry Pearlman, Zenith Electronics Corporation*

F. Jack Pluckhan, Quasar*

Ward Quaal, The Ward L. Quaal Company*

Richard D. Roberts, Telecable Corporation*

Burton Staniar, Westinghouse Broadcasting Corporation*

James Tietjen, SRI International*

Robert Wright, National Broadcasting Company*

EX OFFICIO MEMBERS

Wendell Bailey, National Cable Television Association

Henry L. Baumann, National Association of Broadcasters

Peter Bingham, Philips Laboratories

Joseph Donahue, Thomson Consumer Electronics, Inc.

Brenda L. Fox, Dow, Lohnes & Albertson

Richard Friedland, General Instrument Corporation

Robert Graves, R.K. Graves Associates

Larry Irving, U.S. Department of Commerce, NTIA

Keiichi Kubota, NHK Science & Technical Research Labs

Jae Lim, Massachusetts Institute of Technology

Vonya B. McCann, U.S. Department of State

George Vradenburg III, Latham & Watkins

Margita White, Maximum Service Television Inc. (MSTV)

* Denotes members of the original committee of 25 named in 1987

HDTV Economics

To the mass audience, the difference between NTSC and HDTV is perhaps more akin to the difference between monophonic and stereo sound.
—*W. Russell Neuman, Massachusetts Institute of Technology, 1988*

As noted in Chapter 1, some terrestrial broadcasters in the United States have complained that conversion from NTSC to high-definition television (HDTV)—especially within the prescribed FCC timetable—will place an undue financial burden on stations, especially in small markets, because it will unlikely produce any measurable increase in audience size and advertising revenues. Chris Zell, WETM-TV Chief Engineer, Elmira (New York), put it more bluntly: "HDTV will bankrupt stations. Small stations will be wiped out" (McConnell, 1995a, p. 103). In addition, these broadcasters fear that, even after investing enormous amounts of money into re-equipping their stations, consumers might not follow up with expected acquisitions of HDTV hardware and instead might wait for the availability of a full HDTV programming line-up and cheap sets. For them, consumer diffusion of HDTV could be sluggish, at least in the short term.

Not everyone agrees with this cautious or pessimistic assessment, though. For instance, Pieter Boegels (1988), then President of the European EUREKA HDTV Directorate, delivered an upbeat message about the future of consumer HDTV to the audience of the 1988 International Broadcasting Convention. He stated:

> Will you gamble that HDTV will not make it? Are you prepared to be left out in the cold? People often say to me, "Who wants bigger TV screens? There's no demand." Of course, no-one can predict the future, but keen students of technology and business know that whenever something appears with *clear advantages for the public* [emphasis added]—they flock to buy. (p. 9)

These "clear advantages" generally refer to better pictures, better sound, and larger screen size. Today HD-MAC is defunct, superseded by digital television (DTV), but the fundamental economic question that this statement raises remains as relevant in 1997 as it was in 1988: Will consumers perceive these key features as sufficiently engrossing, as Boegels believes it will, to ensure the successful and rapid diffusion of advanced television (ATV) products paralleling that of the video-cassette recorder (VCR) and the compact disc (CD) player? Will they view them as "clear advantages" over conventional television technologies and "flock to buy" HDTV, DTV, or other ATV variations when introduced in the United States and Europe (Japan is the only country with a regular HDTV service) in the late 1990s? Or will HDTV simply become the next generation of TV receivers—an innovation perceived by consumers to be only marginally better than its predecessor with few or no unique benefits?

This chapter examines three HDTV economic issues—station conversion costs, market potential, and consumer acceptance, although most of it is devoted to the third one. We begin by detailing the results of several NTSC-to-HDTV station conversion cost analyses that have been conducted in the United States between the late 1980s and the mid-1990s. Next, we report estimates about the market size of HDTV sets in the United States and point out some weaknesses of these forecasting studies. Then, we review the body of available HDTV-related consumer studies in Japan, North America, and Europe according to five main variables (awareness, preference, interest, desirability of attributes, and purchase intent and willingness to buy). This third section is guided by diffusion theory and organized to follow Rogers' (1995) steps in the innovation-decision process.[1] The chapter concludes with a discussion of the HDTV versus standard-definition television

(SDTV) implementation issue and the formulation of six propositions about potential consumer acceptance of HDTV.

Station Conversion Cost Studies

Michael Sherlock (1990), NBC President for Operations and Technical Services, has cast the introduction of ATV service in terms of a "chicken and egg" dilemma: "On the one hand, programmers and broadcasters will hesitate to accept the huge cost of producing and broadcasting the new ATV programs until there is a large audience with new ATV sets capable of displaying the improved programs; but, on the other hand, the potential viewers will hesitate to buy expensive new ATV sets until the new programs are on the air" (p. 72). Assuming that broadcasters will first deploy HDTV technology, the next question becomes: How expensive will be the switch to HDTV for them? The answer to that question is quite complex, because not only will these implementation estimates be influenced by the learning curve of the manufacturing process, which will bring about declining costs for HDTV studio and transmission equipment over time, but they will also depend on station size, speed of conversion, and type of conversion.

In 1989, Robert Ross, Director of Broadcast Operations and Engineering for KYW-TV, Philadelphia, issued a first report projecting that the cost of complete station conversion to HDTV could total as much as $38 million ($9 million for "passing the network," i.e., receiving and transmitting network programming; $15 million for studio operations and playback; and $14 million for field operations, which include electronic news gathering equipment). A year later, Ross revised his cost analysis and determined that full conversion costs could range from $1 million for enhanced-definition television (EDTV) in a small market to $38 million for HDTV in a large market, the worst-case scenario. The price for network pass-through could be as low as $250,000 in EDTV and as high as $8 million in HDTV ("Refined HDTV," 1990).

In 1990, CBS and PBS independently predicted that the highest full conversion costs should not exceed $12 million, a 68 percent reduction over previous estimates. CBS reported that the cost for network pass-through could be as low as $741,000 in small markets and as high as $1.5 million in large markets ($1.6 to $3.1 million if we add local commercial insertion capability), while the cost for full conversion could be as low as $5.8 million in small markets and as high as $11.6 million in large markets ("New HDTV Estimates," 1990). Unlike partial conversion, full conversion will enable broadcasters to produce local programs in HDTV format. Among other things, the CBS study assumed that largest markets will begin conversion first, that conversion will be phased in over a period of five to nine years, and that the initial high cost of HDTV equipment will decrease due to economies of scale.

More recently, Weiss and Stow (1993) determined that capital investment for initial conversion to HDTV, including network pass-through and local commercial insertion, would vary from $1.2 million to $2.2 million depending on four different studio and transmission equipment scenarios. In early 1996, PBS updated its 1990-1991 cost analyses and estimated specific implementation costs based on frequency band and level of power (Zou and Kutzner, 1997). Overall these predictions (from $700,000 to $2 million for basic conversion and from $1.2 to $2.5 million for moderate conversion) were similar to those reported in the early 1990s, except perhaps for low-power stations. In a survey conducted in October 1996 with 400 station executives, 46 percent reported that it would cost at least $5 million to fully convert their facilities to DTV. Estimates ranged from $1 million to over $8 million. A majority of the respondents predicted that their stations will convert to DTV within two to five years. Interestingly, only 31 percent stated that the government should not mandate a time line for the conversion to digital television broadcasting. For 72 percent of the respondents, the biggest benefit of DTV was to "be competitive" (Harris Corporation, 1996).

HDTV Forecasting Studies

In what we believe to be the first HDTV market study, a research firm predicted in 1982 that HDTV sets would be available in the United States by 1984 and would number 30 million by 1999 ("Research Firm," 1982)! It is easy to poke fun at these predictions in 1997, but the reality is that, regardless of the forecasting methodology used (e.g., Delphi, scenario writing, new product diffusion models), it is perilous to project the short-run—let alone the long-run—diffusion and adoption of new media products, especially *before* their introduction on the consumer market. Lazarus and McKnight (1984) remind us that marketers and even inventors often misconceived the eventual

consumer use and the diffusion rate of communication technologies—
from the telephone to television to the VCR. "In 1943, *The New York Times* predicted that television might be of value to the military, but didn't expect it to be a commercial success. The *Times* found it hard to believe that American families would want to spend much time sitting still in the same room" (p. 2). The case of the VCR is also illustrative. Of 29 studies forecasting the penetration of home video players (i.e., film cartridge devices, videotape players/recorders, videodisc players) reviewed by Klopfenstein (1985), only one succeeded in predicting accurately the short-term and long-term penetration rates of the VCR in the United States.[2] He found that many forecasts failed to identify obstacles to the diffusion of home video products and drew erroneous comparisons from the development of previous communication technologies.

Japan's pioneering experience with HDTV broadcasting, which began in June 1989 (see Chapter 2), sheds some light on the accuracy of HDTV forecasting studies and offers insights into the evolution of HDTV receiver prices. In 1987, the Ministry of Posts and Telecommunications (MPT) and the Ministry of International Trade and Industry (MITI) independently forecast that HDTV sets would achieve a 30 percent penetration rate by the year 2001 and a 40 percent penetration rate by the year 2000, respectively ("Japanese Government," 1987). Initial prices for a 30-50 inch HDTV receiver were predicted to average $3,000 (Nippon Hoso Kyokai, 1987). But when first introduced on the consumer market in 1989, a 36-inch HDTV receiver cost a whopping ¥4 million (about $29,000; $1 = ¥138) (K. Shoda, Nippon Hoso Kyokai, personal communication, March 17, 1995). In late 1992, the cheapest HDTV set still cost about ¥1.3 million ($10,833) (Schilling, 1992b). In January 1996, the official retail price for a 28-inch Hi-Vision receiver was ¥440,000 (about $4,200; $1 = ¥105), although Japanese consumers could purchase one for ¥290,000 (about $2,800) in electronics discount stores (N. Kumabe, Hi-Vision Promotion Association, personal communication, January 4, 1996). In June 1997, the cheapest 28-inch HDTV receiver in Japan still retailed for ¥360,000 (3,130; $1 = ¥115), although older models could be found for less than ¥200,000 (under $2,000) in some discount stores (N. Kumabe, July 1, 1997). By April 1997, after five and a half years of experimental Hi-Vision satellite broadcasting (see Chapter 2), sales of HDTV sets only totaled 371,000 (HPA, http://www.j-entertain.co.jp/hpa/index.html). In retrospect, then, government forecasts of HDTV sets diffusion have been greatly overoptimistic. Retail prices of HDTV receivers have declined since 1989 but not fast enough to make the purchase of an HDTV set affordable or attractive to a majority of Japanese consumers, hence the growing interest of Japanese consumers in the cheaper widescreen television alternative.

With all these caveats in mind, what do we know about the market potential of consumer HDTV in the United States? In 1988, three reports, American Electronics Association (AEA) (1988), Darby (1988), and Robert R. Nathan Associates (RRNA) (1988), predicted that HDTV set sales in the United States would reach 1.0 million units, 4.1 million units, and 10.4 million units, respectively, by the end of the year 2000. These numbers translate into cumulative household penetrations of about 2 percent, 9 percent, and 43 percent, respectively (based on a projected number of 105,933,000 U.S. households in 2000). A fourth study conducted by the London-based consultancy firm National Economic Research Associates (NERA) in 1992 considered multiple HDTV scenarios and projected the U.S. HDTV receiver sales market to 0.3 million in scenario 1 (sluggish diffusion) and 1.2 million in scenario 3 (moderate diffusion) by the year 2000 (Brown, Cave, Sharma, Shurmer, and Carse, 1992).[3] Household penetration rates would reach 0.5 percent and 2 percent by 2000, respectively.[4] In 1989, the Congressional Budget Office (1989) reviewed the first three studies and concluded that they may underestimate the *rate* of adoption of HDTV and overestimate the *market size* of HDTV, by assuming that this technology will become an instantaneous consumer electronics success. Bayrus (1993) found that of the three studies (AEA, Darby, RRNA), only the HDTV sales forecast by the AEA were congruent with past sales histories of home appliances.

In estimating the market potential of HDTV, these four studies differed in their assumptions about product analogy,[5] beginning of HDTV transmission, impact of EDTV on HDTV, software availability, and retail prices, hence the huge variations observed between these HDTV sales forecasts. These and other forecasting studies have hypothesized that consumer demand for HDTV receivers will depend heavily on five factors: price, availability of programming, buyers' expectations, availability of related video products (e.g., HDTV VCR, EDTV), and buyers' income (AEA, 1988; Brown et al., 1992; Darby, 1988; Lyman, 1985; Mentley and Castellano, 1990; Niblock, 1991; RRNA, 1988; Stow, 1993; Working Party 5, 1988). Because most of this research modeled the diffusion of HDTV after that of color TV, it

would be helpful to look back at the impact of programming availability and price on the diffusion of color television in the United
States:

> When color service was first introduced [in 1954], some six years
> elapsed with a very low rate of market penetration, even though the
> consumer price of color sets fell by a factor of 2.5 during this period.
> Only when full primetime color programming was provided by the
> networks [in 1966] did sales of color sets make significant and rapid
> advances. For HDTV service today, the situation is quite different, and
> much HD programming can be made available at the introduction of
> service. First, 70 percent of all primetime programming is produced on
> 35 mm. film, which is a high definition medium, and intrinsically has
> a wide screen aspect ratio. Such high definition programming can be
> readily transferred to videotape for broadcast, or for delivery by the
> other distribution media of cable, DBS, or home video. ... When full
> primetime programming in color was achieved in 1966, and when
> market penetration "took off", the retail price of a color TV set repre-
> sented 14.7 percent of the average per capita income. The equivalent
> price in 1992 for a similar percentage investment is $3,700, a figure that
> is feasible for a HDTV set. It is also noted that in the four years from
> 1966 to 1970, market penetration rose from 9 percent to 34 percent.
> (Stow, 1993, pp. B-11, B-4, B-5)

Assuming ample programming and initial receiver prices comparable
to those of color TV, Rupert Stow (1993), former CBS Engineering
Director, argued that market acceptance of HDTV will depend on how
consumers perceive the value of the technology, a topic that is covered
below.

HDTV-Related Consumer Acceptance Studies

The third and main section of this chapter reports the findings of 29
quantitative HDTV-related consumer studies that have been conducted in Japan, North America, and Europe between the 1980s and the
mid-1990s (see Table 8.1). About two-thirds were unpublished (i.e.,
reports, conference papers). Not all studies specifically focus on
HDTV's potential diffusion and adoption, but all investigate empiri-
cally at least one aspect of the technology. A brief summary of methodological procedures introduces each reviewed study so that the find-

ings can be interpreted relative to a particular research method. Half of the studies used an experimental design that involves formal manipulation, while the other half relied on survey methodology (mail, phone, or self-administered). Each methodological approach has its strengths and weaknesses. Laboratory experiments offer greater internal validity (i.e., ability to measure what it is supposed to measure) than survey instruments, but they often lack external validity (i.e., ability to generalize sample results to the population) and use small samples. On the other hand, surveys, when administered to random samples, enjoy a higher degree of external validity than laboratory experiments. Unfortunately, most HDTV surveys reviewed here used nonrandom "convenience" samples (e.g., mall intercepts), but about half of them exposed respondents to real-life HDTV pictures prior to questionnaire administration.

Awareness of HDTV

In 1985, the Nikkei Industry Research Institute (NIRI) conducted a mail survey with 1,000 Japanese households and found that, of all respondents ($N = 796$), 58 percent claimed prior knowledge of HDTV (Takahashi, 1985). The Nikkei study was one of the first published consumer surveys on high-definition television and the first to measure HDTV awareness. In May-June 1987, the Broadcasting Technology Association (BTA) surveyed 780 Japanese respondents who had attended demonstrations at the HDTV Fair. Of the respondents, 86 percent reported to have some degree of HDTV awareness—they knew at least the word "high-definition television" (Broadcasting Technology Association [BTA], 1987b). In September-October 1988, the Hi-Vision Promotion Council conducted the largest ever HDTV consumer survey with 88,199 Japanese participants who had viewed Seoul Olympics broadcasts at 81 Hi-Vision (i.e., HDTV) demonstration sites. Only 9 percent of the respondents reported never having heard of Hi-Vision prior to the demonstrations (Hi-Vision Promotion Council [HVPC], 1989).

In the United States, Advanced Television Publishing (ATP) conducted an HDTV marketing survey at the Hecht Company department store, Washington, DC, with 90 shoppers or visitors (Bush, 1987). The ATP study revealed little awareness of HDTV in the United States by early 1987, with less than 25 percent of the respondents claiming having heard of it (Bush, 1987). In the Lupker, Allen, and Hearty (1988) study, described below, North American respondents' HDTV aware-

TABLE 8.1 SUMMARY OF HDTV-RELATED CONSUMER STUDIES

AUTHOR	YEAR	COUNTRY	N	METHOD	VARIABLES
ARDITO	1994	Italy	54	experiment	screen size, viewing distance
BOUWMAN ET AL.	1991	Netherlands	762	survey	awareness, purchase intent, willingness to pay
BOUWMAN ET AL.	1993	Netherlands	753	survey	awareness, attributes, purchase intent, willingness to pay
BTA/OKAI	1987	Japan	780	survey	awareness, interest, attributes, willingness to pay
BUSH	1987	US	90	survey	awareness, preference, interest, attributes
CSP INTERNATIONAL	1984	US	NA	NA	preference, willingness to pay
DETENBER & REEVES	1996	US	132	experiment	screen size
DUPAGNE	1997	US	193	survey	awareness, interest, purchase intent
DUPAGNE & AGOSTINO	1991	Belgium	311	survey	interest, attributes, willingness to pay, purchase intent
HARRIS CORPORATION	1997	US	104	survey	attributes, willingness to pay
HBO	1988	US	820	survey	awareness, preference, purchase intent, willingness to pay
HVPC	1989	Japan	88,199	survey	awareness, interest, attributes, willingness to pay
LOMBARD	1995	US	32	experiment	screen size, viewing distance
LOMBARD ET AL.	1997	US	80	experiment	screen size
LUND	1993	US	50	experiment	screen size, viewing distance
LUPKER ET AL.	1988	Canada/US	6,941	survey	awareness, preference, purchase intent, willingness to pay
McKNIGHT ET AL.	1988	US	356	experiment	screen size, format
NEUMAN	1988	US	613	experiment	preference, viewing distance, willingness to pay
NEUMAN	1990	US	214	experiment	picture quality
NEUMAN ET AL.	1987	US	367	experiment	sound quality
NEUMAN & O'DONNELL	1992	US	500	experiment	picture quality
PITTS & HURST	1989	US	180	experiment	screen format
REEVES ET AL.	1992	US	32	experiment	screen size, viewing distance
REEVES ET AL.	1993	US	40	experiment	picture quality, sound quality, screen size
REEVES ET AL.	1997	US	38	experiment	screen size
SOFRES	1989	France	979	survey	awareness, attributes
TAKAHASHI	1985	Japan	796	survey	awareness, interest, willingness to pay
WOBER	1985	UK	2,304	survey	attributes
WOBER	1989	UK	941	survey	preference, attributes, purchase intent

NOTE: NA = not available; N = number of respondents/participants.

ness varied from 51 percent in Montreal to 65 percent in Toronto at the living room sites and from 36 percent in Ottawa to 59 percent in Toronto at the shopping center sites. In the Home Box Office (HBO) (1988) study, only 27 percent of the surveyed respondents in Danbury, Connecticut, reported prior awareness of HDTV. More recently, Dupagne (1997) conducted a phone survey in March 1995 with a random sample of 193 Miami respondents to determine predictors of HDTV awareness, interest, and purchase intent based on demographics, mass media exposure, ownership of home entertainment products, and importance of television attributes. Results revealed that only 32.1 percent of the respondents were aware of HDTV at the time, a surprisingly low percentage that may be due to the nature of the market but that may also reflect the actual state of HDTV awareness in the *general* American population. Significant predictors of HDTV awareness included income, gender (being male), and importance of picture sharpness.

In Belgium, a 1988 phone survey with 311 randomly selected Brussels respondents indicated that only 17 percent were aware of HDTV (Dupagne and Agostino, 1991). Of those affirming so, only 40 percent were able to articulate what they heard about HDTV. Of the same group, 27 percent reported having learned about HDTV from the print media, 15 percent from interpersonal sources, and 8 percent from television. This result supports the claim that mass communication channels are more effective in creating awareness about an innovation than interpersonal communication channels (see Rogers, 1995). The authors also found that respondents who owned more home entertainment items, watched less television, read a newspaper more frequently, went to a movie theater more regularly, were younger (18-39), had postsecondary education, had higher income, and were male were more likely to be aware of HDTV. In May 1989, SOFRES (1989), a leading French research firm, conducted a survey with a panel of 979 Minitel respondents to assess consumer interest in advanced television systems. Of the respondents, 38 percent were aware of HDTV (but only 17 percent were aware of the D2-MAC standard). In the only HDTV-related longitudinal study, de Jong and Bouwman (1994) reported that the level of HDTV awareness among four different panels of Dutch respondents rose from 36 percent in 1990 to 71 percent in 1993. Significant predictors of HDTV awareness included gender (male), age (negative), and level of education (Bouwman, Hammersam, and Peeters, 1993). These profiles of "early knowers" are consistent with the diffusion literature (Rogers, 1995).

Preference for HDTV versus Conventional Television

In one of the earliest HDTV consumer studies, the CBS study of 1982, 64 percent of the respondents perceived a *major difference in picture quality* between HDTV and NTSC, and 47 percent reported a *major difference in color* (Communications Studies and Planning [CSP] International, 1984).

In February 1987, as noted above, ATP conducted an HDTV consumer survey at a Washington, DC, department store, with 90 shoppers or visitors (Bush, 1987). The purpose of the study was to assess public reactions toward the NHK 1125-line HDTV system. Visitors came to the exhibit as they pleased. Programming segments originated from *Top Gun* (a U.S. theatrical movie), *Dream* (an Italian music video), and *Chasing Rainbows* (a Canadian miniseries). These segments were displayed simultaneously on three monitors of different size (32-inch CRT, 54-inch rear projection, and 120-inch direct projection). After 2-5 minutes of exposure, they were asked to complete a questionnaire. Of the respondents, 87 percent rated the quality of HDTV as *better* than that of conventional television. When asked how much better on a scale from 2 to 10, respondents were split at the two extremes, with 19 percent saying two times better than the picture of their current TV sets and 22 percent saying 10 times ($M = 6.1$).

In October 1987, the Committee for the North American High Definition Television Demonstrations to the Public surveyed 6,941 respondents in three Canadian (Ottawa, Toronto, Montreal) and two American (Seattle [Washington] and Danbury [Connecticut]) cities and produced the most detailed analysis of HDTV reactions ever conducted (see Lupker et al., 1988). The Danbury results will be reported separately below (see Home Box Office [HBO], 1988). The primary purpose of the survey was to determine the extent to which nonexpert viewers discriminate between HDTV and NTSC. At all locations but Seattle there were two types of sites: the shopping center (SC) site and the living room (LR) site. In the SC situation, 4,052 respondents were asked to participate as they passed the displays. Program material comprised a segment of the U.S. movie *Top Gun* and several short music videos. After viewing it on the two HDTV and NTSC sets one at a time for a total of 3-5 minutes, they were asked to complete a questionnaire. In the LR situation, 2,889 respondents were recruited by local advertisements, personal invitations, mall intercept, and telephone survey. These participants were assigned seating positions at viewing distances of 3H (i.e., three picture heights), 5H, 7H, 9H, and

11H. Program material included excerpts from *Oniricon*, an Italian production, *Chasing Rainbows*, a Canadian miniseries, and *Around the World in HDTV*, a Japanese documentary. At both the SC and LR sites, NTSC sets were 25-inch models and HDTV were either 28-inch or 30-inch models.

On average (Danbury excluded), SC and LR respondents perceived a *moderate* to a *considerable* difference between HDTV and NTSC. Of both SC and LR respondents, 51 percent felt that the two systems differed either *greatly* or *considerably*. Eighty percent of the LR respondents judged HDTV to be better than NTSC in overall picture quality, but only 68.5 percent of the SC respondents reported so. Significant predictors of the perceived difference included picture sharpness, sense of depth (i.e., showing the distance of objects from the viewer), motion quality (e.g., lack of blur in moving parts of the image), and set size. In addition, LR respondents preferred HDTV to NTSC in picture sharpness (79 percent), sense of depth (77 percent), screen shape (77 percent), color quality (75 percent), picture brightness (67 percent), motion quality (58 percent), and sound quality (44 percent). Lupker et al. found no significant differences in these technical judgments due to program material (*Oniricon* versus *Around the World in HDTV*) except in sound quality. But respondents seated at 3H distance (1st row) were more likely to indicate preference for HDTV than those at 5H, 7H, or 9H.

As part of a wide North American survey (see Lupker et al., 1988), Home Box Office (HBO) conducted a survey in Danbury, Connecticut, to compare consumer responses to HDTV versus NTSC. A total of 820 American consumers were surveyed at one of two settings: the living room (LR) situation ($n = 507$) and the walk-through (WT) situation ($n = 313$). In the LR situation, recruits were selected randomly by telephone and paid \$15 for their participation. Groups of about 15 respondents entered a room in which rows of seats were arranged according to three different viewing distances (3H, 5H, and 7H). In the WT situation, "intercepted" consumers were asked to watch an HDTV-versus-NTSC demonstration for a few minutes and complete a short questionnaire on the future of television. For both situations, there were two viewing configurations: alternating viewing and side-by-side viewing. In the alternating viewing mode, TV programs were switched back and forth from HDTV to NTSC and consumers were asked to rate the picture quality of each. In the side-by-side viewing mode, TV programs were shown simultaneously on HDTV and NTSC receivers and consumers were asked to rate the picture quality of each. In both situations and configurations, programming was displayed in studio-quality NTSC format on a 25-inch NTSC monitor, and in less-than-studio-quality MUSE HDTV format on a 28-inch HDTV monitor. In the LR situation, programs included *Oniricon* (Italy), *Chasing Rainbows* (Canada), and *Around the World in HDTV* (Japan). The WT situation used the same programs plus four additional materials.

Of the LR respondents, only 39 percent stated that HDTV was better in overall picture quality, while 43 percent preferred the overall picture quality of NTSC. In the WT situation, 55 percent stated that HDTV was better than NTSC, and there were no major differences in preference by viewing angle (52 percent for HDTV in alternating viewing; 56 percent for HDTV in side-by-side viewing). In the LR condition, however, viewing configurations produced different results. In side-by-side viewing, 61 percent of the LR respondents preferred HDTV to NTSC, but in alternating viewing, this number dropped to 39 percent. HBO attributes these conflicting and less-than-positive results to technical difficulties encountered during the demonstrations and other methodological problems (e.g., video quality of HDTV, overrepresentation of housewives in the sample).

The HBO study did not control for variations in program content and therefore did not evaluate the impact of content on overall preference for HDTV versus NTSC, although it suggested, based on subsequent focus group sessions with LR respondents, that "programming content can affect results" (p. 15). Indeed, LR respondents expressed higher interest in *Around the World in HDTV* than in *Oniricon* and *Chasing Rainbows*. Side-by-side results also revealed that male and younger respondents reacted more favorably to HDTV than female and older respondents, respectively. Respondents seated at 3H distance (1st row) were more likely to indicate preference for HDTV than those at 5H or 7H. Side-by-side respondents preferred HDTV for its screen shape (i.e., widescreen) (68 percent), picture brightness (62 percent), color quality (61 percent), and picture sharpness (60 percent).

In December 1987, Neuman (1988) conducted an experiment at the Massachusetts Institute of Technology (MIT)'s Audience Research Facility in the Liberty Tree Mall, Danvers, Massachusetts, to assess viewer preference for HDTV versus NTSC. The 613 participants were recruited by mall intercept and telephone. Each session consisted of two tests. In the first test, the conservative test (also called the single stimulus test), participants were randomly assigned to either the

HDTV sets condition or the NTSC sets condition. They were asked to evaluate the quality of the respective sets. The two screen sizes were 18 inches and 28 inches. In the second test, the comparison test (also called the double or dual stimulus test), participants were asked to view and evaluate programming on two sets (HDTV vs. NTSC) side by side. Participants were seated at three different distances: 1 meter, 2 meters, and 3 meters (18-inch sets: 3H, 7H, 10H; 28-inch sets: 2H, 5H, 7H). Programming included six clips: *Carly Simon* (concert), *Olympics*, *Football*, *Long Gone* (comedy-drama), *Mandela* (drama), and *Lions of Africa* (drama). Viewing exposure totaled about 15 minutes.

The author found that in the side-by-side comparison test, 62 percent preferred the overall picture quality of HDTV to that of NTSC. However, preference for HDTV was highly context-dependent (e.g., program type, set size, viewing distance). For instance, 95 percent of the participants in Group 2—viewing the *Olympics* clip on 28-inch HDTV and NTSC monitors from a distance of one meter (2H)—preferred the quality of HDTV. On the other hand, 89 percent of the participants in Group 1—viewing the *Football* segment on 18-inch HDTV and NTSC sets from a distance of three meters (10H)—preferred NTSC to HDTV. Neuman (1988) found a clear relationship between viewing distance and preference for HDTV: The closer to the screen the viewers sat, the more they preferred HDTV to NTSC. This and other findings (see HBO, 1988; Lupker et al., 1988) support the claim that three picture heights is the optimal viewing distance to discern the attributes of HDTV, although it remains to be seen whether the audience would watch television that close in normal viewing situations (see Ardito, 1994; Lombard, 1995).

In the conservative test, Neuman (1988) found no significant differences between the HDTV and NTSC groups in their evaluations of program liking, program interest, program involvement, and screen quality. In the same vein, there were few evaluation differences based on content, and there were no differences in evaluations of technical characteristics (color, screen shape, picture sharpness, picture brightness, sense of depth, and motion quality). All in all, Neuman (1988) concluded that "HDTV is not the same kind of revolutionary shift in technology as experienced in the transition to color in the 1950s and 1960s in the United States. To the mass audience, the difference between NTSC and HDTV is perhaps more akin to the difference between monophonic and stereo sound" (p. 7).

In July 1989, the British Independent Broadcasting Authority (IBA) surveyed 941 visitors at an exhibition entitled "Lifestyles 2000," which showcased two HDTV sets (38-inch and 54-inch) displaying enhanced-definition television (EDTV) and HDTV program material (Wober, 1989). Seventy-five percent of the respondents reported that the 38-inch set was either *very much* or *quite a bit better* than their home set. This percentage was slightly less for the 54-inch set (67 percent).

Interest in HDTV

In the 1985 NIRI survey, 83 percent of the Japanese respondents stated they were interested in HDTV. When asked what types of programs they would like to see in HDTV, they first selected travelogues, followed by movies and sports (Takahashi, 1985). In the 1987 BTA survey, 62 percent of the respondents claimed to be *very interested* (total interested: 91 percent) in purchasing an HDTV receiver (Okai, 1987). Then in the 1989 HVPC survey, 49 percent of the respondents reported that they were *very interested* (total interested: 94 percent) in Hi-Vision, and when asked whether they wanted it in their home, 76 percent answered affirmatively. Japanese respondents also indicated that the types of programs they would prefer to see in HDTV were sports (39 percent), followed by movies (30 percent), news (28 percent), and dramas (24 percent) (HVPC, 1989).

In the United States, 67 percent of the ATP respondents stated that they were *very interested* (total interested: 79 percent) in having HDTV at home. Reasons for this expressed interest included *quality of the picture* (34 percent), *spend a lot of time watching TV* (23 percent), and *want state of the art* (23 percent). In a more recent U.S. survey, 58.3 percent of the respondents expressed interest in acquiring an HDTV set (somewhat interested: 25.3 percent; very interested: 33.0 percent). Significant predictors of HDTV interest included age (negative), income, frequency of movegoing, frequency of sports viewing, and importance of picture sharpness (Dupagne, 1997).

In contrast, only 27 percent of the Belgian respondents in the Dupagne and Agostino (1991) study felt that is was either *important* or *very important* for them to have an HDTV receiver at home.

Desirability of HDTV Attributes

In Japan, 98 percent of the BTA respondents reported that HDTV had either a *better* or *much better* picture sharpness than their ordinary set. Likewise, almost all respondents (96 percent) favored HDTV color rendition. On the other hand, only 57 percent felt that the screen shape

(16:9) of HDTV was *better* than that of NTSC (4:3). Less than half (48.5 percent) of the Japanese respondents preferred a 30-inch screen size for HDTV to any other size and a mere 13 percent felt that 50 inches and over was the most desirable screen size (Y. Tashima, Ministry of Posts and Telecommunications, personal communication, February 27, 1990). In the 1989 HVPC survey, Japanese respondents were asked to indicate which Hi-Vision feature(s) spurred their interest. Again, picture quality was chosen by an overwhelming majority (71 percent), followed by screen size (47.5 percent) and sound quality (33.4 percent). Of the HVPC respondents, 51 percent stated that they would select a 30-inch screen size should they decide to purchase an HDTV set for home, but only 11 percent indicated that a 50-inch screen size would be suitable for home (HVPC, 1989).

When asked to report their favorable impressions about the HDTV demonstration, American respondents in the 1987 ATP study singled out picture clarity (59 percent of all positive reactions), followed by color rendition (36 percent) and picture brightness (35 percent). Only 18 percent and 11 percent of the favorable reactions referred to the large (i.e., 54-inch) and wide (i.e., looks like a movie screen) picture, respectively. And only 27 percent of the respondents rated the *importance of having HDTV with a 54-inch screen in home* as *very important* (total important: 48 percent). Dislike of the projected picture (26 percent) and size of the set (i.e., too big) (19 percent) topped the list of unfavorable mentions (Bush, 1987). In Ottawa, Canada, the LR respondents who saw *Around the World in HDTV* on a 50-inch HDTV projection monitor judged overall picture quality, color quality, and picture sharpness more favorably than those who saw the program on a 30-inch HDTV CRT monitor (Lupker et al., 1988).

In March 1997, Harris Corporation commissioned a survey involving interviews with 104 participants to assess consumer reactions to the characteristics of American digital HDTV versus those of NTSC. These respondents were recruited from the Washington, DC, area based on screening criteria (e.g., at least six hours of TV viewing per week, VCR ownership). The same material was shown simultaneously on an HDTV monitor and an NTSC monitor. Of the respondents, 98 percent felt that HDTV was superior to traditional television in picture quality; 96 percent stated that they liked the shape of the 16:9 receiver; and 97 percent reported that the 5-channel digital HDTV sound was superior to stereophonic NTSC sound (Harris Corporation, 1997).

In Europe, HDTV surveys have corroborated these Japanese and

North American findings. In 1985, the IBA surveyed 2,304 British respondents to examine their attitudes toward enhanced-screen features (Wober, 1985). The two most important perceived enhancements were sharper pictures and better sound. In addition, an overwhelming majority of British respondents disagreed that a 4 x 3½ foot screen (the size of a window) would *fit in with the size of other things in the room* (84 percent) and would be *a welcome part of the home* (71 percent). Consumer disapproval was even more ostensible when the size of the screen was 7 x 5½ foot (the size of a large double bed): 89 percent and 82 percent respectively.

In the 1988 Belgian study, respondents deemed picture sharpness as the most important television attribute (rated by 77 percent as *very important*), followed by color fidelity (62 percent), sound quality (55 percent), and screen size (33 percent). "Importance of picture sharpness" was *negatively* related to television viewing, *negatively* related to age, and *positively* related to education. Almost three-fourths of the respondents reported that they were either *not too interested* or *not interested at all* in having a TV screen larger than the one they currently own. Only 10 percent of them voiced an interest in a 100 cm × 60 cm set. As we would expect, "interest in a larger screen" was *positively* related to ownership of home entertainment products, *positively* related to moviegoing, *positively* related to income, and *negatively* related to age (Dupagne and Agostino, 1991). In the 1989 SOFRES survey, 91 percent and 82 percent of the French respondents reported to be interested in improved picture quality and in hi-fi stereo sound offered by D2-MAC satellite transmissions, respectively.

In the 1989 IBA survey, British respondents were asked to evaluate the desirability of five HDTV-related attributes (sharper pictures, stereo sound, flat screen, wider pictures, and bigger pictures). Of the respondents, 79 percent *definitely wanted* sharper pictures, 63 percent stereo sound, 52 percent a flat screen, 40 percent wider pictures, and 34 percent bigger pictures. Age was *positively* related to the importance of flat screen, *negatively* related to the importance of stereo sound, and unrelated to the importance of sharper, wider, or bigger pictures (Wober, 1989).

In May 1991, a representative NOS (Nederlandse Omroep Stichting or Netherlands Broadcasting Foundation) panel of 753 Dutch people were asked to rate the importance of 13 potential television improvements. Among HDTV-related features, respondents perceived CD sound to be the most important improvement (*very important*: 55 percent), followed by sharper pictures (50 percent), lighter set (48 percent), and flatter set (48 percent). Only 21 percent felt that a wider screen was a *very important* technological amelioration (Bouwman et al., 1993).

We now turn to a review of the experimental literature to examine more in detail the impact of the three main HDTV characteristics (picture quality, sound quality, and screen size and format) on consumer acceptance. Virtually all those studies were conducted in the United States.

Picture Quality

Neuman (1990) investigated the interactions between three levels of resolution as measured by the number of scanning lines (525-line NTSC, 1125-line HDTV, and 3000-line simulated very high-definition television) and three screen sizes (25-inch, 35-inch, and 180-inch diagonals). The 180-inch screen size was intended to correspond to a flat-screen display. The author set out to determine whether there is "a saturation threshold under normal viewing conditions, beyond which increased resolution is unnecessary and imperceptible" (p. 9). A representative sample of 214 nonexpert adults were recruited from the Greater Boston area. Neuman found the expected interaction between resolution and screen size. At any given resolution, subjects preferred larger screen sizes. For both the 35-inch and 180-inch displays, the higher the resolution, the more positive the overall evaluation of the display. For the 28-inch display, though, participants preferred the 525-line NTSC to the 1125-line HDTV resolution. This finding suggests that consumers may not perceive the incremental value of a higher resolution picture when displayed on a small- or medium-sized TV set. But when asked to rate the *picture quality* of the programs, subjects preferred the 28-inch display to either the 35-inch or 180-inch display for the 1125-line resolution. The author explained these apparent disparities as follows: "The line-structure and artifacts are increasingly evident, and though people express a preference for the larger display, they have to acknowledge that the picture quality is, from a subjective point of view, lower" (p. 23) (see also Lombard, 1995). He concluded that resolution beyond 1125 lines (HDTV) makes a difference but that there is no ideal screen size for advanced television.

Neuman and O'Donnell (1992) compared viewer preference for progressive versus interlaced scanning formats. While interlaced scanning involves two scans (two fields), in which all the odd-numbered television lines (1, 3, 5...) are first scanned followed by all the even-numbered lines (2, 4, 6...), progressive scanning produces a single scan (one complete frame), in which all lines are scanned in one pass. During the development and approval process of the U.S. HDTV transmission standard, the scanning format issue gave rise to one of the most enduring controversies between the members of the Grand Alliance and the computer industry (see Chapter 7) and still remains a subject of heated debate between broadcasters and computer manufacturers. Five hundred participants were assigned to either the "conservative" condition or the "side-by-side" condition. In the conservative test, participants viewed either interlaced or progressive scan pictures and rated the picture quality of the set on a 5-point scale from *bad* to *excellent*. In the side-by-side test, they viewed two monitors simultaneously and rated their technical characteristics on 7-point scales from *lowest* to *highest*. In the conservative test, Neuman and O'Donnell (1992) found no significant differences between the progressive scan ratings and the interlaced scan ratings. In the side-by-side comparison, however, 73 percent of the participants favored NTSC interlaced scanning over progressive scanning.

Reeves, Detenber, and Steuer (1993) examined the effect of video fidelity on attention, memory, and evaluation of content. Forty undergraduate students viewed 16 one-minute clips from four entertainment movies (*Casualties of War, Days of Thunder, Indiana Jones and the Last Crusade*, and *Total Recall*). In this within-subjects experiment, the subjects were exposed to both low-fidelity video (low resolution, low signal-to-noise ratio) and high-fidelity video (high resolution, high S/N ratio). The authors found significant effects for some of the content-based measures, indicating that subjects preferred high-fidelity video to low-fidelity video, but not for attention or memory.

Sound Quality

Neuman, Crigler, Schneider, O'Donnell, and Reynolds (1987) investigated the impact of audio fidelity on viewer perceptions of television programming and technical features. A sample of 367 people were recruited at the Liberty Tree Mall in Danvers, Massachusetts, to participate in this experiment. Each session included a conservative test and a comparison test, performed sequentially. In the conservative test, participants watched a 2½-minute television clip of one of three content types (*Miami Vice* [action-adventure], *Cheers* [situation comedy], *Basketball*) in one of four conditions (mono/low-fidelity;

mono/high-fidelity; stereo/low-fidelity; stereo/high-fidelity). After exposure, participants were asked to rate liking, interest, involvement, picture quality, audio quality, and overall quality of TV set for the clip on evaluation scales (e.g., from *high* to *low*). In the comparison test, participants watched and evaluated two 30-second clips that were identical except in audio quality (e.g., mono/low-fidelity vs. stereo/low-fidelity).

In the conservative test, the authors found significant differences in evaluation scores of program liking, interest, and involvement but only when comparing *extreme* audio conditions. Participants in the stereo/high-fidelity group liked the program content better and found it more involving and more interesting than those in the mono/low-fidelity group. As we would expect, participants in the stereo/high-fidelity condition also rated the sound quality of the program significantly higher than those in the mono/low-fidelity condition. But differing audio quality had little or no impact on evaluations of picture quality and overall quality of TV set. Findings from the comparison tests reflected those of the conservative tests. The authors concluded that subjects were "willing to report the overall quality and sound quality of one set as better than another, but [the results did] not provide clear support for the hypothesis that respondents prefer 'high' quality over 'low' quality sound" (pp. 9-10).

In their study, Reeves et al. (1993) also assessed the impact of audio fidelity and audio spaciousness on attention, memory, and evaluation of content. Using a within-subjects design, 40 participants were exposed to a low-fidelity condition (altered frequency spectrum, lower signal-to-noise ratio) and a high-fidelity condition (unaltered frequency spectrum, higher S/N ratio). The second independent variable, audio spaciousness (i.e., dimensionality), was manipulated between subjects. Participants were assigned either to the surround sound condition (five speakers) or the monophonic sound condition (one speaker). Contrary to expectations, attention was significantly higher, and memory was significantly better, when subjects were exposed to the low-fidelity stimulus rather than to the high-fidelity one. Audio fidelity yielded significant main effects on seven of the eight content items. But on five of these measures, the low-fidelity condition was rated *more highly* than the high-fidelity condition. The results for audio spaciousness were mixed for the three dependent variables, but there were several significant interactions. For instance, women in the mono group responded more slowly to the audio tone and therefore were more attentive than those in the surround group. On the other hand, men were more attentive in the surround condition than in the mono condition.

Screen Size

Reeves, Lombard, and Melwani (1992) conducted a within-subjects experiment with 32 students, manipulating screen size (large = 41-inch; small = 15-inch) and viewing distance (near = 4 feet; far = 10 feet). The dependent variables were attention, memory accuracy, and evaluation. Each subject saw a series of faces. The authors found that subjects were more attentive to large images than to small ones and better remembered images at 4 feet than at 10 feet. On the other hand, screen size and viewing distance had no significant effects on evaluative measures—subjects did not rate large images more positively than they did small images, nor did they evaluate near-distance images more positively than they did far-distance images.

Reeves et al. (1993) also investigated the effects of screen size on attention, memory, and content evaluation, but found results contradicting those of Reeves et al. (1992). Screen size was manipulated using a between-subjects design. The 40 participants were either assigned to the small screen size condition (35 inches) or the large screen size condition (70 inches). Presentations on the larger screen yielded lower attention and memory scores than those on the smaller screen. Of the 10 evaluative items (eight content-oriented, two quality-oriented), only one proved to be statistically significant for screen size. Subjects felt more "a part of the action" when they viewed the clips on a 70-inch set than they did on a 35-inch set.

Lombard (1995) administered an experiment to measure responses to viewing distance and screen size. The design was a within-subjects 2 (close versus normal viewing distance) x 3 (10-inch, 26-inch, and 42-inch screen sizes), yielding six viewing conditions. The dependent variables included emotional responses to and impressions of people on television, both measured by a series of bipolar semantic differential items (e.g., calm/anxious). At each of the six viewing stations, 32 subjects from the Stanford University community watched four excerpts of television news broadcasts featuring an anchor speaking to the camera. Contrary to expectations, subjects' emotional responses to and impressions of people on television were not more positive when they watched television from a close viewing distance (10-inch, 24-inch, and 38-inch for the small, medium, and large screens, respectively) than from a normal viewing distance (30-inch, 72-inch, and 115-inch). But the screen size manipulation revealed a significant effect on the dependent variables. Subjects' emotional responses to and impressions of people on television were significantly higher when the stimuli were displayed on the large screen (42-inch) than on either the small (10-inch) or medium (26-inch) screen. Interestingly, then, while subjects preferred viewing these broadcast news excerpts on a large or medium screen than on a small screen, they favored watching television from a normal viewing distance over a close viewing distance. This finding suggests that just because consumers may prefer a large screen to a small one does not necessarily mean that they are willing to sit much closer to the screen than they do now.

Detenber and Reeves (1996) conducted an experiment to measure the effect of image size (22-inch vs. 90-inch) on three dimensions of emotion (valence, arousal, and dominance). Participants (N = 132) were randomly assigned to a small or large image condition, viewed 60 six-second pictures, and then rated their emotional response after each one. The authors found a significant main effect of image size on arousal and dominance, but not on valence (measured with a pleasing-unpleasing scale). Participants rated large pictures as more arousing and as eliciting more "in control" responses than small pictures.

More recently, Lombard, Ditton, Grabe, and Reich (1997) conducted a between-subjects experiment with 80 undergraduate students to compare evaluative responses to small versus large screen sizes. The first half of the sample (20 males and 20 females) viewed 17 short scenes from current television programming fare on a 12-inch CRT direct view TV set. The other half (20 males and 20 females) watched the same program segments on a 46-inch rear projection TV set. After controlling for the effect of perceived picture quality, the authors found that participants in the large screen condition reported responses of greater intensity than those in the small screen condition. However, there were no differences in reported level of enjoyment between the two groups. For six of the 17 scenes, screen size was found to be a significant factor ($p < .07$). For five of these six scenes (action-adventure, advertising [Coors Lite], advertising [Mobilink], animation, reality), subjects favored the large screen. Again because subjects reported that the picture quality of the small screen was significantly superior to that of the large screen television (see also Lombard, 1995), Lombard et al. (1997) reanalyzed the data adjusting for the effect of perceived picture quality. Again, sign tests indicated that screen size was significant ($p < .05$) for six of the 17 scenes and all significant patterns favored the large screen over the small one. Surprisingly enough, participants were not more likely to report intense responses when viewing two sports clips (football and gymnastics) on a large screen than doing so on a small screen.

Reeves, Lang, Kim, and Tatar (1997) administered a between-subjects experiment with 38 female college students to assess the effects of screen size (picture heights of 56-inch, 13-inch, and 2-inch) on attention (measured by heart rate deceleration) and arousal (measured by skin conductance). All participants viewed 60 six-second video clips taken from television and film. Results revealed that participants were more attentive to, and aroused by, the large screen (56-inch) than to the medium (13-inch) and small (2-inch) screens. The authors also found a significant difference in skin conductance level measures (for attention) between medium and large screens, but not between small and medium screens.

Lund (1993) administered a series of experiments with a total of 50 subjects to evaluate viewing distance preferences as a function of screen size (12-inch; 19-inch; 32-inch; and 60-inch) and resolution (525 lines; 1125 lines). Findings revealed that (1) viewing distance preferences decreased as screen size increased and (2) those preferences were largely unaffected by resolution. Likewise, Ardito (1994) conducted subjective tests at the RAI (Radiotelevisione Italiana) Research Center to evaluate the preferred viewing distance for HDTV programs and examine the interaction between viewing distance and screen size. Of the 54 Italian participants, 90 percent chose a viewing distance between 4H and 6H ($M = 5H$) when viewing HDTV drama pictures on a 54-inch rear projector. Over three-fourths of the subjects found it unacceptable to view television at 3H, the expected optimal viewing distance for HDTV. Of the same participants, 90 percent chose a viewing distance between 4H and 8H ($M = 6.3H$) when viewing HDTV sports pictures on a 38-inch monitor. In this case, 80 percent of the observers reported that 3H was too short a distance to view television. In sum, both Lund (1993) and Ardito (1994) found a *negative* relationship between screen size and preferred viewing distance: The larger the screen size, the shorter the preferred viewing distance from the screen.

Screen Format

In addition to investigating the impact of screen size on evaluative responses and preferred viewing distance, some researchers have looked into the effects of screen format. In 1986, Pitts and Hurst (1989) conducted a side-by-side comparison test with 180 nonexpert RCA employees to determine preference for widescreen (16:9) versus conventional (4:3) television format. Two 35-inch 4:3 NTSC monitors were modified to reproduce video images in 4:3 and 16:9 formats. Ninety percent of the respondents chose widescreen over 4:3. When the 26-inch diagonal was reduced to 20 inches, preference did not change in favor of the 4:3 format. The authors concluded that "people would buy smaller widescreen sets; the market would not need to be limited to the largest displays" (p. 163). Pitts and Hurst also found that content matters: Respondents who either chose widescreen or 4:3 reacted more positively to some program segments than to others.

In early 1988, McKnight, Neuman, Reynolds, O'Donnell, and Schneider (1988) administered an experiment to assess viewer responses to variations of aspect ratio (5:3 versus 4:3) and screen size (35-inch versus 20-inch). A total of 356 subjects were recruited at the Liberty Tree Mall, Danvers, Massachusetts. In the single stimulus test, subjects were randomly assigned to one of the four conditions. They watched two 2-minute clips (*Cheers* [situation comedy] and *1984 Olympics*) from a distance of two meters. In the dual stimulus test, the same subjects were asked to view the same two clips on two NTSC sets side by side, which were varied either in screen size or aspect ratio. The results from the dual stimulus test revealed that 67 percent of the subjects preferred the 5:3 format to the traditional 4:3. They also favored 5:3 in overall picture quality and sense of depth (but not in color quality and picture sharpness). In addition, 87 percent of the subjects exposed to the two screen sizes preferred the 35-inch monitor to the 20-inch monitor. But they rated the smaller screen as superior in color quality, picture sharpness, and picture brightness. In the single stimulus test, there were no significant differences in affective responses (e.g., was the program clip emotionally involving) either between aspect ratio conditions or between screen size conditions.

In recent years, a growing body of research has focused on consumer reactions to letterbox (i.e., a *full-sized* wide picture with black strips at the top and bottom of the screen) versus pan-and-scan (i.e., a *full-screen* picture without the extreme left and right edges of the original wide picture) formats. "Pan and scan" refers to a method of panning back and forth across a widescreen image to convert it into a 4:3 aspect ratio. Of 13 studies conducted in North America, Europe, and Japan (Harvey, 1993; Peeters and van Merwijk, 1995a, 1995b; Pitts, 1992; Russomanno, Trager, and Everett, 1993; Stelmach, 1993), only four (two in Japan, one in the United States, and one in the Netherlands) indicated that participants preferred letterbox to pan and scan (Peeters and van Merwijk, 1995b; Pitts, 1992). More often than not, participants objected to letterbox programs because they felt that they were missing something. "All I could concentrate on were those black bands" (Pitts, 1992, p. xlv) and "I'm not going to pay lots of bucks for a TV set and then not have a full picture on the screen" (p. xlvii) were typical viewers comments.

Purchase Intent and Willingness to Pay

In the 1985 NIRI survey, 38 percent of the Japanese respondents reported to be ready to pay ¥350,000 ($1,458) for a 20-inch HDTV receiver. But only 4 percent expressed the willingness to pay ¥625,000 ($2,604) for such a set (Takahashi, 1985). In the 1987 BTA survey, 42.5 percent of the Japanese respondents were ready to pay ¥300,000-¥399,000 ($2,069-$2,752; $1 = ¥145) for an HDTV home receiver, but only 15 percent would make the same commitment if the price of such a set would range from ¥500,000 to ¥749,000 ($3,448-$5,165) (Okai, 1987). In the 1989 Hi-Vision Promotion Council study, 26 percent of the Japanese respondents were willing to buy an HDTV set for ¥300,000-¥400,000 ($2,400-$3,200), but only 6 percent would do so if the price tag were over ¥600,000 ($4,800) (HVPC, 1989).

In the 1982 CBS study, 80 percent of the American respondents reported willingness to pay as much as 50 percent more for an HDTV set than they would for an ordinary set (CSP International, 1984). In the Lupker et al. (1988) study, LR and SC respondents were significantly more likely to purchase the HDTV set than the NTSC set if it were in their price range. Of the LR respondents, 18.5 percent reported that they *definitely* or *probably would buy* an HDTV set at $3,300 CAN ($2,500 US) within the next two years. Among SC respondents, this percentage rose to 24 percent. Not surprisingly, purchase intention of HDTV went up when its estimated cost fell down to $2,000 CAN ($1,500 US): 34 percent among LR respondents and 36 percent among SC respondents. The authors concluded that "there was little evidence that the *average* viewer would be willing to pay a great deal more to have HDTV" (Lupker et al., 1988, p. 29).

In the HBO study, alternating viewing respondents were asked to express their purchase interest in HDTV. About half reported that they would *definitely* or *probably* buy an HDTV set if it fit their price range, while the other half stated the same intention for NTSC. Both receivers were described as advanced TV sets. Only 16 percent indicated their willingness to purchase an HDTV set at a price of $2,500, but this percentage climbed to 22 percent when the price was dropped to $1,500.

In the 1988 MIT study (conservative test), Neuman (1988) found that 57 percent of the subjects in the HDTV viewing condition would be willing to pay an additional $100 over the price of their current set for the HDTV set on display. But only 6 percent would do so if the increment were $500. By comparison, 41 percent of the subjects in the NTSC viewing condition expressed willingness to pay a premium of $100 for the exhibited NTSC set, and 3 percent would be willing to pay an extra $500.

In March 1995, 15.5 percent of the respondents in the Dupagne (1997) study indicated that they would be likely to purchase an HDTV receiver at $3,000, a rather low-end estimate for the price of the first digital HDTV receivers in the United States. When introduced at the end of 1998, these first sets are expected to cost at least $5,000 (see Chapter 1). In March 1997, only 6 percent of the Harris survey respondents were ready to pay $1,000 or more on top of the price of their current television set for an HDTV set (Harris Corporation, 1997).

In the Dupagne and Agostino (1991) survey, only 1 percent of the Belgian respondents was willing to pay BF100,000 ($2,720; $1 = BF36.77) for an HDTV set (17 percent were ready to pay at least BF50,000 [about $1,360]). Only 6.5 percent stated that it would be either *very likely* or *somewhat likely* they purchase an HDTV set at a price of BF120,000 (about $3,263) to replace their existing television receiver. In the 1989 SOFRES study, 20 percent of the French respondents were willing to acquire a high-end television set at a price of FF8,500-FF9,000 ($1,332-$1,411; $1 = FF6.380) *and* pay an additional FF5,000 ($784) for a built-in D2-MAC decoder and the direct broadcast satellite (DBS) dish (SOFRES, 1989). In the 1989 IBA study, less than one-fifth (16 percent) of the British respondents were *almost certain* to acquire a widescreen TV at a price of £1,000-£1,500 ($1,637-$2,455; $1 = £0.611) (Wober, 1989).

In the Netherlands, Bouwman, Hammersam, and Peeters (1991) found that 43 percent of their 623 respondents were willing to purchase an HDTV set—with cost left unspecified. When cost was specified, willingness to pay decreased dramatically. Only 19 percent indicated that they would be willing to buy an HDTV set if its cost were twice as much as that of an ordinary color TV set (suggested price:

ECU 870; $1,104; $1 = ECU 0.788), and even fewer (6 percent) would do so if HDTV were to cost four times the price of a color TV set. When Bouwman et al. (1993) administered a follow-up survey in 1991 to another panel of 580 respondents, willingness to pay was slightly higher. Of the respondents, 54 percent were willing to purchase HDTV, 24 percent were willing to pay twice as much as color TV for HDTV, and 9 percent were willing to pay three times the price of a color TV set for it. Excluding the "don't know" answers, the percentage of Dutch respondents who reported to be very likely (*zeker wel*) to purchase an HDTV set rose by a mere 7.6 percent from 1990 to 1993 (1990: 13.1 percent; 1991: 16.7 percent; 1992: 13.9 percent; 1993: 14.1 percent). But the percentage of those who stated they would likely (*waarschijnlijk wel*) purchase HDTV climbed by 40.5 percent during these four years (1990: 30.6 percent; 1991: 42.2 percent; 1992: 37.4 percent; 1993: 43.0 percent). The dip in 1992 might be more a function of a change in scaling than a real decline in purchase intent. While in 1991 the response choices included *zeker wel* (very likely), *waarschijnlijk wel* (somewhat likely), *waarschijnlijk niet* (not too likely), and *zeker niet* (not likely at all), *waarschijnlijk niet* was no longer an option in the 1992 survey.

Five of the 29 studies profiled potential HDTV adopters according to demographics and mass media behavior, and their findings are generally consistent with diffusion theory (Rogers, 1995). In the 1987 Japanese BTA study, early adopters of Hi-Vision were expected to be men, in their 40s, with high income, owning new media items, and expressing high interest in HDTV (BTA, 1987b).

In North America, Lupker et al. (1988) ran a series of multiple regression analyses to identify possible early adopters based on demographic variables. The dependent variables were five purchase scenarios of HDTV (e.g., Currently, a high-quality television or a high-quality VCR costs $1,100 CAN [$800 US]. If they were available at $3,300 CAN [$2,500 US] each, how likely would you be to buy each of the following within the next two years?). The only variable that was a significant predictor for all five questions was ownership of a CD player. Three variables predicted responses to four of the five questions: number of TV devices owned (positive), educational level (negative), and employment in a television-related industry (negative). Contrary to adopter characteristics expectations, those with a higher educational level were less inclined to purchase HDTV. Lupker et al. concluded that "the best initial market would be those individuals who already have a reasonably strong orientation towards buying technology (owners of TV and non-TV equipment and pay movie subscribers, but not television industry people)" (p. 50). In the Dupagne (1997) study, the only significant predictor of HDTV purchase intention (from a series of variables covering demographics, media use, ownership of home entertainment products, and television attributes) was the importance of screen size. By comparison, a 1966 survey of 12,604 U.S. adults, at a time when color television was only available in 9 percent of U.S. households, profiled color set owners as having high income, belonging to the 25-49 age group, living in medium-size households, and being venturesome and convenience-oriented (Coffin and Tuchman, 1968). Clearly, these first color TV buyers possessed innovator-type characteristics (see Rogers, 1995).

In Belgium, those who expressed higher purchase intention of HDTV were more likely to be younger, male, with higher income, and owners of home entertainment products (Dupagne and Agostino, 1991). In the Netherlands, Bouwman et al. (1993) found that willingness to purchase HDTV (cost left unspecified) was best predicted by age (negative), followed by the perceived importance of CD sound as a television improvement, HDTV awareness, gender (male), attitude toward technology, the perceived importance of larger pictures, the perceived importance of wider pictures, and television viewing time.

Discussion and Conclusions

In this last section, we first return to Chapter 1 and take a closer look at the potential economic drawbacks of using standard-definition television (SDTV) over HDTV for terrestrial broadcasting delivery. We then summarize and discuss the major findings of the consumer studies about the potential adoption of HDTV by suggesting six preliminary propositions.

ATV Implementation Issues: HDTV versus SDTV

As noted in Chapter 1, American station engineers and managers have griped about the cost of the forthcoming ATV service, troubled by an uncertain return, although by the time of the 1996 NAB convention many had resigned themselves to accepting digital television (DTV) as a competitive necessity (McConnell, 1996a; Harris Corporation, 1996). As was the case with color, it will take some time for all players in the television industry (stations, advertisers, and viewers) to partake of the new ATV environment. CBS's Joseph

Flaherty recalled that "[U.S. broadcasters] had the same problem with color. Remember, everything in the station had to be replaced, except perhaps the audio console. That was a major expense in its day" (West and McConnell, 1996, p. 33). Between 1953 and 1964, neither networks/stations nor advertisers, and not even consumers, flocked to the new color medium (Coleman, 1968; Garvey, 1980). Among other things, local stations objected to monochrome-to-color conversion costs, which averaged about $350,000[6] between 1964 and 1966 (about $1,750,000 in 1996 dollars) ("The Costs," 1967) and saw little justification for raising advertising rates—a necessary step to recoup their investment—provided that color had not increased audience size in the same proportions. Advertisers were not exactly enthused to pay a premium for color commercials because they were able to reach the same audience with cheaper black and white spots. In 1954, Westinghouse's first color receiver cost $1,295 (about $6,500 in 1996 dollars), and although prices fell to $500 by 1963, color remained beyond the financial reach of most Americans (Garvey, 1980). So the economic challenges posed by the introduction of advanced television in the United States (and elsewhere) are not as new as we might first anticipate.

Another economic puzzle that is likely to preoccupy American broadcasters in the years to come is the debate over HDTV versus SDTV delivery (see Chapter 1). In the *Fifth Report and Order*, the Federal Communications Commission (1997a) did not require broadcasters to transmit a certain amount of HDTV programming on their second channel and, instead, left to their discretion the decision to air multiple SDTV programs or a single HDTV program. But does SDTV, a digital EDTV version of NTSC, make sense as an economic strategy? Some broadcasters, including NBC President Robert Wright, have questioned the wisdom of broadcasting in enhanced-definition (EDTV) instead of high-definition, wondering why broadcasters would choose anything less than the best picture quality available at the time in the name of unproven economics (West, 1995a). Other broadcasters, however, have argued that HDTV in and of itself does not legitimize the massive investment required to upgrade NTSC television to advanced television because HDTV signals will not significantly enlarge their pool of viewers, thereby contributing little to advertising revenues.

But what if the reverse were true? What if HDTV attracts more American viewers for longer periods of time, assuming, for instance, that the 16:9 format would become a killer application for ATV service—which is consistent with the growing interest of European and Japanese consumers in widescreen television—and would draw additional viewers and revenues? And what if other technical improvements, such as sharper resolution, digital multichannel sound, and flat-panel display, stimulate a short- or long-term rise in viewership, at the expense of other media, such as cinema? If these advanced television characteristics truly matter, as they seem to according to the empirical literature, then SDTV may not be in the consumer's best interest and delivering programs in SDTV instead of HDTV may not be a wise economic choice for broadcasters. Again, to compare with color TV, a 1964 NBC study found that ratings of NBC evening programs were about 80 percent higher in color TV homes than they were in black and white TV homes (National Broadcasting Company, 1964).

In addition, SDTV will certainly further fragment local and national audiences, it may not receive full must-carry status from the FCC (i.e., requiring cable operators to carry *all* SDTV channels of local broadcasters), and there is no guarantee that it will be a successful revenue-generating vehicle. SDTV technology will be expensive, and these new ancillary services may falter. A. James Ebel (1995), a broadcast consultant involved in broadcasting since 1929, puts it succinctly and ominously:

> The cost of providing SDTV channels will be almost as high as for pass-through HDTV. New transmitter, feed line and antenna will be required along with digital SDTV processing equipment. ... The addition of extra SDTV channels will not necessarily bring in large amounts of revenue. Advertising or pay-per-view revenue is not inexhaustible. The problem of getting additional software (programs) for the new channels will be substantial. (p. 76)

Jay Fine, CBS Senior Vice President of East Coast Operations, echoes these concerns and points out the logistical problems that will derive from a multichannel DTV environment:

> What do you label the other channels? Is it Channel 33A, B, C, D and F? What do you put on these channels? You are going to have to pay for program services. You are going to have to have additional production. (Beacham, 1997, p. 28)

Consumer Acceptance of HDTV: Empirical Propositions
■ PROPOSITION 1:
The level of HDTV awareness has risen in Japan, the United States, and Europe since the mid-1980s.

The literature review clearly points out that the degree of consumer HDTV awareness varies by geographic area. Japanese consumers have been more aware of HDTV and sooner than their American and European counterparts, which is hardly surprising since Nippon Hoso Kyokai (NHK) began research on HDTV in the mid-1960s. Surveys indicated that virtually all Japanese respondents (91 percent) were aware of HDTV by the late 1980s, while HDTV awareness averaged about 50 percent in the United States and less than 50 percent in European countries at the time. More importantly, the literature generally suggests that HDTV awareness has increased in each of the triad countries since the mid-1980s. In the Netherlands, where the only longitudinal study has been conducted, level of HDTV awareness soared by almost 100 percent in four years, from 36 percent in 1990 to 71 percent in 1993.

Awareness-knowledge is the first step in Rogers' (1995) innovation-decision process, and therefore it is paramount for consumer electronics manufacturers to expose the general public to HDTV technology as early as possible. Indeed for consumers to go to the persuasion stage and react favorably or unfavorably to HDTV, they first must be made aware of its existence. In the case of HDTV, as well as many consumer electronics products, awareness-knowledge is used to create a need for the innovation that did not preexist in the mind of most individuals— rather than the reverse, i.e., an existing need prompting individuals to seek knowledge about a new idea (Rogers, 1995). Consumers rarely embrace new electronics items by themselves and may perceive little need for acquiring yet a new television set—no matter how advanced its features are. Whereas Takahashi (1985) reported that 90 percent of his Japanese respondents complained about various technical aspects of NTSC technology, Dupagne and Agostino (1991) found that 97.5 percent of their Belgian respondents were either *very satisfied* or *satisfied* with their current PAL set (see also Dupagne, 1997; HBO, 1988; SOFRES, 1989; Wober, 1985). In addition, a majority (55 percent) reported that they would change *nothing* about their TV set. Some of the Belgian respondents preferred programming improvements instead of technological advancements (see also the MIT focus group study in Neuman and O'Donnell, 1992).

All in all, it will be up to consumer electronics companies to devise effective marketing campaigns to bring HDTV to the attention of the public and stimulate demand. It has been argued that consumers' lack of exposure adversely affected initial sales of color TV sets in France and the United States (Dupagne and Agostino, 1991). One way to prod HDTV awareness is to schedule public demonstrations not only at consumer electronics fairs but also in department stores and other "strategic" locations. The Japanese, under the aegis of the Ministry of Posts and Telecommunications, organized such demonstrations throughout the country in the late 1980s and managed to create a high level of awareness for Hi-Vision (see Chapter 3). Interestingly, while this success stimulated interest in the technology (BTA, 1987; HVPC, 1989), it did not translate into significant actual purchase behavior (see below). Before the demise of HD-MAC in the early 1990s, the Commission of the European Communities had recommended the pursuit of four promotional activities to introduce HDTV technology in Europe: scheduling of public demonstrations, participation in exhi-bitions/fairs, distribution of high-quality brochures, and lobbying of key European, Japanese, and U.S. decision makers (Commission of the European Communities, COM (88)299, 1988). But apparently little consumer promotion has ever been undertaken, although there were some "EUREKA" HD-MAC receivers distributed throughout Europe for the 1992 winter and summer Olympics. In the United States, the City of Denver and Tele-Communications Inc. opened an HDTV the-ater in the Denver International Airport in November 1995 (Dickson, 1995c). This joint venture between a municipality and a cable TV oper-ator represented one of the first opportunities (a Hitachi HDTV set has been on display at the Smithsonian since 1993) for the U.S. public to see HDTV on a regular basis. In addition, both WRAL-TV, Raleigh, NC, and WETA-TV, Washington, DC, plan to install HDTV receivers in public venues, such as malls and libraries, to acquaint the public with the new format when their HDTV facilities become operational (see Chapter 1).

■ PROPOSITION 2:
A majority of viewers prefer HDTV to conventional television.

Despite supporting evidence presented in this chapter, this state-ment is contentious. Although 62 percent to 87 percent of North

American respondents (excluding Danbury) felt that HDTV was better than NTSC, one can argue that these results do not prove an unconditional consumer endorsement of HDTV technology. If surveyed viewers had really perceived HDTV attributes as revolutionary, nearly 100 percent of them would have favored HDTV over NTSC. On the other hand, one can conclude that these real-life demonstrations, more often than not, evoked positive reactions toward HDTV and that preferences are situation-bound attitudinal measures anyway. Furthermore, preference for HDTV may not necessarily reflect intended or actual purchase behavior. Over the years, the precise nature of attitude-behavior relations has generated one of the most enduring controversies in social science research. While some early research indicated that attitudes are often negligibly related to overt behavior, later studies found that attitudes can have better predictive value by improving attitude measurement and incorporating situational factors (Severin and Tankard, 1992; see also Kim and Hunter, 1993). Similarly, buyer-intent surveys have been notoriously unreliable for predicting actual purchase behavior (see Klopfenstein, 1989), although the predictive value of intentions has improved over the years since Juster (1966) declared in his classic study that "neither intentions nor attitudes reduce unexplained variance to the extent that consistently reliable forecasts are obtainable either from survey variables alone or from survey variables in conjunction with observable financial variables" (p. 661).

Even for color television, now available in 98 percent of U.S. households, the initial relationship between attitudes and actual purchase was not a simple one because of lack of programming and high receiver prices. According to NBC and A.C. Nielsen estimates, it took color TV 13 years (until 1966) and 20 years (until 1973) to reach household penetration rates of 10 percent and 50 percent, respectively (Television Bureau of Advertising, 1987), and this despite *highly positive* consumer attitudes toward color TV in the early 1950s. For instance, in October 1951, the Opinion Research Corporation (ORC) surveyed 2,776 nonindustry visitors on behalf of the Radio Corporation of America (RCA) to assess public reaction to RCA color television (Opinion Research Corporation [ORC], 1951). Respondents were asked to view a 26-minute program (a musical variety show followed by a short band performance) and complete a questionnaire. Ninety-three percent of them found color television more enjoyable than black and white television; 77 percent rated the overall quality of color TV pictures as either *excellent* or *very good*; 76 percent found the clearness of detail in the color TV pictures *excellent* or *very good*; and 64 percent considered the trueness-to-life of the colors *excellent* or *very good*. In 1953, when the Federal Communications Commission (FCC) approved the NTSC color TV standard, the ORC replicated the survey, which yielded even more favorable reactions (ORC, 1953). Of the 671 respondents, 98 percent found color television more enjoyable than black and white television; 87 percent felt that the overall quality of the color pictures was either *excellent* or *very good*; 85 percent considered the clearness of detail in the color pictures *excellent* or *very good*; and 77 percent stated that the trueness-to-life of the colors was either *excellent* or *very good*. In sum, while HDTV may not knock "the socks off everyone who sees it" (Neuman, 1988, p. 10) at least initially, there is little reason to believe that HDTV would not diffuse as rapidly as color TV in the United States, which by all accounts had a long gestation period before gaining wide consumer acceptance.

A major concern for broadcasters is whether program type affects attitudes toward HDTV. The general consensus in this review is that content matters and can affect evaluative responses (HBO, 1988; Lombard et al., 1997; Lombard, Grabe, Reich, Campanella, and Ditton, 1996; McKnight et al., 1988; Neuman, 1988; Neuman and O'Donnell, 1992), although it remains unclear to what degree. For instance, in an HBO focus group session, a participant admitted: "I was so enthralled by the movie that I didn't see any difference in the two TV's" (HBO, 1988, p. 15). If it is indeed the case, the next logical question becomes: What program genres are most likely to elicit positive responses toward HDTV? There is some preliminary evidence that sports programming aired in HDTV could produce highly favorable attitudes (HVPC, 1989; Neuman, 1988; see also Chapter 2), which if confirmed could in turn increase existing viewership. In his U.S. survey, Dupagne (1997) found that sports viewing was positively correlated with interest in HDTV ($p < .07$) and purchase intent. On the other hand, it may well be that these attitudes are conditioned on perceptions of HDTV attributes and the viewing configuration itself (e.g., preferred viewing distance). Clearly, this is one aspect of HDTV economics that deserves more research.

■ PROPOSITION 3:
The most important perceived benefit of HDTV is picture sharpness.

Picture sharpness, and by extension picture quality in general, was singled out as the most important/desirable HDTV or television attribute by an overwhelming percentage of respondents across regions—from 50 percent to 98 percent. Furthermore, experimental studies indicated that subjects preferred high-resolution/fidelity pictures to low-resolution/fidelity pictures when asked to evaluate program material (Neuman, 1990; Reeves et al., 1993).

■ PROPOSITION 4:
Improved sound quality is not perceived as important by viewers as improved picture quality.

Of all surveyed respondents, 33 percent to 97 percent deemed sound quality as being an important/desirable HDTV or television feature. The experimental evidence was more ambiguous, though. While subjects in the Neuman et al. (1987) study preferred stereo/ high-quality sound to mono/low-quality sound, they were unable to discriminate between more subtle changes in audio quality (e.g., stereo vs. mono). Contrary to expectations, Reeves et al. (1993) found that subjects rated more highly the low-fidelity condition than they did the high-fidelity condition for five of the eight content-based measures.

But these findings might change with the current boom in sales of home theater systems, which offer viewers a movie theater-like experience at home with a minimum of a large-screen TV set, five or more speakers, a video player, and a multichannel receiver (see Harris Corporation, 1997). In the United States, the sound component of the Digital Television Standard is the 5.1 channel Dolby AC-3 system (see Chapter 1). Prices of home theater systems range from less than $1,000 to over $200,000 (Wallace, 1995).

■ PROPOSITION 5:
Consumer acceptance of large-screen television depends on its size and weight.

Of the three main HDTV attributes, screen size and format ranks last. Overall, less than half of the surveyed respondents (11 percent to 96 percent) felt that a larger/wider screen was an important/desirable quality for HDTV or regular television. Few respondents, especially in Japan where space is a premium, would acquire a 50-inch HDTV receiver for their home (see BTA, 1987b; Bush, 1987; HVPC, 1989). Dupagne and Agostino (1991) found that while a majority of Belgian respondents owned a 24-inch or larger television set—Belgians are more inclined to own larger sets than some of their European neighbors—nearly 75 percent expressed little or no interest in having a larger set than the one they currently own. Understandably most consumers would be indifferent to a 50-inch CRT weighing some 250 pounds (about 125 kg)! In the Dupagne and Agostino (1991) study, some respondents voiced concerns that a large-screen TV set would not fit in their living room (see also Wober, 1985).

The experimental findings support the use of wider (16:9) screen format, but are less positive for the use of larger screen sizes, especially in regard to evaluative responses (i.e., attitudes). Subjects in the Lombard et al. (1997) study favored the large screen (46-inch) over the small one (12-inch) for only *some* of the scenes (see also Detenber and Reeves, 1996; Reeves et al., 1993). Methodologically we can wonder what is the adequate size for a small versus large screen and whether the inclusion of a medium-size set would not throw off the results. In other words, how *big* is big? In a subsequent study, Lombard (1995) found that emotional responses to and impressions of the anchors were more positive when subjects viewed these broadcast news excerpts on a large screen (42-inch) than either on a small (10-inch) or a medium (26-inch) screen. However, subjects *liked* viewing on the medium-size screen almost as much as they did on the large one.

On the other hand, subjects in the Neuman (1988, 1990) studies opted for NTSC-type picture quality when asked to compare HDTV versus NTSC on smaller screens. So viewers may discriminate better between HDTV and conventional television on a large screen than on a small screen. Furthermore, the growing popularity of large-screen TVs should alleviate fears that consumers will automatically discount screen sizes greater than 30-inch. For instance, the share of color TV sets in the United States with screens 27-inch and larger skyrocketed by 400 percent in eight years, from the EIA-estimated 5 percent in the first 39 weeks of 1988 to 25 percent in 1995 ("Giant Screens," 1996). Maybe screen size is becoming a relative term: Consumers today may not necessarily regard a 25-inch television set as a *large-screen* TV, although they probably would have a decade ago.

But perhaps more than anything else, it will be flat-screen technology that will solve the issue of receiver bulkiness and convince consumers to embrace larger screen sizes. In mid-1995, several Japanese consumer electronics companies, including Sony, NEC, and Matsushita, announced that they developed color display panels using liquid crystal, plasma, or a combination of both and would introduce them on the Japanese consumer market sometime in 1996. NEC priced its 40-inch color panel to about ¥450,000 (about $5,300) (Mitsusada, 1995). While flat-screen TVs will gradually replace the cumbersome large-screen CRT sets, they will also require viewers to rearrange their living room or their viewing area. In their experiment, Tanton and Stone (1988) found that most subjects were willing to make some minor adjustments to their domestic viewing habits to accommodate HDTV (e.g., sitting closer).

In the final analysis, the optimal display size for HDTV will depend on the domestic viewing environment. Tanton and Stone (1988) concluded that the most practical and desirable screen size for HDTV hovers between 1 and 1.25 meters in diagonal (see also Westerink and Roufs, 1989). The challenge for consumer electronics manufacturers will be to sell large-screen HDTV receivers that optimize the impact of such HDTV attributes as picture sharpness and wide aspect ratio while being suitable in size and weight for most consumers' homes.

■ PROPOSITION 6:
Consumers are unwilling to pay a high premium for HDTV.

Aside from validity issues of willingness-to-pay questions, few respondents (ranging from 1 percent to 26 percent) were ready to pay $2,500 for HDTV—a low-end figure for the initial cost of an HDTV receiver. This finding held across the triad. Price is a cardinal consideration when consumers purchase television sets and is likely to be the main objection to HDTV (see Dupagne and Agostino, 1991; Lupker at al., 1988; "The Market," 1995). It is a twist of irony that despite all the public demonstrations and all the positive reactions toward HDTV in Japan, sales of Hi-Vision sets have failed to achieve any significant dent in the Japanese consumer electronics market (less than 400,000 by April 1997). The culprit is undoubtedly the high price of the technology. In June 1989, when Hi-Vision satellite broadcasting debuted with one hour a day, an HDTV receiver cost about $29,000; eight years later in June 1997, the official retail price was still over $3,000. The main lesson we can draw from the Japanese experience thus far is that programming availability in and of itself does not guarantee the successful diffusion and adoption of HDTV if receiver prices remain high too long. So it is clear that price will be the key determinant of HDTV/DTV adoption in the United States and Europe.

9 · A Global Model of HDTV Policymaking

By refusing in 1986 to allow the imposition of a true technological "diktat" [by the Japanese] on the international community, Europe allowed the debate to be put back on its healthier and much more favorable foundations, to a true cooperation between nations and regions. Everyone agrees today to recognize the diversity of technological and economic contexts, which will not permit, at least until the middle term, a unified distribution standard on a global level.

—*Patrick Samuel, CEO, International HD, 1990*

One of the overarching themes of this book is that telecommunications technology policymaking needs to be examined from both a national and a global perspective. Since national models vary according to political and regulatory imperatives, observers have rarely attempted to create an international model that incorporates the common elements found in respective national systems. In this concluding chapter we will propose such a model *specifically for advanced television policymaking* that can be utilized to better comprehend the linkages that connect international telecommunication service providers and regulatory agencies. In an era when nations are routinely forming political and economic linkages such as the European Union (EU) and the North American Free Trade Agreement (NAFTA), the creation of such global models is important in comprehending the importance of technology policymaking for national and regional economies. Multinational corporations (MNCs) based in the G-8 nations have extended their already prodigious global reach with the transformation of formerly-socialist economies into capitalist systems in Eastern Europe and Asia. Many of these corporations such as Philips, Thomson, AT&T, and Sony have been active players in global HDTV standardization battles.

We also wish to emphasize the fundamental point that modern telecommunications technology has facilitated the "control at a distance" that makes such organizations functional on a global basis. The medium is the method, to paraphrase Marshall McLuhan. MNCs could not operate without communication technologies such as undersea telephone links, computer-assisted design (CAD) data lines, and satellite videoconferencing. Engineers and managers in Osaka, Sunnyvale, and Stuttgart need these links to transmit product design and marketing information from corporate planning centers to factories located around the globe.

The explosive growth of the Internet since 1990 has rapidly expanded the number of data linkages between all industrialized nations, not just the G-8 economic powerhouses. National telecommunication regulators are wrestling with a number of new digital standardization issues related to the expansion of the Internet, such as the adoption of protocols for the exchange of audio and video data over the Internet. Other pressing regulatory issues concern transnational censorship, privacy, and the security of financial transactions. For instance, in December 1995, the U.S.-based online service CompuServe bowed to the threat of prosecution under German law by blocking access to more than 200 sexually oriented computer bulletin boards and picture databases by its subscribers in the United States and around the world (Markoff, 1995; Nash, 1996). But in a surprising reversal in February 1996, CompuServe restored most of these services and announced that it would provide filter software to customers— software that can be used by individual subscribers to block material that they deem offensive (Lewis, 1996).[1] Not only is the world still shrinking electronically, communications policy issues that were previously resolved on a national basis are increasingly topics for transnational negotiation.

This chapter will build upon previous nationally and regionally oriented models (e.g., Japan, the United States, and Europe) to create a global model of HDTV technology policymaking. Prior to presenting the international model, we will provide a brief recapitulation of developments from earlier chapters.

The International Development of HDTV

On December 24, 1996, the U.S. Federal Communications Commission (FCC) finally established a standard for advanced television in the United States. From 1991 to 1995, the Commission, through its Advisory Committee for Advanced Television Service (ACATS), conducted tests of various alternative ATV systems, and the FCC approved a digital transmission system developed by a consortium of former competitors known as the Grand Alliance (GA). The Commission has articulated a proposed 10-year period during which American broadcasters would simulcast high-definition programming contemporaneously with NTSC (National Television System Committee) programs. The DTV Standard would then replace NTSC for terrestrial television transmission at the end of that period (FCC, 1997a).

The FCC's decision on advanced television has multibillion-dollar implications for North American broadcasters and consumers. The DTV Standard is digital and is incompatible with existing cameras, switchers, and receivers—it would require eventual replacement of all NTSC hardware. The new standard is also incompatible with both the Japanese analog Hi-Vision/MUSE HDTV[2] system and the analog HD-MAC system developed by the European EUREKA EU-95 consortium, although European broadcasters have since established digital transmission standards through the Digital Video Broadcasting (DVB) group (see Chapter 4). Policymakers and television manufacturers in Europe and Japan have been closely following U.S. advanced television standardization efforts (Powell and Itoi, 1994).

U.S. policymaking concerning ATV standardization cannot be accurately examined without placing it in the context of research and policy initiatives in Japan and Europe. The Japanese began research on high-definition television at NHK in 1964, and developed the world's first working prototypes in the early 1980s (Carbonara, 1992). Direct broadcast satellite transmissions of the Hi-Vision/MUSE system began in 1989, and 14 hours of HDTV programming are presently transmitted daily (see Chapter 2). NHK attempted to have the Hi-Vision/MUSE system adopted as a world HDTV standard in 1986, but were rebuffed by representatives of the European Community at the quadrennial CCIR Plenary Assembly in Dubrovnik (Carbonara, 1992). The "Dubrovnik rejection" was a watershed event that dashed Japanese hopes of having its HDTV technology promptly adopted as a worldwide production standard and led to European and American efforts to develop alternative systems. Because of the prospect of potential Japanese domination of HDTV technology, government officials in North America and Europe were drawn into the debate over the role of national governments in the creation of industrial policies that promoted selected high-technology industries. This debate continues today with the proponents and opponents of national industrial policies citing the case study of the development of HDTV technology to support their positions (see Chapters 3, 4, and 5; see also Beltz, 1991; Tyson, 1993).

The purpose of this chapter is to analyze two significant interrelated questions:

1. How is the standardization of new communication technologies, such as HDTV, influencing the role of international telecommunication policymakers and regulators?

2. What type of international HDTV policymaking model incorporates the changing roles of telecommunication regulators, broadcasters, manufacturers, and consumers in an era when the number and type of transmission technologies are rapidly expanding?

International HDTV policymaking makes an ideal case study in examining these issues because of the active involvement of Japanese, American, and European governments in creating policies designed to promote national champions and hinder foreign competitors. At stake are thousands of jobs in high-technology industries and billions of dollars in potential international trade income (or deficits) if one part of the industrialized world can dominate the design and production of a new generation of digital televisions. The creation of incompatible national standards for ATV systems in Japan, North America, and Europe provides an ideal case study for investigating these two research questions.

Telecommunications Policymaking Models

Models of telecommunications policymaking systems illustrate distinctive national (and supranational in the case of the EU) political and economic attributes.[3] These symbolic models are useful in conceptualizing very complex politico-economic systems comprised of hundreds or even thousands of individual participants and the organizations they represent. A comparative analysis of these systems is instructive in seeking to understand the behavior of participants and the technology policies that emerge from the interrelated systems. As the HDTV case study will demonstrate, these formerly-distinctive systems are increasingly interconnected as multinational telecommunications corporations extend their global reach, and as international standard-setting bodies such as the International Telecommunication Union (ITU) decline in stature while regional organizations such as ACATS (United States), BTA/ARIB (Japan), and DVB (Europe) assume a larger standard-setting role (see Chapters 3 and 4). Besen and Farrell (1991) made the point that regional standards organizations (RSOs) are making inroads into the "turf" of the ITU due to the latter organization's slow responsiveness to global standardization needs. The RSOs have moved into this vacuum by being more responsive to regional companies and governments. In the case of HDTV, European and American RSOs have also acted to protect regional industries by erecting standards barriers to Japanese HDTV technology. Thus, we see an increasing global fragmentation of technology policymaking (including standard-setting), with increasing influence being gained by regional standards organizations and the multinational corporations that are key constituents of the RSOs. The MNCs are active on a global basis in a number of regional standards organizations—for example, in the HDTV case study, Philips and Thomson are active in both the European Digital Video Broadcasting group and the American Grand Alliance consortium. In light of these trends, we will first revisit each of the national and regional HDTV policymaking systems with the intention of seeking a global model that would connect them.

The U.S. HDTV Policymaking Model

The Krasnow, Longley, and Terry (1982) [KLT] broadcast policymaking model explained in Chapter 6 is the most relevant and inclusive framework for studying HDTV standardization in the United States. The model describes three primary categories of participants: the FCC, Congress, and the "regulated industries." Three additional groups—the White House, the courts, and citizen groups—are less influential but have played key roles in national telecommunications policymaking since the FCC was created by the Communications Act of 1934.

We modified the KLT model in the case of HDTV standardization by splitting the "regulated industries" entity into two distinct camps—

broadcasters and television hardware manufacturers—since they have very divergent objectives in the standardization game (see Figure 9.1). One of the basic problems with the KLT model is that it lumps the two entities together when their policymaking goals are often diametrically opposed, as was the case when the White House suggested shortening the planned 10-year transition period for conversion to an advanced television system. Manufacturers supported the idea while broadcasters were adamantly opposed (McConnell, 1995a).

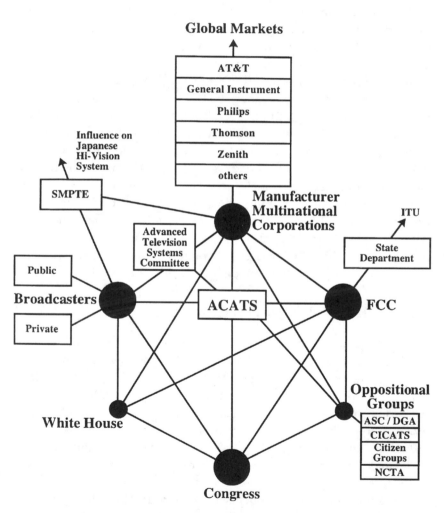

FIGURE 9.1. United States HDTV policymaking model.

One rationale for this bifurcation is clear—while the enthusiasm of American broadcasters for an incompatible HDTV standard that would render obsolete their entire physical plant has been ambivalent at best, the manufacturers group has supported the transition for obvious economic reasons. They have often been at odds during the work of the FCC's Advisory Committee, but broadcasters have also recently (post-1995) advocated the proposed transition to digital broadcasting (McConnell, 1997b).

The other rationale is to highlight the roles that multinational corporations such as Philips and Thomson have played in both U.S. HDTV standardization and also their roles as key participants in the EUREKA 95 and DVB initiatives in Europe. This linchpin role will become clearer as we develop the global HDTV model below. The U.S. HDTV model has four primary participants: (1) broadcasters; (2) manufacturers; (3) the FCC; and (4) Congress. The FCC's Advisory Committee on Advanced Television Service (ACATS) is placed at the center of the model since it was the nexus of standardization activity in the United States.

Secondary-level players in the model include the White House, the Advanced Television Systems Committee (ATSC), the Society of Motion Picture and Television Engineers (SMPTE), the U.S. State Department, and an "oppositional groups" entity. The White House has played a limited role, but has become more assertive in recent months concerning the setting of a "date certain" for the reversion of present NTSC spectrum for auction. The White House would like to auction the present NTSC channels 60-69 and dedicate the estimated $5 billion in income to U.S. school construction (McConnell, 1996d). The ATSC is included along the broadcaster-industry axis as the industry-sponsored consortium that codified the Grand Alliance HDTV transmission system into what is officially known as the "ATSC Digital Television Standard" that the FCC modified as the American HDTV standard in December 1996 (FCC, 1996c). SMPTE has been included in the model due to its influential role in shaping the Japanese Hi-Vision HDTV standard into the SMPTE 240M high-definition television production standard presently used in Japan and the United States. The State Department is included as the official link between the U.S. government (via the FCC) and the International Telecommunication Union concerning telecommunication policy. However, since the 1986 Dubrovnik rejection of the Japanese 1125/60 system, the State Department and the ITU have taken a back seat relative to the activity of regional standards organizations such as ACATS, the ATSC, and the European Telecommunications Standards Institute (ETSI).

Oppositional Groups to the U.S. Digitial Television Standard

The model also includes a node for parties opposed to the FCC's adoption of the ATSC Digital Television Standard. Chapters 6 and 7 have outlined the opposition of elements of the U.S. computer industry—especially Microsoft, Intel, and Apple—to the inclusion of interlaced scanning in the proposed standard. Organized as the Computer Industry Coalition on Advanced Television Service (CICATS), they were able to force changes in the standard (dropping all mandated scanning parameters) to make it more hospitable for hybrid PC-TVs in the future.

Other oppositional groups were less successful in their policymaking goals. The American Society of Cinematographers (ASC) and the Directors Guild of America (DGA) working together as the Coalition of Film Makers were unsuccessful in having the FCC mandate a provision that would have required broadcasters to transmit a film in the same aspect ratio in which it was photographed. They were bucking a trend in which the Commission was seeking a DTV standard that was as open as possible, rather than one that was overly prescriptive. The National Cable Television Association (NCTA) fruitlessly asked to FCC to not set *any* mandated DTV standard, but to leave it completely to the marketplace. They were concerned about the costs that cable systems would incur in carrying DTV signals to subscribers, and with potential competition from multichannel SDTV broadcast stations.

Citizen groups such as the Media Access Project (MAP) accused the FCC and Congress of "giving away" the digital spectrum to broadcasters. They argued that this spectrum should be auctioned to support public broadcasting and public interest programming. The FCC's counterarguement was that the spectrum allocated to broadcasters is solely for DTV transmission—allowing the government to reclaim the NTSC spectrum at some future date (now set at 2006) for auction and reallocation.

The dominant voices amongst the opposition groups were not Steven Spielberg and the filmmakers group, the cable TV industry, and not the ad hoc citizen groups, but were rather the politically powerful stars of the U.S. computer industry—Intel and Microsoft. They were able to forge a compromise on the DTV Standard that created a place for the computer industry in the digital television marketplace. As detailed in Chapter 6, enhancing the international competitiveness of these U.S.-based multinationals was a factor in the Commission's decision to modify the standard.

Multinational manufacturing corporations such as Philips, Thomson, and General Instrument have dominated the ATV standardization process in the United States. The national commercial television networks and the cable television industry have also been active participants in the work of the FCC's Advisory Committee, but the group with the most to gain financially from the transition to a new broadcasting system is the manufacturing contingent.

Both Thomson of France and Philips of the Netherlands have been influential players in both the United States and Europe. These companies were instrumental in blocking the international adoption of the Japanese Hi-Vision/MUSE standard in Dubrovnik in 1986. They championed the HD-MAC European alternative system, but are also part of the U.S. Grand Alliance consortium. The government of France is presently seeking to privatize state-owned Thomson, which purchased RCA's consumer electronics division from General Electric in 1987,[4] and both Thomson and Philips are major manufacturers of television sets in the United States. Prior to the formation of the Grand Alliance, Thomson and Philips were allied with NBC (owned by General Electric) and the Sarnoff Research Center (formerly RCA's research arm) in the Advanced Television Research Consortium (ATRC). The ATRC was a distinctively Euro-American organization, built on the fragmented foundations of the Radio Corporation of America.[5] In light of these transnational linkages, it is instructive to compare the U.S. advanced television policymaking system with a similar model that describes HDTV policymaking in Europe.

A European HDTV Policymaking Model

The close relationships between European governments and their national high-technology industries are hallmarks of their policymaking systems. The intraregional trade conflicts that led to the development of the incompatible PAL (Anglo-German) and SECAM (French) television standards have given way to regional cooperation under the auspices of the European Community/ European Union.

The European model is a triad comprised of broadcasters (public and private), EU institutions, and manufacturer multinationals. These

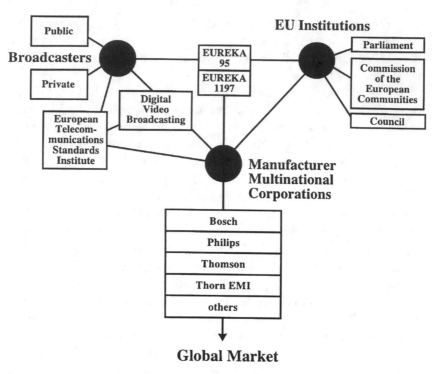

FIGURE 9.2. European HDTV policymaking model.

groups were active partners in the creation of the EUREKA 95 advanced television consortium and EUREKA 1197 digital research group that succeeded it. Both projects were created as a direct result of formal EU industrial policies that were examined in detail in Chapter 4. Although EUREKA 95 failed to create a viable HDTV technology, it indirectly laid down the foundations for research in digital technologies that led to the industry-dominated Digital Video Broadcasting (DVB) group. For this reason, in the model the HDTV EUREKA initiatives are located in the nexus of policymaking system while DVB is solely on the broadcaster-industry axis. The European Telecommunications Standards Institute (ETSI) is also included as a key player in the development and dissemination of the DVB standards and is a prime example of the regional standards organizations (RSOs) described by Besen and Farrell (1991) that are challenging the ITU's monopoly on telecommunications standardization.

The model is also useful in that it outlines the varied levels of EU institutions that played a role in creating industrial policies to protect the European consumer electronics industry. As noted in the quote from Patrick Samuel at the start of this chapter, the European Union consciously rejected the Japanese Hi-Vision "diktat" in 1986 in Dubrovnik and that "everyone agrees today to recognize the diversity of technological and economic contexts, which will not permit, at least until the middle term, a unified distribution standard on a global level" (Samuel, 1990, p. 25). While this comment was made in 1990 in the midst of the development of the 1250/50 European ATV standard, the EU member states are still not willing to concede to either Japanese or American dominance of their "audiovisual" software or hardware markets. The growth of the innovative DVB group is evidence of this. It is anticipated that the development and adoption of the DVB-T (terrestrial) and DVB-S (satellite) digital television standards will prompt a flurry of digital television activity in the near future.

National cultural issues are also an important (and often divisive) issue within the EU. One rationale in the United States for support of the Japanese Hi-Vision/MUSE system in Dubrovnik in 1986 was that a single world standard for HDTV would facilitate the international exchange of television programming, and implicitly would enhance the export of American television programming in the PAL/SECAM world—especially Europe (U.S. Department of State, 1985). The proliferation of U.S.-produced media programs on European cable, satellite, and broadcast television networks is a major irritant in Euro-American relations and was a significant roadblock to passage of the 1994 General Agreement on Tariffs and Trade (GATT).

Thus, the development of European telecommunications standards is inextricably linked to policies designed to protect regional economic and cultural interests. The 1986 European rejection of Hi-Vision/MUSE and the subsequent development of HD-MAC must be analyzed from the multiple perspectives provided by the model. For the EU it was more than just an issue of standardization—important cultural and national economic competitiveness issues were at stake.

The Japanese HDTV Policymaking Model

The Japanese equivalent of the Advanced Television Systems Committee (ATSC) is the High Definition Television Committee of the Broadcasting Technology Association (BTA) (see Figure 3.1 in Chapter 3). HDTV standards policy formation in Japan has been guided primarily by the Ministry of Posts and Telecommunications (MPT) and the BTA. The chart of the Japanese HDTV policymaking system below permits direct comparisons with the European and American models.

As noted in Chapter 3, much of the work of the BTA committee was largely a rubber-stamp approval of NHK's 15 years of research on the Hi-Vision/MUSE technologies. No significant changes in the standard were made by the High Definition Television Committee, but important modifications (such as image aspect ratio) were made based upon input from U.S. organizations such as the Society of Motion Picture and Television Engineers. For this reason we have included a link between NHK's technical laboratories and SMPTE in the United States.

Our model of the Japanese HDTV policymaking system places the Broadcasting Technology Association (BTA)[6] at the nexus formed by

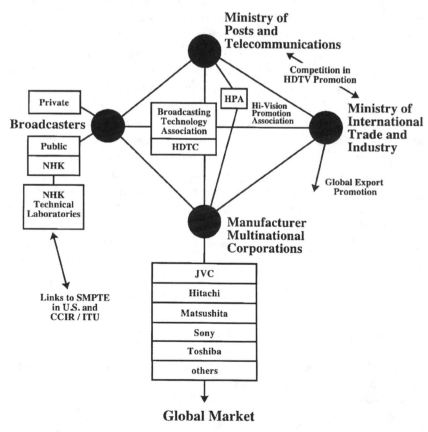

FIGURE 9.3. Japanese HDTV policymaking model.

the four key participants: broadcasters (dominated by NHK), manufacturer multinationals, the Ministry of Posts and Telecommunications (MPT), and the Ministry of International Trade and Industry (MITI). It is interesting to compare this diagram with Figure 3.1 in Chapter 3 as MITI is nowhere to be seen in the BTA/MPT chart. As we pointed out in that chapter, the MPT and MITI went head-to-head in their competitive efforts to promote HDTV technology in Japan and abroad, but only the MPT was involved in HDTV standardization matters. MITI, on the other hand, has been a significant player in guiding manufacturing efforts by Japanese multinationals such as Sony in creating 1125/60 HDTV hardware for export markets. The Japanese have dominated global markets in television sets, VCRs, and camcorders since the 1980s and it is inconceivable that MITI would not be involved in developing a potential global market for HDTV hardware.

The Japanese system gives government regulators in the MPT significant influence over internal policy, and those in MITI direction over external telecommunications trade policy. Since World War II, the Japanese government has supported high-technology communication manufacturing industries with direct and indirect subsidies, and by developing aggressive industrial policies that promote the growth of export markets for their high-technology products (U.S. Congress, 1990).

Characteristics of the Japanese economic model are centralized export-trade planning by the government with multimillion-yen research subsidies for targeted export industries. The goal of the regulators is not to harass the big trading companies with antitrust action but rather to *coordinate* their activities with the goal of making each multinational company into a world leader in sales and technology in their respective specialties. To this end, MITI officials subdivide national R&D efforts to avoid duplication of effort and assign specific manufacturing efforts to avoid duplication of effort among industries (Okimoto, 1989).

Much of the support provided by the MPT was an elaborate promotion campaign for Hi-Vision (U.S. Congress, 1990). The Hi-Vision Promotion Council and its successor, the Hi-Vision Promotion Association (HPA), have underwritten a series of HDTV demonstration broadcasts, including all of the Olympic Games since 1988. It is difficult to imagine the FCC acting in such a promotional role in the United States. The government's role in Hi-Vision promotion underscores the unique level of Japanese governmental-industrial cooperation in meeting mutually beneficial national technology objectives.

The Japanese HDTV policymaking system is to a large degree a triadic export-oriented system with broadcasters, manufacturers, and government agencies as key players. It is interesting that NHK, a public broadcaster, started the development of HDTV for the technical, social, and economic reasons outlined in Chapter 3. However, once the technology was perfected, manufacturers led by Sony created a formidable lead in HDTV technology that both European and American manufacturers have endeavored to close. The development of an innovative American digital transmission system may actually prove to be a Pyrrhic victory for the United States. The Japanese are world leaders in the development of digital television hardware (D1, D2, D5 video formats etc.), and as one Sony executive noted, "We're manufacturers. If there is sufficient demand, we'll build anything" (Tyson, 1993, p. 242).

A Global Model

In studying the test case of international efforts to standardize HDTV, it is apparent that the national and regional models outlined above need to be placed in a systemic global framework. While the national/regional models are useful for intrasystem analysis, they omit the important influences of multinational corporations and international regulatory agencies. What is needed is an international HDTV policymaking model that can link a series of interrelated national systems.

Multinational Manufacturers at the Center

The multinational corporation (MNC) manufacturers are the linchpin of the model. The economic superpowers of Japan, the United States, and the European Union are tied to each other and the rest of the world by cable and satellite linkages established by multinational corporations that operate throughout the world system. A number of scholars have criticized the growing power and global influence of MNCs in exporting the mass-consumer cultures of the industrial superpowers (especially that of the United States) to the rest of the world (e.g., Hopkins and Wallerstein, 1982; Schiller, 1992). There is also a reverse flow of consumer goods (such as television sets) manufactured by MNCs in low-wage nations and marketed in the more affluent ones.

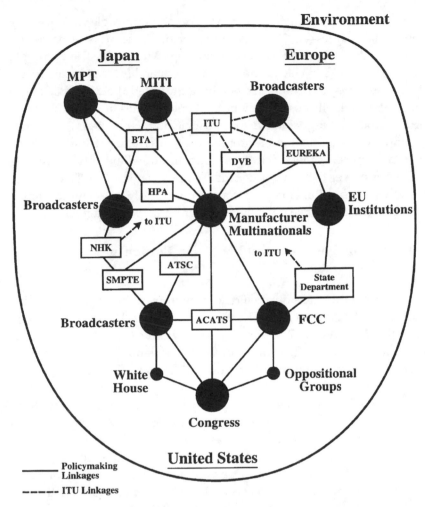

FIGURE 9.4. Global HDTV policymaking model.

Like it or not, the multinational corporation is an intrinsic linking element in the global telecommunications model. Consumers and regulators vary between nation-states, but MNCs do not. The pervasive global reach of the MNCs is based upon the telecommunications linkages developed by corporate giants such as AT&T, Sony, and Philips. The control-at-a-distance that is the sine qua non of the multinational corporation would not be possible without instantaneous electronic communication.

As noted above, the advent of digital television broadcasting has introduced a new category of MNC competitors to global consumer electronics battles—U.S.-based computer manufacturers of hardware (e.g., Intel and Compaq), and software (e.g., Microsoft). Their struggle to force the FCC to consider an open ATV standard that incorporates both television and computer scanning indicated that they wished to be active competitors not only in the United States but in any market that introduced digital television broadcasting. How the U.S. computer industry fares in the environment of international high-volume, low-margin consumer electronics competition will be very interesting to observe over the coming decade (see Brinkley, 1997b). The point to be made is that the emerging competition in digital television hardware/software will be global in scope and intensely waged.

The Role of Thomson and Philips in Europe and the United States

On the hardware side, the French government has injected massive amounts of capital to keep Thomson afloat since the parent company purchased RCA in 1987. Thomson is the world's second largest television manufacturer after Philips of the Netherlands, and the two companies have made multimillion-dollar investments in HDTV research in Europe and the United States. Both companies were betting that HDTV would lead to significant profits by the late 1990s, but millions were squandered on the MAC system, and U.S. HDTV royalties will now have to be shared among other Grand Alliance partners.

The big change in the wind for Thomson is impending privatization. A deal in late 1996 with Daewoo Electronics Company of South Korea fell through after protests by French labor unions over the sale of the "national patrimony" to foreign interests (Whitney, 1996, p. C9). These privatization efforts mirror similar efforts in other quasi-socialist states in Scandinavia and Eastern Europe. As these governments divest their state-managed software and hardware monopolies, a more balanced regulatory structure is emerging with greater separation between the state and manufacturers/broadcasters.

France may divest Thomson and encourage the growth of commercial broadcasters, but the nation will probably continue to guard its technological and cultural independence within the EU and with other OECD nations. French willingness to go to the mat over agricultural subsidies and media quotas (limiting American film and TV imports) during the 1994 GATT talks demonstrates that national cultural issues can have a profound impact on continental and global policy formation.

Cultural issues notwithstanding, the trend for French and European communications policy is for greater privatization of national media monopolies. The growing dominance of commercial TV over public broadcasting is a subject of intense debate within European nations, especially as they face the prospect of a flood of American-produced programming on the new commercial networks. Despite funding restrictions, most EU member states are committed to maintaining their public service broadcasting operations.

In the area of industrial policy, Cynthia Beltz (1991) argued that EU and Thomson/Philips efforts to develop an analog HD-MAC standard have been an expensive and fruitless example of the pitfalls of governmental efforts to make such policies. Her thesis is that the European experience with HDTV industrial policy is a textbook illustration of the dangers of placing government bureaucrats in charge of technology planning.

Laura Tyson (1993) countered that if the Europeans had not made the effort to block the Japanese Hi-Vision/MUSE systems in Dubrovnik, the analog formats would have become world standards. She argued that if it were not for EU industrial policies, the FCC's ATV competition might not have taken place, and European and American manufacturers would now be paying HDTV royalties to NHK and Sony—a repeat of Japanese hegemony in VCR technology.

Tyson is assuming that consumers will be lining up in droves to buy new HDTV sets, which is by no means assured, but her basic point is well-taken. Nations have an obligation to their economies to promote the growth of industries that create high-skill, high-wage jobs, and most of the G-8 nations have created industrial policies to this end.

One of the key tenets of this book is that it is impossible to isolate national telecommunications policy by nation or even geographic affiliation (the EU or NAFTA). All national and continental regulatory systems must be seen as nested in the global network, not as independent and unrelated parts.

Summary and Conclusions

The global HDTV policymaking case demonstrates the lengths to which national and regional governments are prepared to go to develop and protect their high-technology industries. Japan presents a remarkable success story as a nation rising from the ashes of the Second World War to economic superpower status primarily on the strength of its manufacturing and export prowess. The nations of Europe, many equally devastated, transcended centuries of discord to forge an economic and political union that provides competitive advantages to Member States by erecting trade barriers to outsiders. The United States, long an open market to these competitors, has belatedly forged regional ties with Mexico and Canada as part of NAFTA. In this politico-economic environment linked by the growing influence of multinational corporations, there are several important trends that have influenced global telecommunications policymaking:

1. The rising influence of regional standards organizations (RSOs).
RSOs such as the Advanced Television Systems Committee (ATSC) (backed by a decade of work by the FCC and ACATS), and European Telecommunications Standards Institute (ETSI) have emerged in response to governmental and industrial initiatives to create standards that provide a competitive advantage to national industries. Their growth at the expense of international groups such as the ITU has been attributed to their greater speed in resolving standards issues (Besen and Farrell, 1991), but we think that regional economic protectionism is equally important as a rationale. The European EUREKA project, the DVB group, and the FCC's Advisory Committee were each influenced by the threat of potential Japanese dominance of advanced television technology. The RSOs may be more efficient than their international counterparts, but the balkanization of global communication standards does not bode well for future international program exchange.

2. The creation of international technology standards by MNCs.
In nonregulated industries such as consumer electronics and computer manufacturing, standardization agreements are being forged by multinational corporations without government intervention. The best example of this is the 1995 agreement by Sony/Philips and Toshiba to merge their competing digital video disk systems into a single Digital Versatile Disk (DVD) standard (Roberts, 1996). Since this was not a telecommunications standard it did not require approval by any government agency, but such agreements point to the speed that MNCs can move when they see an economic incentive for doing so. While speed and efficiency are hallmarks of the emerging global "technopoly" (Postman, 1992), the danger with delegating telecommunications standard-setting to the multinationals is that consumer and national

cultural interests are usually ignored. World governments should not abdicate their regulatory mandates in the sole interest of efficiency.

3. *The explosive growth of digital telecommunications technologies.*

The third trend is linked to the development of new communication technologies such as DVD, the Internet, and digital compression tools such as MPEG. The world is presently in the midst of a historic transition from analog communication technologies to a host of digital modalities. Books, video programs, brochures, films, photographs, children's games are gradually being converted into digital forms that can be transmitted anywhere in the world in a nanosecond. This is why the global battles over the creation of digital standards for advanced television are significant—much more so than sharper pictures or better audio. The ultimate convergence of the disparate worlds of computers and television is at the heart of the process and is one reason for the hotly contested debate over interlaced and progressive scanning. The victory of elements of the U.S. computer industry in deleting any mandated scanning parameters from the American ATV standard solidified the impending convergence of these formerly divergent worlds in consumer electronics.

A Postscript on Media Convergence

The most significant impact of the HDTV set may not be the ability to display the entire widescreen frame from a film such as *Dances With Wolves*, but rather that it could also "deliver" the morning newspaper. Until the advent of closed-captioning for the hearing-impaired, television was rarely thought of as a device for communicating text on screen. Computer screens were originally designed for textual display and television for moving images. These distinctions have already become outdated and will become more so in an era of digital telecomputers.

In the Smithsonian Museum of American History in Washington, D.C., there is an elaborate exhibit on the development of communication and computer technology. The display winds through five large rooms as dual parallel exhibits—one side of each room contains communication technology such as Morse's telegraph key and the other half displays computer technology from the same era (e.g., Jacard's punch cards for a power loom). The exhibits flow in parallel until the advent of the personal computer where they merge together. Perhaps it is no accident that at the point of merger, the display case contains a working Hitachi HDTV set showing widescreen rodeo footage. The point is not that the content was uniquely American and the display technology Japanese, but rather that an analog interlaced display system was sitting amidst all the digital systems around it.

By late 1998 such HDTV television displays will be processing sound and picture through digital circuits, presaging a struggle between global computer manufacturers, television set builders, and content providers for consumer attention. After decades of advanced television development fueled by multibillion-dollar investments on three continents, consumers making basic purchasing decisions will finally get to cast their decisive vote on the success or failure of these technologies. It is possible that widescreen HDTV will fail to catch on with consumers in the face of competition from lower-cost PC-TV systems. It is equally possible that computer manufacturers may misjudge the level of interactivity that viewers want with their television sets.

What is most likely is there will be a multiplicity of digital displays in all sizes and shapes available to consumers. The ability to interact with others through e-mail and videoconferencing will be there if needed, but so will be the ability to immerse the viewer in a multisensory environment similar to a movie theater. Films shown in high resolution on a widescreen HDTV set are dazzling, but some viewers would be just as happy seeing their granddaughter take her first steps hundreds of miles away via low-resolution Internet-delivered video. The technology for both scenarios is available today, but it is the global consumer who will decide what the ultimate level of interactivity in this digital communication environment will be.

■ Notes

1. In Japan, HDTV is generically known as "Hi-Vision." But to be quite accurate, Hi-Vision refers to the 1125/60 production standard.

2. Actually, NTSC operates at 59.94 fields and 29.97 frames per second.

3. Hi-Vision's aspect ratio was initially 5:3 but was changed to 16:9 after the U.S. Advanced Television Systems Committee adopted a 16:9 format to facilitate conversion from film to HDTV and vice versa (NHK Science and Technical Research Laboratories, 1993) (see Chapter 2).

4. Although there is little consensus on what is the exact optimal ratio of viewing distance to screen height for watching NTSC television, the modal prediction revolves around 7H (Lund, 1993).

5. Japan's direct broadcast satellite (DBS) channels have a bandwidth of 27 MHz.

6. The CCIR began work on color television in earnest in 1955, soon after the United States adopted its NTSC system. At that time, many countries were not even considering color television, instead focusing their efforts on planning and implementing monochrome television technology. Between 1955 and 1961, Study Group 11 (television broadcasting) made some progress by unanimously agreeing on several color television parameters. But from 1961 on, the Administrations decided to go their separate ways. In 1961, both France and the United Kingdom disclosed their intent of broadcasting color programs in 625 lines, and at a 1964 CCIR meeting in London, the Europeans appeared poised to endorse a 625-line color standard (Herbstreit and Pouliquen, 1967). In March 1965, shortly before a scheduled CCIR meeting in Vienna, France and the Soviet Union dropped a bombshell by announcing that both nations will cooperate on the adoption of a SECAM-based color television standard (Crane, 1979). The Franco-Soviet accord exacerbated already existing tensions and rivalries between CCIR participants, antagonizing many delegations, including some French delegates, who criticized this kind of "politicking" and the inopportune timing of the communiqué. "Coming as it did, prior to the Vienna discussions, the announcement of the accord was interpreted as a rebuff to the C.C.I.R. itself. In effect, France and the Soviet Union had proclaimed intentions of proceeding with SECAM, regardless of the C.C.I.R. decision" (Crane, 1979, p. 73). By the time of the Vienna conference, CCIR members were irremediably deadlocked, and likewise the CCIR Plenary Assembly in 1966 ended without an agreement. The outcome of the Oslo CCIR Plenary Assembly formally divided the world into three main incompatible color television standards, which complicated program exchange and raised the cost of the receivers.

7. A number of sobriquets have been applied to NTSC, SECAM, and PAL systems since the 1950s or 1960s. The Europeans nicknamed NTSC "Never Twice the Same Colour," for the reason explained below. The Americans, on the other hand, derided SECAM as "Something Essentially Contrary to America" or "System Essentially Contrary to the American Method." As to PAL, it stood for "Peace (or Perfect) at Least"! Although the three systems have much in common (e.g., 4:3 aspect ratio, interlaced scan-

ning structure), they differ from each other in number of scanning lines, frame rate, field rate, and color modulation. Both PAL and SECAM transmit 625 lines, while NTSC uses 525 lines. They operate at 25 frames and 50 fields per second, while NTSC scans a full frame every 1/30th a second and a field every 1/60th a second. For historical reasons associated with the frequency of the public electricity supply, 625-line systems generally have a 50 Hz field rate while 525-line systems generally have a 60 Hz field rate. NTSC is susceptible to chroma phase shift, causing noticeable changes of hue (color), particularly in the skin tones (for a review of NTSC defects, see FCC, 1987a). PAL eliminates this problem, by reversing "the subcarrier phase on *each scan line,* nullifying any hue shift problems by comparing the phase of one scan line with the phase of the next" (Anderson, 1984, p. 27). Both NTSC and PAL transmit the two chrominance signals (A = R - Y and B = B - Y) simultaneously on each scanned line according to the following pattern: A and B (line 1), A and B (line 2), and so on. On the other hand, SECAM sends the R-Y signal on one line and the B-Y signal on another line according to the following sequential pattern: A (line 1), B (line 2), A (line 3), B (line 4), and so on (Crane, 1979). Most American engineers would now recognize the technical advantages of PAL and SECAM systems over NTSC (McKnight and Neil, 1987), but to the layperson these differences in picture quality are largely undetectable, especially when viewing cable television channels.

8. The ATSC membership counts about 50 companies, including the five founding organizations, from all spheres of the television industry.

9. This was not the first time that multiple advanced television systems were being studied. In November 1982, a SMPTE study group considered three alternatives to an HDTV format: enhanced television, extended NTSC, and a "translatable" format, which would display twice as many lines as NTSC and yet would remain compatible with NTSC through the use of a converter ("HDTV Alternatives," 1982).

10. CCIR Recommendation 601 specifies "sampling rates for video signals and is often followed in many high end digital effects systems. For main digital studio equipment, this document recommends sampling the luminance (Y) signal at 13.5 MHz and the color difference signals R-Y and B-Y at 6.75 MHz each. The ratios between sampling frequencies give rise to the term 4:2:2" (CasaBianca, 1992, pp. 143-144).

11. The 80 Hz field rate was regarded as the lowest rate to downconvert fields to either 60 Hz or 50 Hz easily and satisfactorily ("ATSC Comes Up," 1984).

12. While some engineers still complain about the flicker problem in television systems operating at 50 Hz, others claim that this is now a moot point—there is no necessary relationship between television and electrical current. The latter group would advise to abandon 50 and 60 Hz and select a higher field rate (McKnight and Neil, 1987).

13. European consumer electronics firms often alluded to the demise of the U.S. consumer electronics industry (see Chapter 5) as a likely prospect in Europe should the European authorities elect to endorse the Japanese/U.S. 1125/60 standard.

14. At that time, EUREKA members included: Austria, Belgium, Denmark, Finland, France, Germany, Greece, Iceland, Ireland, Italy, Luxembourg, the Netherlands, Norway, Portugal, Spain, Sweden, Switzerland, Turkey, and the United Kingdom.

15. The objective of EUREKA is to encourage European-wide collaboration on advanced technology projects that will strengthen the productivity and competitiveness of European products in European and world markets (EUREKA EU95, 1988).

16. Progressive scanning is a process whereby all lines are scanned in one pass. If

NTSC used progressive scanning, all 525 lines would be displayed in one frame (i.e., 1, 2, 3 … up to 525) instead of two fields as in interlaced scanning, thereby producing a picture with less flicker but doubling the bandwidth.

17. Disagreement between broadcasters and computer hardware/software manufacturers on selection of a scanning structure, which snagged the formation of the Grand Alliance (see Chapter 7) and almost derailed the FCC approval of the digital television standard (see Chapter 6), was already apparent in the late 1980s when some participants challenged the adoption of the 1125/60 standard. While most broadcasters and the Electronic Industries Association (EIA) favored an interlaced scanning structure, the computer industry promoted progressive scanning. Just prior to the 1989 Extraordinary Meeting of Study Group 11, all involved parties convened for a marathon session at the NAB to resolve this controversy and present a unified U.S. front in Geneva. Such an agreement did not happen, however, and the U.S. position on scanning structure was left open (J. W. Reiser, Federal Communications Commission, personal communication, April 22, 1996).

18. FCC rulemaking consists of at least four steps: (1) initiation of rulemaking procedure (by the public, Congress, the executive branch, the judicial branch, or the FCC itself); (2) initial Commission action (*Memorandum Opinion and Order* denying petition, *Notice of Inquiry* [*NOI*], *Notice of Proposed Rule Making* [*NPRM*], or *Report and Order* [*R&O*] adopting minor editorial changes); (3) evaluation of comments and replies; and (4) *Report and Order* (decision). If a party is not satisfied with an FCC *Report and Order*, it can petition the Commission for reconsideration. After reviewing the issues, the FCC adopts a *Memorandum Opinion and Order* (*MO&O*) either denying the petition or modifying its initial decision. The FCC issues an *NOI* when it simply seeks information on a given topic and an *NPRM* when it considers specific changes to an already existing rule. An *NOI* must be followed by either an *NPRM* or a *Memorandum Opinion and Order*. FCC rulemaking can become quite complex, involving numerous *NOIs*, *NPRMs*, and *R&Os*, as is the case for HDTV. By April 1996, the FCC's file on ATV (MM Docket No. 87-268) had encompassed 32 three-inch-thick binders!

19. The Electronic Industries Association joined in as the eighth member in 1989, but announced its resignation in 1995.

20. The ATTC was the largest of three facilities used to test the earlier six ATV systems and the digital HDTV Grand Alliance system. It was renamed the Advanced Television Technology Center in 1996. Cable Television Laboratories, Inc. (CableLabs), an R&D program established by a consortium of American cable television operators, conducted cable-related laboratory and field testing on the proposed systems for ACATS. The Advanced Television Evaluation Laboratory, an Ottawa-based facility of Canada's Department of Communications (now Industry Canada), performed subjective assessment tests to determine how typical viewers rated the transmission interference characteristics and picture quality of ATV systems (ACATS, 1995).

21. The members—with the support of the TV stations in the case of INTV, MST, and the NAB—contributed approximately $14.5 million for the construction and operation of the Test Center, with the remainder coming from HDTV developers in fees for having their systems tested ($7.5 million) and other income from contracts for service and equipment (e.g., $2.8 million paid by CableLabs to access ATTC facilities) (P. Fannon, Advanced Television Test Center, personal communication, December 29, 1995).

22. An augmentation channel, an alternative to simulcasting, would offer an extra

full or half channel to supplement the main 6 MHz NTSC channel and carry the ATV information (e.g., additional resolution, wider aspect ratio).

23. Cornwell, Thompson, Trager, and Hatfield (1993) have questioned the feasibility and even the soundness of the FCC's proposed 15-year NTSC phase-out window. They point out that it took 20 years for color television to pass the 50 percent household penetration mark since color TV sets were first introduced, and almost 10 years to do so when full-time color television programming was available. If the penetration of HDTV resembles that of color TV, they argue, 50 percent of U.S. viewers will end up without television service if they do not elect to purchase an HDTV set by the end of the transitional period.

24. In its *Fourth Interim Report*, and even earlier (see ACATS, 1989), ACATS (1991) had urged "the establishment of voluntary agreements among proponents to synthesize their designs ... in the unlikely event that each system proves to be inadequate" (p. 19).

25. Active lines are those lines that convey picture information.

26. MPEG-2 (Moving Picture Experts Group) is a family of flexible international video coding and data transport (or packaging) standards that specify the data-stream "syntax" of coding systems. It is best viewed as a "tool box" enabling a service provider to design systems at greater or lesser levels of sophistication. MPEG-2 video offers four source formats, or Levels, from low VCR-type definition to high definition, each with a range of bit rates. Generally, the higher the bit rate, the higher the picture quality. It also allows five sets of compression tools, or Profiles, that together make up the coding system. For instance, the first profile (i.e., Simple Profile) has the fewest tools, whereas the final one (i.e., High Profile) provides for "'a super-system,' designed for the most sophisticated applications where there is no constraint on bit rate" (Digital Video Broadcasting, 1996, p. 17).

27. In March 1994, broadcast television networks and stations pledged $1.2 million to design and build an alternate modulation scheme, known as COFDM (coded orthogonal frequency division multiplexing) (Foisie, 1994). Broadcasters believed that COFDM might improve station coverage, reduce or eliminate the effects of ghosting, and offer greater flexibility for ancillary nonbroadcast services. But in July 1995, the Certification Experts Group of ACATS unanimously agreed not to approve COFDM technology for testing, concluding that the COFDM-Limited Liability Corporation had not demonstrated "the superiority of COFDM over VSB for the majority of markets" (ACATS, 1995, p. 18).

28. In simple terms, the U.S. digital HDTV transmission system works as follows: A charge-coupled HDTV camera produces images that are converted to some 1.2 gigabits (1.2 billion bits) per second of information (occupying approximately a 30 MHz bandwidth). This massive amount of information is then compressed using MPEG-2 compression technology. The compression ratio is up to 60:1, yielding a compressed set of information of some 20 million bits per second. MPEG-2 tools are also used to encode the bits into packets of data, each of which contains "labels" as to its content and use in reconstructing the image. The MPEG-2 video encoding and transport system is analogous to other packetized data systems, in which a message at the originating site is broken into digital code and put into packets of data, which are then routed on various links—in this case broadcast, cable/fiber, or satellite TV—and eventually reassembled at the receiving site. At that stage, video, audio, and ancillary (e.g., for closed-captioning) data packets are multiplexed (i.e., combined) into a single data stream. The next step is

to modulate/transmit those packets. Analog NTSC television uses amplitude modulation (AM) for video and frequency modulation (FM) for audio. That is, a modulator in the transmitter varies the amplitude or the frequency of a carrier wave according to the amplitude of the video intelligence and frequency of the audio intelligence, respectively. In digital HDTV, vestigial sideband (VSB) modulation carries audio, video, and ancillary data packets on a radio wave at a rate of about 19.3 megabits per second (Mbps) into the 6 MHz TV channel bandwidth—the same size as today's channel. But because coaxial or fiber cable is a noise-free and clean transmission path (i.e., closed and controllable), whereas over-the-air broadcasting is subject to certain uncontrollable forces, cable operators will be able to carry two HDTV signals on a single 6 MHz cable channel ("HDTV Field," 1995). Finally, in the home TV receiver the HDTV signal is demodulated (stripped of its radio wave) and demultiplexed, and the digital information is decoded and reassembled. While the new HDTV receiver would be required to receive and display the new HDTV signal in full HDTV quality, it is expected—and some manufacturers are already presenting some prototypes—that set-top (or in-NTSC-set) adaptors would also be available to downconvert the incoming HDTV signal to an NTSC-type signal that is viewable on an NTSC receiver. This output derived from the digital HDTV (or an SDTV) signal will look "cleaner" (i.e., no noise, snow, or ghosts), but *not* sharper (i.e., higher resolution), than a normal NTSC signal because the artifacts caused by analog NTSC transmission would be eliminated (P. Fannon, Advanced Television Test Center, personal communication, December 29, 1995).

29. Central to the Grand Alliance HDTV system's principle of interoperability is a layered digital architecture that can interface with other applications (e.g., computer technology) at any stage in the coding/transmission process. The four primary layers of the Grand Alliance system are: the picture layer, the audio and video compression layer, the transport (packetization) layer, and the transmission (modulation) layer.

30. Yet, in 1995, widescreen TV set sales only represented 1 percent of all TV set sales in Europe (Johnson, 1996).

31. Costello (1995) reported that the major reason why widescreen receivers are quasi-inexistent in the United States and why manufacturers are unlikely to introduce any before the beginning of HDTV broadcasting is the lack of available programming in this format.

32. At that time, EUREKA members included: Austria, Belgium, Denmark, Finland, France, Germany, Greece, Hungary, Iceland, Ireland, Italy, Luxembourg, the Netherlands, Norway, Portugal, Russia, Slovenia, Spain, Sweden, Switzerland, Turkey, and the United Kingdom.

33. In June 1989, NHK began airing HDTV programming for one hour a day from the BS-2b satellite. This experimental HDTV schedule was increased to eight hours a day in November 1991 and was further extended to nine hours a day in January 1994, to 10 hours a day in April 1994, to 11 hours a day in April 1995, and to 13 hours on weekdays and 14 hours on weekends in April 1996. As of June 1997, the number of Hi-Vision hours aired by NHK and the associated commerical broadcasters totaled 14 hours a day (98 hours a week).

34. In its *Second Report and Order*, the FCC explicitly forbade broadcasters to use their second channel for multiple NTSC channels. "The reason we are awarding broadcasters a second channel is to permit them to move to an improved technology without service disruption. If a broadcaster chooses not to broadcast in ATV, there is no reason for

awarding that broadcaster an additional license" (FCC, 1992a, pp. 1108-1109).

35. A 1994 study estimated that revenues from video data broadcasting could exceed $250 million by 1998 (McConnell, 1995b).

36. If one defines Clear-Vision as broadcasting using a ghost canceller reference signal, the proportion of programming aired in Clear-Vision is almost 100 percent (Y. Osumi, Ministry of Posts and Telecommunications, personal communication, February 6, 1996).

37. In June 1997, the official retail price for a 28-inch Hi-Vision receiver was ¥360,000 ($3,130; $1 = ¥115), although Japanese consumers could purchase an older model for just under ¥200,000 ($1,739) in electronics discount stores (N. Kumabe, Hi-Vision Promotion Association, personal communication, July 1, 1997).

38. Those who have expressed the greatest interest in PALplus programming include ARD, ZDF, 3sat, and Premiere in Germany; MCM in France; and RTBF and CANAL+ Belgique in Belgium. ZDF and 3sat broadcast about 10% of their total programming in PALplus in 1996.

39. On March 1, 1993, the ITU was revamped into three sectors, the Radiocommunication Sector (ITU-R), the Telecommunication Standardization Sector (ITU-T), and the Telecommunication Development Sector (ITU-D). The standard-setting functions of the CCIR and those of the International Consultative Telegraph and Telephone Committee were merged to form the ITU-T. It would be logical to assign HDTV activities to the ITU-T, but this was not the case. To date, *only* radiocommunication activities involving radio systems that interconnect with public telecommunication networks have been transferred to the Telecommunication Standardization Study Groups (International Telecommunication Union [ITU], 1994). In the case of HDTV as well as all broadcast activities, this work therefore remained within the Radiocommunication Sector. The ITU-R operates very much in the same way as its predecessor. The World Radiocommunication Conferences (WRC) are held every two years in conjunction with the Radiocommunication Assemblies (RA) that provide technical support for the work of the WRCs. So the WRCs and RAs replaced the World Administrative Radio Conferences (WARC) and the CCIR Plenary Assemblies, respectively. These changes notwithstanding, the Radiocommunication Assembly carries the same duties and involves the same participants as the former CCIR Plenary Assembly; there are only a few statutory differences between the two entities. Similarly, the WRC is very close to the former WARC (P. Amarsingh, International Telecommunication Union, personal communication, November 27, 1995). At the WRCs, ITU members review, and if necessary, adopt radio regulations. The RAs establish study groups and approve their program of work. These Radiocommunication Study Groups examine technical and operational Questions related to radiocommunication issues, develop draft ITU-R Recommendations, and prepare Reports for future WRCs. The RAs also approve, modify, or reject draft Recommendations submitted by the Study Groups. As was the case in the ex-CCIR, Study Groups can set up Working Parties (WP) and Task Groups (TG) to study specific technical Questions (see ITU, 1994; http://www.itu.ch/itudoc/itu-r).

40. Both European and Japanese proposals had 1920 pixels per line and a 16:9 aspect ratio. The United States took the position that square pixels is a key picture characteristic for the digital age and therefore recommended that the number of active lines be 1080

[1920 x (9 ÷ 16)]. As drafted for the RA-95, the Japanese and European proposals did not have square pixels (R. Hopkins, Sony Pictures High Definition Center, personal communication, April 5, 1996).

CHAPTER 2

1. The Hi-Vision Promotion Association is an organization of Japanese consumer electronics manufacturers, broadcasters, and government agencies such as the Ministry of Posts and Telecommunications and the Ministry of International Trade and Industry who seek to promote the diffusion of the Japanese 1125/60 Hi-Vision HDTV system.

2. The qualifier "full-time," "full-scale," or "regular" does not imply that the schedule of HDTV programming is currently irregular or part-time. As of June 1997, the number of broadcast Hi-Vision hours totaled 14 a day. Rather, what it means is that the type of license granted by the Ministry of Posts and Telecommunications to NHK and associated commercial broadcasters will become a *regular* commercial license, instead of an experimental one (Y. Ogaki, Nippon Hoso Kyokai, personal communication, July 1, 1997).

3. The five key members of Vision 1250 are Thomson, Philips, BTS, the British Broadcasting Corporation, and RTP broadcast networks. There are an additional 36 "participant" members from throughout Europe, in addition to the REBO Group from the United States.

4. The American Society of Cinematographers and the Directors Guild of America protested the adoption by the Federal Communications Commission of an HDTV system with an aspect ratio of 1.78:1. They advocated a wider aspect ratio of 2:1. As noted above, the aspect ratio of the NHK Hi-Vision format was widened from 1.67:1 to 1.78:1 at the behest of the Society of Motion Picture and Television Engineers in 1985 and further modifications at this date are unlikely. The 1.78:1 (16:9) aspect ratio has emerged as a global standard for widescreen television in both analog and digital formats.

CHAPTER 3

1. NTSC is an acronym for National Television System Committee, the broadcasting industry group that designed the monochrome television standard approved for U.S. service in 1941, as well as the color standard adopted in 1953. The NTSC color standard is used throughout North America and much of Central America, parts of South America, Asia (including Japan), and the Pacific.

2. Nippon Hoso Kyokai (NHK) is Japan's public service broadcasting organization. Modeled after the British Broadcasting Corporation, it is required by law to provide nationwide service and derives almost all of its funding (97%) from monthly fees paid by viewers (see Browne, 1989; Ito, 1978; NHK, 1995).

3. Although Hi-Vision is used as a generic term for HDTV in Japan, it refers specifically to the 1125/60 studio standard.

4. Research into high-definition telecasting had been previously undertaken in the United States during the late 1950s and early 1960s. It never reached a commercial stage, though. As American firms increasingly dropped out of television manufacturing or were sold to foreign interests, domestic HDTV research dwindled (U.S. Congress, 1990).

5. Television images are composed of scanning lines which run from left to right and top to bottom across the face of the TV picture tube. With interlaced scanning, all the even-numbered lines are displayed first, and then the electron gun inside the TV set returns to scan the odd-numbered lines. It takes two of these rapidly-scanned "fields" to create a complete, 525-line television frame.

6. The field rate refers to the number of times an image is vertically scanned in one second. With NTSC interlaced scanning, there are approximately 60 total fields per second, or 30 complete frames.

7. The flicker effect, which consists of visible alternations of light and dark, is heavily influenced by the field rate.

8. Aspect ratio is the ratio of the television screen's width to its height.

9. Line structure refers to how the scanning lines of a television image are designed to make a complete frame. Interlaced scanning, which has already been mentioned, is one type of line structure that combines odd and even fields (made up of odd and even lines, respectively) to create a complete television frame.

10. With progressive scanning, all horizontal lines are scanned in one pass (one field) instead of two passes (two fields) as in interlaced scanning (see Donow, 1988).

11. MUSE stands for multiple sub-nyquist sampling encoding. "Simply stated, MUSE divides the [HDTV] signals corresponding to a single image into four parts through image sampling (the decomposition of an image into pixels) and subsampling (the reduction of a sample by discarding redundant pixels to reduce the volume of transmitted information). These four parts are then put back together at the receiving end in a form nearly identical to the original. This compression process enables an 8 MHz basebandwidth to transmit a 27 MHz FM HDTV signal and permits HDTV broadcasting over a single 27 MHz channel from a satellite" (MPT, 1989, p. 10).

12. "'Simulcast' is a contraction of 'simultaneous broadcast' and means the broadcast of one program over two channels to the same area at the same time" (FCC, 1990, p. 5629).

13. Each satellite has three transponders. Unfortunately, the third transponder of BS-3a became inoperational because of power failure due to a solar panel malfunction. The third transponder of BS-3b serves as a backup channel for the other four channels (H. Sakamaki, Nippon Hoso Kyokai, personal communication, August 26, 1993).

14. The consensus model holds that "The bureaucracy, the majority party, and big business have so many interests in common, and dominate the governmental system to such an extent, that policymaking is basically cooperative rather than conflictual" (Campbell, 1984, p. 294).

15. The Association of Radio Industries and Businesses (ARIB) was created through the merger of the Broadcasting Technology Association (BTA) and the Research and Development Center for Radio Systems (RCR) (M. Wakao, ARIB, personal communication, September 10, 1996).

16. ATV, or advanced television, is an umbrella term which refers to any of the newer systems used to improve television image reproduction, including HDTV and EDTV (as well as IDTV, or improved-definition television).

17. Figure 3.1 reflects the BTA nomenclature as of November 1987. In June 1996, there were five subcommittees: the HDTV Emission Standards Subcommittee, the HDTV Studio Standards Subcommittee, the HDTV Program Transmission Standards Subcommittee, the HDTV Production Technology Subcommittee, and the Picture Quality Assessment Subcommittee (M. Wakao, Association of Radio Industries and

Businesses, personal communication, September 10, 1996).

18. In June 1996, BTA/ARIB's High Definition Television Committee consisted of NHK, 19 manufacturing companies, 8 commerical broadcasters, and 1 telecommunication company (M. Wakao, Association of Radio Industries and Businesses, personal communication, September 10, 1996).

19. On October 15, 1991, the Hi-Vision Promotion Association (HPA) was established as a not-for-profit corporation by 117 industry organizations, including broadcasters, electronics manufacturers, and production companies. From November 25, 1991 to November 24, 1994, the primary functions of HPA were to operate the Hi-Vision channel on the BS-3b satellite (BS-9) and promote Hi-Vision applications in Japan. But on November 25, 1994, the task of managing the channel was transferred to NHK and seven commercial broadcasters. So HPA's role is now limited to promoting Hi-Vision broadcasts and applications (N. Kumabe, Hi-Vision Promotion Association, personal communication, August 30, 1996).

20. Japan has five commercial networks: Nippon Television Network (NTV), Tokyo Broadcasting System (TBS), Asahi National Broadcasting (ANB), Fuji Television Network (CX), and Television Tokyo (TX).

21. The distinction between an EDTV system and an HDTV system is not always clear. Both commonly provide higher picture resolution, a wider-screen aspect ratio, more dynamic digital audio, and better color performance, in some combination. It is generally agreed, however, that HDTV equipment must provide twice the horizontal and vertical resolution—or picture sharpness—of today's television receivers if it is to be considered truly high-definition. EDTV systems do not meet this criterion, placing them on a lower level of performance than HDTV.

22. Until March 1995, there were six BTA/ARIB committees that dealt with EDTV: the EDTV Planning Committee, the EDTV System Development Committee, the EDTV International Investigation Committee, the Ghost Canceller Committee, the EDTV Studio Facilities Committee, and the Transmission Facilities Committee (M. Wakao, Association of Radio Industries and Businesses, personal communication, September 10, 1996).

CHAPTER 4

1. The EU is a blanket term for the European Community (EC) and its member states. The EC refers to the central institutions of the EU: the Council, the Commission, the European Parliament (EP), the European Court of Justice (ECJ), and the Economic and Social Committee (ESC). Policymaking is carried out by the EC institutions and implemented throughout the EU. In this chapter we shall refer to the EU/EC as the EU for the sake of simplicity.

2. The CCIR was a permanent organ of the Geneva-based International Telecommunication Union (ITU) in charge of recommending global telecommunications standards. With the restructuring of the ITU in 1993, it was replaced with the Radiocommunication Sector, which handles HDTV and broadcast standardization activities (see Chapter 1).

3. In May 1993, Richard Wiley, the chairman of the U.S. FCC's Advisory Committee on Advanced Television Service invited the EU to assume a collaborative stance over HDTV, which was ranked among the world's greatest technological rivalries ("HDTV: All Together Now," 1993).

4. In the context of political economy, "international capital integration" refers to the process of capital concentration and economic policy convergence of national economies, resulting in a global economy whose boundaries are more blurry and which in turn makes local or regional policy inputs more irrelevant or more contingent on further uncontrollable factors. Capital becomes intermeshed and concentrated via mergers, acquisitions, alliances, and take-overs.

5. The institutions of the EU include the Commission (CEC), the Council, the European Parliament (EP), the European Council, the European Court of Justice (ECJ), the Economic and Social Committee (ESC), and the Auditors Council. There is a tangle of links between these institutions which provide certain checks and balances between them. The EU being an evolving economic and political entity entails that its institutional decisionmaking processes have been and are still under change and ongoing revision. Thus in the course of the decade that this study reviews, three different legislative modes (with regard to Directives) have affected the relationships between the three main institutional actors, the CEC, the Council, and the EP. The first MAC Directive of 1986 was plainly the outcome of a single-reading consultation procedure, whereby the Commission submitted a proposal to the Council, who then solicited comments from the EP and the ESC. Those opinions had a mere nonbinding consultative nature. The second MAC Directive was adopted according to the two-reading cooperation procedure, introduced by the Single European Act (SEA) of 1987, which established the EP as a significant legislative agent (see Dupagne, 1992). However, although the EP policymaking prerogatives were set, a resolved Council could still bypass the EP views. The third Directive on the use of transmission standards (1995) was adopted under the codecision procedure, which for the first time provides the EP with a veto power over the legislative process of the EU (see Nugent, 1994). A brief description of the EU institutions relevant to this case study follows.

The *Commission* has a combination of political, administrative, legislative, and implementational powers. It selects and elaborates on specific policy areas and proceeds to presenting policy proposals to the Council and to Parliament. The Commission is a powerful institution because it can initiate and hold the agenda. Because of variable levels of power between the Directorates-General (DG), rivalries may grow, particularly when roles are overlapping or antagonistic. The DG X (Information, Communication, Culture and Audiovisual) and DG XIII (Telecommunications, Information Market and Exploitation of Research) whose priorities overlap and whose respective cultural versus industrial priorities clash is a case in point.

The *European Parliament* (EP) consists of 518 members, who sit by political group. It exercises legislative and supervisory powers which are conferred upon it particularly by the recent treaty revisions. Legislative functions are strongest on instruments requiring the codecision procedure. Moreover, its consent is required for the ratification of EC Treaties as well as for accession and budgetary issues. Optional/Formal opinions are delivered following debate and voting in the plenary session. The debates draw on reports by specialist committees assigned to review proposed legislation. A "rapporteur," who is designated by a system of allocation of report tasks according to party political proportionality, does the preparatory work for the Committee.

The *Council of Ministers* is composed of the ministers of the member states. The ministerial Council is "thematic," i.e., its composition changes depending on the type of issues on the board. Ministers of culture and/or of communications and ministers of telecommunications are in charge of putting forward and adopting policies on broadcasting issues. Majority voting is proportional to the population of each member state.

The *European Council*, instituted by the SEA in 1986, is made up of the leaders of member states and the President of the Commission. It convenes at least twice a year. Its prerogative is to draw up the future strategic lines and goals of the EC.

Finally, the *Economic and Social Committee*, composed of representatives of employer organizations, trade unions, and interest groups, is consulted on most legislative proposals but its views are not necessarily incorporated into legislation.

6. Teleconferencing and, more recently, point-to-point casting (e.g., telephone call) are examples of non-broadcast network-based applications where the use of high-definition screens loomed large. Home video is also a major and lucrative market segment of non-broadcast, non-network-based applications for HDTV quality screens.

7. Without a common approach to deregulation, the objective of a large homogeneous telecommunications market "would be seriously jeopardised and a great deal of fragmentation would be inevitable if the member states were to legislate without previous agreement," considered Commissioner Heinz Narjes (Hughes, 1988, p. 73).

8. The 1986 Semiconductor Trade Agreement between the two powers was seen as an example of a global cartel denounced by the EU at the General Agreement on Tariffs and Trade (GATT) negotiations. This was compounded by the fact that policymaking in Japan was a function of centralistic cooperation between an oligopoly of firms and the Ministry of International Trade and Industry (MITI) and that industrial cooperation and collusion between the United States and Japan formed a special rapprochement and intertwining of policies in the 1980s. As a result "the West Europeans came to regard the United States and Japan as linked not merely via the alliance portfolios of companies but via a series of tacit and formal agreements at industry and government level" (Tunstall and Palmer, 1990, p. 53; see also Sharp and Pavitt, 1993).

9. Since the last internal restructuring of the Commission, Directorate-General XIII (DG XIII) is responsible for Telecommunications, Information Market and Exploitation of Research, while Directorate-General III (DG III) is responsible for Industry. Overlapping of competences is not unusual in the EU Commission.

10. SAT 1 and RTL-Plus, operating German commercial broadcasters, reacted furiously because their viewers would suddenly need new equipment or D2-MAC decoders (Humphreys, 1988). For a more thorough account of this policy battle from the point of view of the leading member states, see also Kuhn (1988) and Goodfriend (1988). On the other hand, the Luxembourg-based channel CLT was the spearhead of every movement toward deregulation and commercialization. In 1996 CLT merged with the Bertelsmann subsidiary called Ufa in Germany. As Stüdeman (1996) put it: "This is basically about clearing up commercial television in Germany. There used to be three main players: Bertelsmann, Kirch, and CLT. With the merger Bertelsmann has narrowed that down to two and has made life easier for itself" (p. 19). The combination of these developments points to the strengthening and further consolidation of those leading media market forces that throughout the last decade fought against MAC transmission standards and also against the open and commonly controlled MultiCrypt CA system.

11. Understandably France was one of the leading advocates of an aggressive HDTV strategy, partly because state-controlled Thomson had invested heavily in the original MAC family.

12. HD-MAC disposed of the same basic characteristics as D2-MAC, but had the

advantage of being able to deliver pictures with four times greater definition. This was achieved by doubling the number of horizontal lines from 625 to 1250 as well as doubling the number of pixels per line (CEC, 1992b). Both D2-MAC and HD-MAC standards were designed for satellite delivery, because terrestrial transmission systems did not have the bandwidth capacity required for analog HDTV signal transmission (see Slaa, 1991).

13. The World Administrative Radio Conference (WARC) of 1977 planned broadcasting satellite services (BSS), including DBS, in the high-frequency band of 12 GHz, because they require a broader spectrum capacity and could be received by small diameter antennas (one meter). By contrast it planned the fixed satellite services (FSS) in the low-frequency bands requiring medium- and low-power satellites aimed for telecommunications. These required a large diameter (nondomestic) antenna for reception by Ministries of Posts, Telephones, and Telegraphs (PTT) and cable operators. Improvements in reception capacity from the FSS band enabled the circumvention of the 1986 Standards Directive.

14. At the time the MEDIA program was an EU pilot project designed to promote European audiovisual programming.

15. The European Economic Interest Grouping (EEIG) created "a flexible mechanism working transnationally to allow partners from different member states to pool their resources to achieve objectives in their common interest" (Lalor, 1990, p. 27).

16. HD-DIVINE (Digital Video Narrow-band Emission) is a terrestrial digital HDTV system that was developed in 1991 by a Scandinavian consortium.

17. "The politicians betrayed us ... the broadcasters were promised money by a certain time and that money was repeatedly postponed. Several made programs, they were going to buy equipment, and they budgeted for it ... if there is no money, nothing will happen" (Piete Boegels, Head of the EUREKA 95 HDTV project, quoted in Flynn, 1993a, p. 1).

18. The Moving Picture Experts Group (MPEG-2) was established by the International Standards Organization (ISO) to provide the basis for a picture coding and compression system (Morgan, 1993). In addition, MPEG-2 embodied a case of competition in standard-setting, whereby small groups of market forces vied to demonstrate and have their product approved by ISO and ETSI (European Telecommunications Standards Institute) (Morgan, 1992). This was the manner in which an international standard for compression of digital signals, which is the most complicated and expensive element in digital broadcasting systems, had evolved by April 1993 ("HDTV: All Together Now," 1993).

19. The Commission invited the DVB group to develop an industrial position on conditional access for digital TV systems particularly in connection with this Directive (Schoof and Brown, 1995).

20. The conditional access (CA) controversy involved two competing encryption systems, SimulCrypt and MultiCrypt. CANAL+ and BSkyB, two major pay-TV broadcasters, proposed their own proprietary, exclusively-accessed SimulCrypt system, while the "traditional" commercial and public broadcasters favored the publicly managed and openly accessed MultiCrypt approach. The latter group opposed SimulCrypt on vertical integration grounds—it would allow programmers to own and control a CA system ("Has Europe," 1994). In the face of such polarization, the DVB Steering Board gave joint approval to both systems in May 1994. The DVB Steering Board, however, rejected Simulcrypt's code of conduct for not being open on commercial issues. The SimulCrypt

strategy relied on different proprietary conditional access systems for country or linguistic zones (Leclerq and Flynn, 1994). By July 1994, DVB members were still deadlocked. Thus these unresolved issues were deferred to a Board meeting in August 1994 but even then the DVB group was unable to reach a consensus on a voluntary basis. While there was a skew toward SimulCrypt, DVB did not unanimously agree on its selection. In sum, Simulcrypt came out as a de facto but not properly voted winner. Problems within DVB were inevitable as stakes and conflicts of interest were high. Certain members even feared that the only reason for CANAL+'s membership was to block agreements that would have threatened its pay-TV monopoly (Flynn, 1993b). The MultiCrypt proponents managed, however, to impose certain checks and balances, hoping that SimulCrypt operators will not abuse their dominant position. Thus DVB imposed a different line from that laid out by the EU, stepping closer to laissez faire and away from the consensual and voluntary agreement on standards.

21. Following the DVB input to the Commission for the approval of both MultiCrypt and SimulCrypt, MultiCrypt proponents predicted that "MEPs and ministers would tend to back only the MultiCrypt proposal" for conditional access (CA) systems (Leclerq and Flynn, 1994, p. 7). But the EP and the Commission proposed neither the "dual system" nor the mandatory inclusion of the open interface MultiCrypt but instead the proprietary "smart card" interface SimulCrypt. The new climate of agreement between leading market (DVB) and political (EP, Commission) actors enabled the EU Council to adopt unanimously the amended draft Directive on October 24, 1995. The Council accepted all the amendments proposed by the EP, including those on the controversial issue of conditional access.

22. In September 1996, French pay-TV giant CANAL+ acquired Dutch-based Nethold for 6.1 million new CANAL+ shares and $45 million in cash. The new group will offer pay-TV service to over 8.5 million subscribers in France, Italy, Spain, Scandinavia, Greece, Benelux, Germany, as well as in several Central European countries ("Nethold/Canal+ Deal," 1996).

CHAPTER 5

1. These average annual rates of change were computed by dividing the 1990 index value by the 1977 index value and raising the quotient to the 1/13th power (because the period covered 13 years). For instance, for the United States, we first divided 95.7 by 75.5 and then raised 1.267 to the exponent value of 1/13 (0.077), yielding 1.018 or a 1.8 percent average annual change (1.018 - 1 × 100).

CHAPTER 6

1. Diagram by Peter Seel. This version of Krasnow, Longley, and Terry's system diagram shows only the six categories of policymaking actors. It does not attempt to characterize the types of communication paths between them. See their more comprehensive representation on Krasnow et al. (1982), page 136 of the 3rd edition.

2. "Contemporary" indicates it was after the Japanese development of the 1125/60 format. See Carbonara (1992) for details on earlier RCA and CBS high-definition research.

3. The Association of Maximum Service Telecasters (MST) is now called the Association for Maximum Service Television (MSTV).

4. The first U.S. DBS satellite was launched in December 1993. It is transmitting in NTSC, but has the capability to relay HDTV signals as well. DBS broadcaster DIRECTV announced in April 1994 that it will transmit NTSC signals in 16:9 widescreen proportions, indicating a likely upgrade to full HDTV transmission when the U.S. standard begins DTV service in 1998.

5. The chairs of the subcommittees were Joseph Flaherty of CBS for Planning; Irwin Dorros of Bellcore for Systems; and James Tietjen, then of Sarnoff for Implementation. See Appendix in Chapter 7 for a list of Committee members.

6. In addition to Wiley who represents CBS through his law firm, the other representatives were Ward Quaal, a broadcast industry consultant; William Henry, the Chairman of the industry-sponsored ATSC at the time; Joseph Flaherty of CBS; and James Tietjen of Sarnoff Labs, a division of RCA in 1987.

7. Each frame of an NTSC television program is scanned in two interlaced halves. However, computer displays are scanned in distinct progressive frames with less flicker than TV sets. This fundamental display distinction is addressed in greater detail in Chapter 7.

8. Zenith Electronics was sold to LG Electronics (Goldstar) of Korea in July 1995.

9. This was later modified to three years for application and three years for construction in the Third R & O (Sept. 1992).

10. COFDM stands for Coded Orthogonal Frequency Division Multiplexing, a type of television transmission technology developed in Europe. The Broadcasters Caucus of the Advanced Television Systems Committee (ATSC) asked the Advisory Committee on Advanced Television Service to investigate this technology for HDTV transmission in the United States. COFDM was ultimately rejected by ACATS as an HDTV transmission option in the United States.

CHAPTER 7

1. Interviews with Richard Wiley concerning the Advisory Committee and his role within it were conducted in Washington, DC, on November 3, 1993, and May 23, 1994.

2. The process of media convergence is incremental and gradual, but a good benchmark was the 1993 introduction of the Mosaic browser (at the University of Illinois) for the presentation of multimedia content over the Internet.

3. Due to quirks in the NTSC standard that resulted from the introduction of color in the early 1950s, the actual rates are 59.94 fields per second and 29.97 frames per second. These are often rounded off, but the true rates are short of 60 and 30. The U.S. ATV standard will use 24, 30, and 60 true frames per second.

4. The original members of the ATTC were ABC, CBS, INTV, MSTV, NAB, NBC, and PBS.

5. The advertisement appeared in *Broadcasting*, April 6, 1992, p. 32.

6. An international Common Image Format would be 1080 pixels high by 1920 pixels wide or 2,073,600 total pixels in a 16:9 aspect ratio. This may be the foundation for a group of easily transcodeable global HDTV standards in the future. Square pixels refer to a design in which the vertical and horizontal spacing of pixels is the same.

7. An earlier deadline of March 15, 1993 for the start of retesting had been extended until May 24.

8. FCC Chairman Reed Hundt added Craig Mundie of Microsoft and Samuel Fuller of Digital Equipment Corporation to the Advisory Committee five weeks prior to the

delivery of the ACATS final report on November 28, 1995.

9. The ATSC or Advanced Television Systems Committee is a broadcast industry group chaired by James McKinney that documented the standard based on the Grand Alliance system. The ATSC performed a more limited role in the standard-setting process than its NTSC predecessor. See Chapter 1 for details.

10. One causative factor that Braun (1995) attributes to this collaboration amongst prior competitors in the United States was the 1984 passage of the National Cooperative Research Act (NCRA). The NCRA made it possible for U.S.-based companies to work together during the research phase of the R&D process by removing the threat of anti-trust action by the Justice Department. The NCRA was passed by Congress in response to a perceived lack of international competitiveness by American companies, particularly in the consumer electronics industry. See Chapter 5 on industrial policy for details.

11. Instead of asking for a ballot vote of ACATS on the acceptance of the Grand Alliance standard at their final meeting on November 28, 1995, Richard Wiley took a voice vote. After a number of yeas were shouted out, he asked for "all opposed" and silence followed. He gaveled the meeting to a close and later said that the vote was unanimous (see Brinkley, 1997a). He had not called for any abstentions in the vote and later was informed by a Microsoft employee that their representative on the ACATS panel, Craig Mundie, would have voted to abstain if given the chance. (Mundie was still unhappy with the allowance of interlaced scanning in the standard.) Wiley said later that he had forgotten to call for any abstentions. "I didn't think about it. I was so pleased to wrap this up. Craig Mundie says that he abstained from the vote and I have accepted that. He didn't say anything at the time. The decision was unanimous except for that abstention. I have told Craig that I would accept that" (R. Wiley, personal communication, July 1, 1997). This is a minor footnote in the ACATS history as the vote of the 25 members was close to unanimous, but it is important to set the record straight.

12. Sikes, as then-head of the National Telecommunications and Information Administration, was also an *ex officio* member of the Advisory Committee on Advanced Television Service at its inception.

CHAPTER 8

1. Rogers (1995) defines "diffusion" as "the process by which an *innovation* is *communicated through certain channels* over *time* among the members of *a social system* [emphases added]" (p. 5). The first element of this definition presupposes the existence of an innovation, which refers to an idea that is perceived as new by an individual. People evaluate an innovation in terms of five main attributes, relative advantage, compatibility, complexity, trialability, and observability, which account for 49-87% of the variance in rate of adoption. For instance, if consumers view HDTV's (sharper) pictures, (crisper) sound, and (wider) screen size as superior to conventional television's, then HDTV would enjoy a relative advantage over the existing color television systems (NTSC/PAL/SECAM) and the perceived importance of these HDTV attributes might exert a powerful impact on purchase intention and rate of adoption. The second element, communication channels, involves both interpersonal (e.g., word of mouth) and mass media (e.g., television) channels. While mass media channels offer the most effective means to create awareness and knowledge, i.e., to inform the widest possible audience of individuals about the existence of an innovation, interpersonal channels are best to persuade potential adopters about the merits of an innovation. The third element is

time, which is an important dimension in determining the innovation-decision process and measuring the adopters' degree of innovativeness. The innovation-decision process is "the process through which an individual (or other decision-making unit) passes from first knowledge of an innovation to forming an attitude toward the innovation, to a decision to adopt or reject, to implementation and use of the new idea, and to confirmation of this decision" (p. 20). Therefore, it contains five steps: knowledge (e.g., HDTV awareness), persuasion (e.g., reactions to HDTV and its attributes), decision (e.g., purchase intent and willingness to pay for HDTV), implementation, and confirmation. Implementation and confirmation stages do not really apply to this review of the literature since none of the studies were conducted at a time when HDTV sets were available on the consumer market. Rogers (1995) classified adopters into five groups according to their level of innovativeness: innovators (2.5% of adopters), early adopters (13.5%), early majority (34%), late majority (34%), and laggards (16%). Earlier adopters (innovators, early adopters, early majority) differ from later adopters (late majority, laggards) in terms of socioeconomic status, personality values, and communication behavior. Among other things, they are more highly educated, have higher income, are less dogmatic, are better able to cope with uncertainty, are more cosmopolite, and are more exposed to mass media and interpersonal communication channels than later adopters (Rogers, 1995). Finally, the fourth element of diffusion is the social system, which is "a set of interrelated units that are engaged in joint problem-solving to accomplish a common goal" (p. 23). Members of a social system can be individuals, groups, or organizations.

2. Management Horizons Inc. forecast that less than 2 percent of U.S. households would own a VCR by 1980; 4-9 percent by 1983; and 55-70 percent by the 1990s (Klopfenstein, 1985). These predicted rates were close to the actual ones. *Screen Digest* reported that the VCR penetration in the United States was 2.5 percent in 1980, 11.1 percent in 1983, and 70.3 percent in 1990 ("Patterns," 1990).

3. Scenario 1 was referred to as the niche outcome, whereby HDTV would develop as a limited service delivered primarily by cable and satellite to leisure businesses and high-income households. The authors deemed this scenario the most likely outcome. Scenario 2, a more optimistic scenario, assumed that the United States would successfully develop HDTV as a digital terrestrial system, which would then prompt European and Japanese broadcasters to follow suit and adopt the technology. This was regarded as the least probable outcome. Finally, scenario 3, a combination of scenarios 1 and 2, posited a successful digital-based HDTV diffusion in the United States but also assumed that Europeans and Japanese would initially use analog HDTV systems before adopting the U.S. digital system. This was judged fairly likely. Interestingly, the scenario (#2) that was deemed the least likely in 1992 might turn out to be the most likely in the late 1990s, although Japan does not plan to replace MUSE before the year 2000 (see Chapter 1).

4. NERA further envisioned that by 2001 HDTV set ownership would reach 5 percent in Japan, 2 percent in the United States, and 2 percent in Europe under scenario 1; 8 percent in Japan, 6 percent in the United States, and 6 percent in Europe under scenario 2 (rapid diffusion); and 4 percent in Japan, 6 percent in the United States, and 2 percent in Europe under scenario 3. Worldwide annual estimates by the year 2000 ranged from 0.75 million to 2.95 million to 10.95 million HDTV receiver units (see AEA, 1988; Brown et al., 1992).

5. One of the thorniest issues in new media forecasting is product analogy. Most HDTV forecasts have relied on the market history of color TV, implicitly or explicitly

assuming that the diffusion of HDTV will parallel that of color TV (AEA, 1988; Brown et al., 1992; Darby, 1988; RRNA, 1988; Stow, 1993; Working Party 5, 1988). The unanswered question is, of course, whether this is an adequate historical analogy. For HDTV to mimic the penetration of color TV, consumers would have to perceive that the qualities of HDTV (wider and larger screen; CD-quality sound; sharper pictures) are as important or appealing as the advent of color in television.

6. $350,000 represented the average cost of fully equipping a station for color. Interestingly, the Katz study found no relationship "between color set penetration and either station investment in color equipment or the number of commericals broadcast in color. Stations in markets with low color penetration [were] often fully equipped and [carried] a high percentage of color advertising. The converse also [was] true in some instances" ("The Costs," 1967, p. 36). Color set penetration averaged 12 percent in spring 1966.

CHAPTER 9

1. German prosecutors have denied that they asked for CompuServe's outright ban on the offending sexually-oriented services, and asked only that access to them be denied to the company's German customers. CompuServe stated that they did not want to be placed in a position of providing content access on a country-by-country basis and thus decided to uniformly block access to every customer. This is an interesting regulatory question in an era of porous electronic national boundaries, and one that will not be resolved soon.

2. Hi-Vision is a 1125-line analog production standard and MUSE is the companion transmission compression system.

3. "Telecommunications" is defined here in the contemporary sense that includes all electronic modes of communication, wired or not, rather than the traditional definition of solely telephonic communication.

4. The French government planned to privatize Thomson in 1996 by selling the company to a French-South Korean consortium, with the consumer electronics division Thomson Multimedia becoming a division of Daewoo Electronics Company. The deal fell through due to intense political opposition in France (see Whitney, 1996).

5. RCA developed the "compatible" NTSC color television standard presently in use in the Western Hemisphere and Japan.

6. The Broadcasting Technology Association (BTA) was renamed the Association of Radio Industries and Businesses (ARIB) in 1995.

■ References

ABC Broadcast Operations and Engineering. (1988, December 23). *Summary statement of Capital Cities/ABC, Inc.: Appeal of ANSI approval of ANSI/SMPTE 240M-1988-for television-signal parameters-1125/60-high-definition production system.* New York: Author.

Adam, J. A. (1990, September). Industries transcend national boundaries. *IEEE Spectrum,* pp. 26-31.

Advanced TV brain trust. (1987, October 12). *Broadcasting,* p. 37.

Advisory Committee on Advanced Television Service. (1988). *Interim report of the FCC Advisory Committee on Advanced Television Service.* Washington, DC: Author.

Advisory Committee on Advanced Television Service. (1989). *Second interim report.* Washington, DC: Author.

Advisory Committee on Advanced Television Service. (1991). *Fourth interim report.* Washington, DC: Author.

Advisory Committee on Advanced Television Service. (1992a). *Fifth interim report.* Washington, DC: Author.

Advisory Committee on Advanced Television Service. (1992b). *Minutes of the seventh meeting.* Washington, DC: Author.

Advisory Committee on Advanced Television Service. (1993). *Minutes of the eighth meeting.* Washington, DC: Author.

Advisory Committee on Advanced Television Service. (1995). *Final report and recommendation.* Washington, DC: Author.

Alliot, J. (1996, April). Many are called, few are chosen. *Cable and Satellite Europe,* pp. 58-61.

Amdur, M. (1992, October). EC overview: Sizing up fortress Europe. *Broadcasting Abroad,* pp. 30-31.

American Electronics Association. (1988). *High definition television (HDTV): Economic analysis of impact.* Santa Clara, CA: Author.

American Electronics Association. (1989). *Development of a U.S.-based ATV industry.* Santa Clara, CA: Author.

American National Standards Institute. (1989, April 18). *ANSI Appeals Board hearing of appeal of BSR action to approve SMPTE 240-1989.* New York: Author.

American Technology Preeminence Act of 1991, Pub. L. 102-245, 106 Stat. 7 (1992).

An approach for development of a world standard for high-definition television compatible with existing systems. (1986). Paris: Ministry of Industry, Division of Electronics Industries and Computer Science.

Anderson, G. H. (1984). *Video editing and post-production: A professional guide.* White Plains, NY: Knowledge Industry Publications.

Andreassen, J. (1994, April). The parade of children. *Diffusion,* pp. 29-33.

Andrews, E. L. (1996, January 11). Dole steps up criticism of telecommunications bill. *The New York Times,* p. C4.

An open message to the nation's broadcasters. (1996, April 15). *Broadcasting & Cable,* p. 53.

Ardito, M. (1994). Studies of the influence of display size and picture brightness on the preferred viewing distance for HDTV programs. *SMPTE Journal, 103,* 517-522.

A rosy hue for color TV in Europe. (1966, October 22). *Business Week,* pp. 94-98.

Ashbacker Radio Corp. v. FCC, 326 U.S. 327 (1945).

Ashworth, S. (1997, July 3). Border stations face unique DTV concerns. *TV Technology,* pp. 1, 14.

ATSC comes up with parameters for HDTV system. (1984, March 19). *Broadcasting,* p. 86.

ATSC: Looking at the better picture. (1985, November 25). *Broadcasting,* pp. 68-69.

ATSC recommends single worldwide HDTV standard. (1985, March 25). *NAB Radio/TV Highlights,* pp. 1-2.

Baron, S. (1997). Report on ITU Task Group 11/3, digital terrestrial television broadcasting (DTTB) final meeting, Sydney, Australia, November 11 to 15, 1996. *SMPTE Journal, 106,* 180-183.

Bayrus, B. L. (1993). High-definition television: Assessing demand forecasts for a next generation consumer durable. *Management Science, 39,* 1319-1333.

Beacham, F. (1996a, August 23). Digital TV airs grand soap opera. *TV Technology,* pp. 1, 8.

Beacham, F. (1996b, October 25). Pressure builds for DTV compromise. *TV Technology,* pp. 1, 27.

Beacham, F. (1997, June 19). CBS demos HDTV, addresses key issues. *TV Technology,* pp. 1, 28.

Behrens, S. (1986, May). The fight for high-def. *Channels,* pp. 42-47.

Beltz, C. A. (1991). *High-tech maneuvers.* Washington, DC: AEI Press.

Beltz, C. A. (1994). Lessons from the cutting edge: The HDTV experience. *Regulation, 16*(4), 29-37.

Besen, S. M., and Farrell, J. (1991). The role of the ITU in standardization: Pre-eminence, impotence or rubber-stamp? *Telecommunications Policy, 15,* 311-321.

Blue ribboners go to work on TV's future. (1987, November 23). *Broadcasting,* pp. 35-37.

Boegels, P. W. (1988). The EUREKA HDTV project—Philosophy and practice. *High definition television* (pp. 3-10). Selected papers presented by EUREKA 95 participants at the International Broadcasting Convention, Brighton, England.

Bouwman, H., Hammersam, M., and Peeters, A. (1991). Is HDTV alleen maar interessant voor voetbal-liefhebbers. *Massacommunicatie, 19,* 50-61.

Bouwman, H., Hammersam, M., and Peeters, A. (1993, May). *The demand for a better television and acceptance of HDTV.* Paper presented at the annual conference of the International Communication Association, Washington, DC.

Braun, M. (1994). *AM stereo and the FCC: Case study of a marketplace shibboleth.* Norwood, NJ: Ablex.

Braun, M. (1995). Research joint ventures and the development of digital HDTV. *Journal of Broadcasting & Electronic Media, 39,* 390-407.

Brinkley, J. (1996a, December 2). Defining TVs and computers for a future of high definition. *The New York Times,* pp. C1, C11.

Brinkley, J. (1996b, December 25). FCC clears new standard for digital TV. *The New York Times,* pp. C1, C15.

Brinkley, J. (1997a). *Defining vision: The battle for the future of television.* New York: Harcourt Brace.

Brinkley, J. (1997b, July 7). PC industry calls for truce in TV wars. *The New York Times,* p. C2.

Broadcasting Technology Association. (1987a, November 16). Comments to the *Notice of Inquiry* (NOI) of the Federal Communications Commission, MM Docket No. 87-268.

Broadcasting Technology Association. (1987b). *A survey of audience reaction at the HDTV fair* (in Japanese). Tokyo: Author.

Brown, A. W. (1987). The campaign for high definition television: A case study in triad power. *Euro-Asia Business Review, 6*(2), 3-11.

Brown, A. W. (1994). Advanced television between the lines. *I & T Magazine*, No. 14, pp. 5-9.

Brown, A., Cave, M., Sharma, Y., Shurmer, M., and Carse, P. (1992). *HDTV: High definition, high stakes, high risk.* London: National Economic Research Associates.

Browne, D. R. (1989). *Comparing broadcast systems: The experiences of six industrialized nations.* Ames, IA: Iowa State University Press.

Burgess, J. (1989, October 26). HDTV forces renew call for federal aid. *The Washington Post*, p. E3.

Burgess, J., and Richards, E. (1989, September 13). Commerce to drop role in HDTV. *The Washington Post*, pp. C1, C4.

Burrows, P. (1994, March 14). Craig Fields's not-so-excellent adventure. *Business Week*, p. 32.

Burton, F. N., and Saelens, F. H. (1987a). Japanese strategies for serving overseas markets: The case of electronics. *Management International Review, 27*(4), 13-18.

Burton, F. N., and Saelens. F. H. (1987b). Trade barriers and Japanese foreign direct investment in the colour television industry. *Managerial and Decision Economics, 8*, 285-293.

Bush, S. (1987). *A survey of audience reaction to NHK 1125 line color television.* Portland, OR: Advanced Television Publishing.

Bylinsky, G. (1991, Spring-Summer). DARPA: A big pot of unrestricted money. *Fortune* (special issue), p. 65.

Campbell, J. C. (1984). Policy conflict and its resolution within the governmental system. In E. S. Krauss, T. P. Rohlen, and P. G. Steinhoff (Eds.), *Conflict in Japan* (pp. 294-334). Honolulu: University of Hawaii Press.

Can the global information infrastructure exist without standards? (1996). *ITU News*, No. 3, pp. 10-14.

Carbonara, C. P. (1990a). HDTV: An historical perspective. Part 2: 1940 to the present. *HDTV World Review, 1*(2), 52-55.

Carbonara, C. P. (1990b). A current history of high-definition television: Part three. *HDTV World Review, 1*(4), 4-9.

Carbonara, C. P. (1992). HDTV: An historical perspective. In L. CasaBianca (Ed.), *The new TV: A comprehensive survey of high definition television* (pp. 3-26). Westport, CT: Meckler.

Carey, J., and Bartimo, J. (1990, July 23). "If you control … computers, you control the world." *Business Week*, p. 31.

Carey, J., and Harbrecht, D. (1990, February 5). Holding the line against an industrial policy. *Business Week*, p. 60.

Carter, A. (1989, April). Japanese HDTV plans in full swing. *TV Technology*, pp. 1, 50.

CasaBianca, L. (Ed.). (1992). *The new TV: A comprehensive survey of high definition television.* Westport, CT: Meckler.

CBS breakthrough on HDTV compatibility. (1983, September 26). *Broadcasting*, p. 77.

CCIR. (1986). The present state of high-definition television (report 801-2). *Recommendations and reports of the CCIR: Vol. XI-I* (pp. 43-55). Geneva: Author.

CCIR. (1990, January 15). *Draft recommendation XA/11 (Mod F): Basic parameter values for the HDTV standard for the studio and for the international programme exchange* (Doc. 11/1007-E).

CCIR puts an end to hope for HDTV standard. (1986, May 19). *Broadcasting*, p. 70.

CCIR sets some HDTV parameters, world production standard fails to emerge. (1990, June 4). *Broadcasting*, pp. 64, 66.

CCIR Study Group. (1989, January 9). *HDTV activities in the United States* (Doc. IWP 11/6-2016).

CCIR Study Groups. (1985, April 9). *Proposal for the draft recommendation for worldwide HDTV studio standard* (Doc. 11A-85/10).

CCIR Study Groups. (1987, October 30). *Draft recommendation: Parameter values for a single world-wide high definition television standard for programme production and for the international exchange of HDTV programmes* (Doc. 11/161-E).

CCIR Study Groups. (1989, June 30). *Conclusions of the Extraordinary Meeting of Study Group 11 on high-definition television* (Doc. 11/410-E).

Choate, P. (1990). *Agents of influence*. New York: Alfred A. Knopf.

Choy, J. (1989, January 13). Developing advanced television: Industrial policy revisited. *Japan Economic Institute Report*, pp. 1-9.

Choy, J. (1994, March 4). Future of Japan's advanced television system debated. *Japan Economic Institute Report*, pp. 5-7.

Clark, B. (1988, April 22). Transcript from the Advanced Television Seminar. Seattle, WA: KCTS-TV.

Clear advantages to high resolution. (1981, February 16). *Broadcasting*, pp. 30, 32.

Clifford, J. (1992). HDTV from a business perspective. In L. CasaBianca (Ed.), *The new TV: A comprehensive survey of high definition television* (pp. 65-69). Westport, CT: Meckler.

Closing in on a new FCC. (1989, June 19). *Broadcasting*, p. 28.

Cloud, D. S. (1989, May 13). Washington policy on R&D proving divisive issue. *Congressional Quarterly Weekly Report*, pp. 1107-1111.

Coalition to exploit land-based digital. (1995, September). *Cable and Satellite Europe*, p. 10.

Coffin, T. E., and Tuchman, S. (1968). The impact of color—A profile of color TV set owners: Television's "class" audience. In H. W. Coleman (Ed.), *Color television: The business of colorcasting* (pp. 121-135). New York: Hastings House.

Cole, A. (1992, June). Zenith/AT&T chips fail. *TV Technology*, p. 8.

Cole, A. (1995, February). HDTV development forges ahead. *TV Technology*, p. 66.

Coleman, H. W. (Ed.). (1968). *Color television: The business of colorcasting*. New York: Hastings House.

Commission of the European Communities. COM(88)299 final, 13.6.1988, *Report on high definition television* (Communication from the Commission).

Commission of the European Communities. (1992a). *Europe and the technologies of information and communication*. Luxembourg: Office for Official Publications of the European Communities.

Commission of the European Communities. (1992b). *A television for tomorrow*. Luxembourg: Office for Official Publications of the European Communities.

Commission of the European Communities. COM(93)557 final, 17.11.1993,

Communication from the Commission to the Council and the European Parliament, a framework for Community policy on digital video broadcasting.

Commission of the European Communities. COM(94)347 final, 19.7.1994, *Europe's way to the information society. An Action Plan* (Communication from the Commission to the Council and the European Parliament and to the Economic and Social Committee and the Committee of Regions).

Commission of the European Communities. COM(94)455 final—94/476 COD, 25.10.1994, *Amended Directive proposal on the use of standards for the transmission of television signals (including repeal of Directive 92/38/EEC).*

Commission of the European Communities. COM(96)346 final, 26.7.1996, *Second annual report on progress in implementing the Action Plan for the introduction of advanced television services in Europe* (Report from the Commission to the Council, the European Parliament and the Economic and Social Committee).

Commission spells out position on HDTV. (1993, May). *XIII Magazine*, No. 10, p. 4.

"Common image" sought as solution to HDTV standard. (1989, February 13). *Broadcasting*, p. 84.

Common Position adopted by the Council on 10 February 1992 with view to adopting a Directive on the adoption of standards for satellite broadcasting of television signals, No. 4160/92.

Communications Act of 1934, 47 U.S.C. 303.

Communications Studies and Planning International Inc. (1984). *Prospects for high definition television*. New York: Author.

Compromise letter to Commissioner Susan Ness. (1996, November 27). In Federal Communications Commission Docket MM 87-268, Advanced Television Systems and Their Impact Upon the Existing Television Broadcast Service. Washington, DC: Federal Communications Commission.

Computer Industry Coalition on Advanced Television Service. (1996, July 11). Comments of the Computer Industry Coalition on Advanced Television Service (Volume 1 of 2: Comments), In the Matter of Advanced Television Systems and Their Impact Upon the Existing Television Broadcast Service, MM Docket No. 87-268.

Congress has a strange idea. (1996, March 12). *San Francisco Chronicle*, p. C2.

Congressional Budget Office. (1989). *The scope of the high-definition television market and its implications for competitiveness.* Washington, DC: Author.

Cook, W. J. (1990, September 10). Spoils of a good air war. *U.S. News and World Report*, pp. 75-80.

Cookson, C. (1995). Intorduction of wide screen to television series production. *Symposium record of the 19th international television symposium and technical exhibition* (pp. 369-373). Montreux, Switzerland: The Symposium.

Coopers & Lybrand. (1996). *Review of Action Plan for advanced television services.* London: Author.

Coppola, F. F. (1992). HDTV and electronic filmmaking. In L. CasaBianca (Ed.), *The new TV: A comprehensive survey of high definition television* (pp. 93-97). Westport, CT: Meckler.

Cornwell, N. C., Thompson, W. B., Trager, R., and Hatfield, D. N. (1993, August). *"Colorizing" HDTV: Is consumer adoption of color television an appropriate comparison for acceptance of high-definition television?* Paper presented at the annual convention of the Association for Education in Journalism and Mass Communication, Kansas City,

MO.

Costello, M. (1995, June). The big picture for consumer television. *Broadcast Engineering,* pp. 26, 28.

Crane, R. J. (1979). *The politics of international standards: France and the color TV war.* Norwood, NJ: Ablex.

Curtis, P. J. (1994). *The fall of the U.S. consumer electronics industry.* Westport, CT: Quorum Books.

Dahl, R. A. (1961). *Who governs? Democracy and power in an American city.* New Haven, CT: Yale University Press.

Darby, L. F. (1988). *Economic potential of advanced television products.* Washington, DC: National Telecommunications and Information Administration.

DARPA director move draws Hill criticism. (1990, April 30). *Broadcasting,* p. 70.

Davis, B. (1990). Fading picture: High-definition TV, once a capital idea, wanes in Washington. In J. F. Rice (Ed.), *HDTV: The politics, policies, and economics of tomorrow's television* (pp. 225-232). New York: Union Square Press.

Defense Department wants in the HDTV picture. (1988, December 26). *Broadcasting,* pp. 32-33.

de Jong, A., and Bouwman, H. (1994). *Consumenten en nieuwe media: 1990-1993.* Amsterdam: Het Persinstituut.

Delesalle, B. (1994, April). The Winter Olympic Games: Sport as a team operation. *Diffusion,* p. 34.

Dertouzos, M. L. (1991, January). Building the information marketplace. *Technology Review,* pp. 28-40.

Detenber, B. H., and Reeves, B. (1996). A bio-informational theory of emotion: Motion and image size effects on viewers. *Journal of Communication, 46*(3), 66-84.

Dickson, G. (1995a, September 25). Sony/Philips, Toshiba agree on DVD format. *Broadcasting & Cable,* p. 56.

Dickson, G. (1995b, October 2). HD Vision takes the long view. *Broadcasting & Cable,* pp. 52-53.

Dickson, G. (1995c, November 6). TCI brings HDTV to the public. *Broadcasting & Cable,* p. 111.

Dickson, G. (1996, April 1). WETA-TV to build HDTV station. *Broadcasting & Cable,* pp. 60-61.

Digital era could mean 18 channels. (1995, December 16). *The Times,* p. 4.

Digital Video Broadcasting. (1995). *Going ahead with digital television.* Grand-Saconnex, Switzerland: Author.

Digital Video Broadcasting. (1996). *The new age of television: DVB.* Grand-Saconnex, Switzerland: Author.

"Digitally assisted TV." (1986, May 21). *Communications Daily,* p. 4.

Does Japan have the will to lead? (1989, January). *HDTV Newsletter,* pp. 11-15.

Do not adjust your set. (1993, February 27). *The Economist,* pp. 65-66.

Donow, K. R. (1988). *HDTV: Planning for action.* Washington, DC: National Association of Broadcasters.

Down to the wire. (1991, July). *Cable and Satellite Europe,* p. 16.

Doyle, S. (1993, June 18). Wiley urges HDTV peace. *Broadcast,* p. 20.

Drake, W. J. (1994). The transformation of international telecommunications standardization: European and global dimensions. In C. Steinfield, J. M. Bauer, and L. Caby

(Eds.), *Telecommunications in transition: Policies, services and technologies in the European Community* (pp. 71-96). Thousand Oaks, CA: Sage.

Drumare, X. (1995). France television and the 16:9 from 1991 to 1995. *Symposium record of the 19th international television symposium and technical exhibition* (pp. 335-350). Montreux, Switzerland: The Symposium.

D2-MAC: The transmission standard for consumer satellite television. (1988). Eindhoven, Netherlands: Euromac, D2-MAC Transmission and Conditional Access System Consortium.

Dupagne, M. (1990). High-definition television: A policy framework to revive U.S. leadership in consumer electronics. *The Information Society, 7,* 53-76.

Dupagne, M. (1992). EC policymaking: The case of the "Television Without Frontiers" directive. *Gazette, 49,* 99-120.

Dupagne, M. (1997, July). "A profile of potential high-definition television adopters in the United States." Paper presented at the annual convention of the Association for Education in Journalism and Mass Communication, Chicago, IL.

Dupagne, M., and Agostino, D. E. (1991). High-definition television: A survey of potential adopters in Belgium. *Telematics and Informatics, 8,* 9-30.

Dutton, W. H. (1992). The ecology of games shaping telecommunications policy. *Communication Theory, 2,* 303-328.

Easton, D. (1965). *A framework for political analysis.* Englewood Cliffs, NJ: Prentice-Hall.

Easton, D. (1968). A systems analysis of political life. In W. Buckley (Ed.), *Modern systems research for the behavioral scientist* (pp. 428-436). Chicago: Aldine.

Ebel, A. J. (1995, September 18). SDTV no answer. *Broadcasting & Cable,* p. 76.

EC drafts new MAC directive. (1991, March). *Cable and Satellite Europe,* p. 6.

Editorial. (1995, June). *DVB News,* p. 2.

EIA opposes TV tax. (1989, May 19). *Communications Daily,* p. 1.

Engineers search for HDTV standards. (1987, September 28). *Broadcasting,* pp. 75-76.

EUREKA EU95. (1988). *The road to high definition television.* Eindhoven, Netherlands: Author.

European Audiovisual Observatory. (1994). *Statistical yearbook.* Strasbourg: Council of Europe.

European counterattack on HDTV. (1988, February 1). *Broadcasting,* p. 67.

Faltermayer, E. (1991, Spring-Summer). The thaw in Washington. *Fortune* (special issue), pp. 46-51.

Fannon, P. M. (1994). HDTV update: The US Grand Alliance sets a standard for world harmonisation. *Intermedia, 22*(2), 37-38.

Farnsworth, C. H. (1989, January 25). Commerce choice vows to work with the industry. *The New York Times,* p. A17.

FCC gains a deregulatory diplomat. (1989, August 14). *Broadcasting,* pp. 29-30.

FCC to take simulcast route. (1990, March 26). *Broadcasting,* pp. 38-40.

Feder, B. J. (1995, July 18). Last U.S. TV maker will sell control to Koreans. *The New York Times,* pp. A1, D2.

Federal Communications Commission. (1970). 41 F.C.C. (Television Matters, September 1, 1950 to June 30, 1965).

Federal Communications Commission. (1987a). Advanced Television Systems and Their Impact on the Existing Television Broadcast Service (*Notice of Inquiry*), 2 FCC Rcd. 5125.

Federal Communications Commission. (1987b). Freeze on Applications to Amend TV Table of Allotments, 52 Fed. Reg. 28346.

Federal Communications Commission. (1987c). Formation of Advisory Committee on Advanced Television Service and Announcement of First Meeting, 52 Fed. Reg. 38523.

Federal Communications Commission. (1988). Advanced Television Systems and Their Impact Upon the Existing Television Broadcast Service (*Tentative Decision and Further Notice of Inquiry*), 3 FCC Rcd. 6520.

Federal Communications Commission. (1990). Advanced Television Systems and Their Impact on the Existing Television Broadcast Service (*First Report and Order*), 5 FCC Rcd. 5627.

Federal Communications Commission. (1991). Advanced Television Systems and Their Impact Upon the Existing Broadcast Service (*Notice of Proposed Rule Making*), 6 FCC Rcd. 7024.

Federal Communications Commission. (1992a). Advanced Television Systems and Their Impact on the Existing Television Broadcast Service (*Second Report and Order/Further Notice of Proposed Rule Making*), 7 FCC Rcd. 3340.

Federal Communications Commission. (1992b). Advanced Television Systems and Their Impact Upon the Existing Television Broadcast Service (*Second Further Notice of Proposed Rule Making*), 7 FCC Rcd. 5376.

Federal Communications Commission. (1992c). Advanced Television Systems and Their Impact Upon the Existing Television Broadcast Service (*Memorandum Opinion and Order/Third Report and Order/Third Further Notice of Proposed Rule Making*), 7 FCC Rcd. 6924.

Federal Communications Commission. (1995). Advanced Television Systems and Their Impact Upon the Existing Television Broadcast Service (*Fourth Further Notice of Proposed Rule Making and Third Notice of Inquiry*), 10 FCC Rcd. 10540.

Federal Communications Commission. (1996a). Advanced Television Systems and Their Impact Upon the Existing Television Broadcast Service (*Fifth Further Notice of Proposed Rule Making*), 11 FCC Rcd. 6235.

Federal Communications Commission. (1996b). Advanced Television Systems and Their Impact Upon the Existing Television Broadcast Service (*Sixth Further Notice of Proposed Rule Making*), 11 FCC Rcd. 10968.

Federal Communications Commission. (1996c). Advanced Television Systems and Their Impact Upon the Existing Television Broadcast Service (*Fourth Report and Order*), 11 FCC Rcd. 17771.

Federal Communications Commission. (1997a). Advanced Television Systems and Their Impact Upon the Existing Television Broadcast Service (*Fifth Report and Order*), MM Docket No. 87-268.

Federal Communications Commission. (1997b). Advanced Television Systems and Their Impact Upon the Existing Television Broadcast Service (*Sixth Report and Order*), MM Docket No. 87-268.

Fink, D. G. (1980a). The future of high-definition television: First portion of a report of the SMPTE Study Group on High-Definition Television. *SMPTE Journal, 89*, 89-94.

Fink, D. G. (1980b). The future of high-definition television: Conclusion of a report of the SMPTE Study Group on High-Definition Television. *SMPTE Journal, 89*, 153-161.

Five U.S. companies picked to receive Defense funds for HDTV displays. (1989, June 19).

Broadcasting, p. 42.

Flint J. (1992, April 13). HDTV: Too close for comfort? *Broadcasting*, pp. 4, 14.

Flynn, B. (1992a, October). Canal+ bets on digital TV. *Advanced Television Markets*, p. 2.

Flynn, B. (1992b, November). KPMG puts MAC case. *Advanced Television Markets*, p. 2.

Flynn, B. (1993a, July-August). HD-MAC: Montreux swansong. *Advanced Television Markets*, pp. 1-2.

Flynn, B. (1993b, September). Digital MoU sealed. *Advanced Television Markets*, pp. 1-2.

Foisie, G. (1994, March 28). Road to digital HDTV takes detour. *Broadcasting & Cable*, p. 40.

Fox, R. (1995, April). The digital dawn in Europe. *IEEE Spectrum*, pp. 50-53.

Freeman, J. P. (1984). The evolution of high-definition television. *SMPTE Journal, 93*, 492-501.

Fujio, T. (1985). High-definition television systems. *Proceedings of the IEEE, 73*, 646-655.

Fujio, T., et al. (1982). High-definition television. *NHK Technical Monograph*, No. 32.

Fuller, C. (1993, June 4). European giants back HDTV move. *Broadcast*, p. 10.

Garvey, D. E. (1980). Introducing color television: The audience and programming problem. *Journal of Broadcasting, 24*, 515-525.

Gatski, J. (1990, January). HDTV legislation loses momentum. *TV Technology*, pp. 1, 10.

Giant screens saved 1995 color TV sales. (1996, February 5). *Television Digest with Consumer Electronics*, pp. 11-12.

Gilder, G. (1989). *Microcosm: The quantum revolution in economics and technology.* New York: Simon and Schuster.

Gilder, G. (1994). *Life after television.* (Rev. ed.) New York: Norton.

Glenn, W. E. (1988). *High definition television.* New York: Society for Information Display.

Goodfriend, A. (1988). Satellite broadcasting in the UK. In R. Negrine (Ed.), *Satellite broadcasting: The politics and implications of the new media* (pp. 144-175). London: Routledge.

Gordon, C., and REBO Studio. (1996). *The guide to high definition video production.* Boston, MA: Focal Press.

Government support for HDTV. (1987, November 2). *Broadcasting*, p. 53.

Grindley, P., Mowery, D. C., and Silverman, B. (1994). SEMATECH and collaborative research: Lessons in the design of high-technology consortia. *Journal of Policy Analysis and Management, 13*, 723-758.

Grout, F. (1988). Actions et promotion du MPT et du MITI. *La télévision haute définition au Japon* (pp. 20-22). Tokyo: Ambassade de France au Japon.

Grout, F. (1989). *Budgets publics japonais relatifs à la télévision à haute définition et aux satellites de télévision directe.* Tokyo: Ambassade de France au Japon.

Guistiniani, A. C. (1994, April). The winter games: Sport as a team operation. *Diffusion*, p. 37.

Habermann, W., and Wood, D. (1986). Images of the future: The EBU's part to date in HDTV system standardization. *EBU Review: Technical*, No. 219, 267-280.

Hamamatsu City in Shizuoka prefecture designated as "Hi-Vision City." (1994, September 5). *MPT News*, p. 3.

Hansen, G. (1966). Colour television in Europe. *EBU Review: Technical*, No. 98, 138-141.

Harris Corporation. (1996). *Digital TV survey findings.* Melbourne, FL: Author.

Harris Corporation. (1997). *Consumer DTV screening survey.* Melbourne, FL: Author.

Harvey, S. (1993). *Audience reaction to letter box pictures on TV.* London: British

Broadcasting Corporation, Broadcasting Research Department.

Has Europe solved its digital problems? (1994, October 12). *Broadcasting & Cable's TV International*, p. 8.

HDTV: All together now. (1993, May 29). *The Economist*, p. 74.

HDTV alternatives. (1982, November 15). *Broadcasting*, p. 41.

HDTV competitor goes through hoop. (1992, September 4). *Broadcasting*, p. 17.

HDTV Directorate. (1994). *EUREKA 95 HDTV project: Final report*. Eindhoven, Netherlands: Philips Consumer Electronics B.V.

HDTV: Efforts to redefine TV on display in Washington. (1987, January 12). *Broadcasting*, pp. 134-135.

HDTV field tests. (1995, July 31). *Cable World*, p. 35.

HDTV production standard debated at NTIA. (1989, March 13). *Broadcasting*, p. 67.

HDTV: Progress report on television's next quantum leap. (1987, October 26). *Broadcasting*, p. 74.

HDTV transmission tests set to begin next April. (1990, November 19). *Broadcasting*, pp. 52-53.

Healy, M. (1993, November). Operators speak on digital TV. *Advanced Television Markets*, p. 12.

Herbstreit, J. W., and Pouliquen, H. (1967). International standards for colour television. *Telecommunication Journal, 34*, 16-23.

High definition dominates Montreux. (1989, June 26). *Broadcasting*, pp. 47-51.

High definition TV: So close and yet so far away. (1986, April 7). *Broadcasting*, pp. 134-138.

Hill, A. (1993a, February 19). Europe will follow US lead over high-definition TV. *Financial Times*, p. 16.

Hill, A. (1993b, February 26). French bid to win UK TV accord. *Financial Times*, p. 3.

Hill, A. (1993c, March 12). Call for action on digital HDTV. *Financial Times*, p. 3.

Hills, J., and Papathanassopoulos, S. (1991). *The democracy gap*. New York: Greenwood.

Hi-Vision Promotion Council. (1989). *HDTV audience survey report*. Tokyo: Author.

Home Box Office. (1988). *Consumer response to high definition television*. New York: Author.

Homer, S. (1993, April). Europe's HDTV trek takes a sharp turn toward digital. *Broadcasting Abroad*, pp. 15-16.

Homer, S. (1996, February). Digital TV storms Europe. *Broadcasting & Cable International*, pp. 28-29.

Hopkins, T. K., and Wallerstein, I. M. (1982). *World-systems analysis: Theory and methodology*. Beverly Hills, CA: Sage.

Hora, J. (1995, April). Past is prologue. *American Cinematographer*, pp. 119-120.

H.R. 1267, 101st Cong., 1st Sess. (1989).

H.R. 1516, 101st Cong., 1st Sess. (1989).

H.R. 2287, 101st Cong., 1st Sess. (1989).

H.R. 4764, 101st Cong., 2d Sess. (1990).

H.R. 4933, 101st Cong., 2d Sess. (1990).

Hughes, D. (1986, June). HDTV draws much attention. *TV Technology*, p. 9.

Hughes, R. (1988). Satellite broadcasting: The regulatory issues in Europe. In R. Negrine (Ed.), *Satellite broadcasting: The politics and implications of the new media* (pp. 49-74). London: Routledge.

Humbert, M. (1984). Le typhon electronique japonais. *Chroniques d'Actualités de la*

SEDEIS, 30, 150-162.

Humphreys, P. (1988). Satellite broadcasting policy in West Germany: Political conflict and regional competition in a decentralised system. In R. Negrine (Ed.), *Satellite broadcasting: The politics and implications of the new media* (pp. 107-143). London: Routledge.

Improving the look of television through HDTV. (1984, April 30). *Broadcasting,* pp. 142, 144.

In re Japanese Electronic Products, 723 F.2d 238 (3rd Cir. 1983).

In re Japanese Electronic Products Antitrust Lit., 807 F.2d 44 (3rd Cir. 1986).

International Telecommunication Union. (1994). *The International Telecommunication Union.* Geneva: Author.

Ito, M. (1978). *Broadcasting in Japan.* London: Routledge and Kegan Paul.

Jackson, T. (1993, September 10). How to stand out in a crowd. *Financial Times,* p. 19.

Jacobi, F. (1995). High definition television revisited. *Television Quarterly,* 27(4), 17-23.

Japanese government high on HDTV. (1987, September 7). *Japanese Industry Newsletter,* pp. 12-13.

Japanese legislation of telecommunications: Vol. 5. Broadcast law of Japan. (1991). Tokyo: Communications Study Group.

Japanese official suggests replacing HDTV system. (1994, February 23). *The Wall Street Journal,* p. B11.

Jessell, H. A. (1992, September 7). HDTV option: Multiple and be fruitful. *Broadcasting,* p. 28.

Johnson, C. (1982). *MITI and the Japanese miracle: The growth of industrial policy, 1925-1975.* Stanford, CA: Stanford University Press.

Johnson, C. (1984). Introduction: The idea of industrial policy. In C. Johnson (Ed.), *The industrial policy debate* (pp. 3-26). San Francisco, CA: ICS Press.

Johnson, D. (1996, September). Widescreen expands slowly. *Broadcasting & Cable International,* pp. 20-24.

Johnson, D., and Davies, S. T. (1996, November 25). Japan prepares for DTH crush. *Broadcasting & Cable,* pp. 28-29.

Johnstone, B. (1993, March 25). Keeping an eye out. *Far Eastern Economic Review,* pp. 59-62.

Joosten, M. (1992). *High and digital? Convergences and collisions in old and new HDTV.* London: Westminster University, Centre for Communication and Information Studies.

Jurgen, R. K. (1988, April). High-definition television update. *IEEE Spectrum,* pp. 56-62.

Jurgen, R. K. (1989a, January). Consumer electronics. *IEEE Spectrum,* pp. 59-60.

Jurgen, R. K. (1989b, October). Chasing Japan in the HDTV race. *IEEE Spectrum,* pp. 26-30.

Juster, F. T. (1966). Consumer buying intentions and purchase probability: An experiment in survey design. *Journal of the American Statistical Association, 61,* 658-696.

Kageki, N. (1994, February 28). HDTV shift stuns industry. *The Nikkei Weekly,* pp. 1, 27.

Kaitatzi-Whitlock, S. (1994). European HDTV strategy: Muddling through or muddling up? *European Journal of Communication, 9,* 173-192.

Kaitatzi-Whitlock, S. (1997). The privatizing of conditional access control in the European Union. *Communications & Strategies,* no. 25, 91-122.

Karr, A. R. (1995, December 13). Computer firms contest standard for digital TV. *The Wall*

Street Journal, p. B7.

Kehoe, L. (1993, March 4). A blurred vision of the future. *Financial Times,* p. 16.

Kim, H. (1992). Theorizing deregulation: An exploration of the utility of the "Broadcast Policy-Making System" model. *Journal of Broadcasting & Electronic Media, 36,* 154-172.

Kim, M. S., and Hunter, J. E. (1993). Attitude-behavior relations: A meta-analysis of attitudinal relevance and topic. *Journal of Communication, 43*(1), 101-142.

Klopfenstein, B. (1985). Forecasting the market for home video players: A retrospective analysis (Doctoral dissertation, The Ohio State University, 1985). *Dissertation Abstracts International, 46,* 546A.

Klopfenstein, B. (1989). Problems and potential of forecasting the adoption of new media. In J. L. Salvaggio and J. Bryant (Eds.), *Media use in the information age: Emerging patterns of adoption and consumer use* (pp. 21-41). Hillsdale, NJ: Lawrence Erlbaum.

Kodaira, S. I. (1986). Japan. In R. Paterson (Ed.), *International TV and video guide* (pp. 125-128). London: Tantivy Press.

Krasnow, E. G., Longley, L. D., and Terry H. A. (1982). *The politics of broadcast regulation* (3rd ed.). New York: St. Martin's Press.

Krivocheev, M. I. (1991). Current CCIR activities in HDTV. *Telecommunication Journal, 58,* 699-709.

Krivocheev, M. I. (1993). The first twenty years of HDTV: 1972-1992. *SMPTE Journal, 102,* 913-930.

Kuhn, R. (1988). Satellite broadcasting in France. In R. Negrine (Ed.), *Satellite broadcasting: The politics and implications of the new media* (pp. 176-195). London: Routledge.

Kumabe, N. (1988, April 22). Transcript from the Advanced Television Seminar. Seattle, WA: KCTS-TV.

Kumada, J. (1997, June). *The introduction of digital HDTV in Japan.* Paper presented at the "HDTV '97" workshop, Montreux, Switzerland.

Kupfer, A. (1991, April 8). The U.S. wins one in high-tech TV. *Fortune,* pp. 60-64.

LaFrance, V. A. (1985). The United States television receiver industry: United States versus Japan, 1960-1980 (Doctoral dissertation, The Pennsylvania State University, 1985). *Dissertation Abstracts International, 46,* 2763A.

Lalor, E. (1990). HDTV and the European Economic Community. *HDTV World Review, 1*(2), 24-28.

Lambert, P. (1992a, March 10). By HDTV's early light. *Broadcasting,* pp. 10-11.

Lambert, P. (1992b, May 11). HDTV test analysis off to sticky start. *Broadcasting,* p. 32.

Lambert, P. (1992c, May 25). NBC, Thomson confident despite HDTV snags. *Broadcasting,* p. 9.

Lambert, P. (1992d, October 5). FCC and broadcasters battle toward flexible HDTV conversion. *Broadcasting,* pp. 4, 14.

Landler, M. (1996a, November 26). Industries agree on U.S. standards for TV of future. *The New York Times,* pp. A1, C6.

Landler, M. (1996b, November 28). Film makers fail to win concession in digital TV standard. *The New York Times,* p. C2.

Laroche, C. (1992, February). Astra 1D: Is SES banking on digital HDTV? *Advanced Television Markets,* p. 3.

Lartigue, J. P., and Oudin, M. (1996). *The emergence of 16/9 and digital markets.* Brussels: Vision 1250.

Lazarus, W., and McKnight, L. (1984). *Forecasting the new media*. Cambridge, MA: Massachusetts Institute of Technology, The Media Laboratory.

Leclercq, T. (1994, May). MEPs oppose Simulcrypt. *Advanced Television Markets*, p. 4.

Leclercq, T., and Flynn, B. (1994, June). CA stalls digital directive. *Advanced Television Markets*, p. 7.

Le Duc, D. R. (1973). *Cable television and the FCC: A crisis in media control*. Philadelphia: Temple University Press.

Lessing, L. (1949, November). The television freeze. *Fortune*, pp. 124-125.

Lewis, P. H. (1996, February 14). On-line service ending its ban on sexual materials on the Internet. *The New York Times*, pp. A1, C2.

L'industrie electronique japonaise. (1985, November 27). *Problèmes Economiques*, pp. 28-31.

Lombard, M. (1995). Direct responses to people on the screen: Television and personal space. *Communication Research, 22,* 288-324.

Lombard, M., Grabe, M. E., Reich, R. D., Campanella, C. M., and Ditton, T. B. (1996, August). *Screen size and viewer responses to television: A review of research*. Paper presented at the annual convention of the Association for Education in Journalism and Mass Communication, Anaheim, CA.

Lombard, M., Ditton, T. B., Grabe, M. E., and Reich, R. D. (1997). The role of screen size in viewer responses to television fare. *Communication Reports, 10,* 95-106.

Long, T., and Stenger, L. (1986). The broadcasting of HDTV programmes. *EBU Review: Technical*, No. 219, 297-314.

Lukes, S. (1974). *Power: A radical view*. London: Macmillan.

Lund, A. M. (1993). The influence of video image size and resolution on viewing-distance preferences. *SMPTE Journal, 102,* 406-415.

Lupker, S. J., Allen, N. J., and Hearty, P. J. (1988). *The North American high definition television demonstrations to the public: The detailed survey results*. Montreal, Canada: Committee for the North American High Definition Television Demonstrations to the Public.

Lyman, P. (1985). "HDTV: Who pays for the dream." *Second International Colloquium on New Television Systems: HDTV '85*. Montreal, Canada: Canadian Broadcasting Corporation.

Machut, J. L. (1994, April). The winter games: Sport as a team operation. *Diffusion*, p. 36.

Macpherson, A. (1990). *International telecommunication standards organizations*. Boston, MA: Artech.

Mahan, E., and Schement, J. R. (1984). The broadcast regulatory process: Toward a new analytical framework. In B. Dervin and M. J. Voight (Eds.), *Progress in communication sciences: Vol. IV* (pp. 1-21). Norwood, NJ: Ablex.

Markoff, J. (1989, November 16). Cuts are expected for U.S. financing in high-tech area. *The New York Times*, pp. A1, D9.

Markoff, J. (1990, April 6). High-detail TV faces fund cuts. *The New York Times*, pp. D1, D13.

Markoff, J. (1995, December 29). On-line service blocks access to topics called pornographic. *The New York Times*, pp. A1, C4.

Markoff, J. (1996, March 11). New Xerox company for flat-panel displays. *The New York Times*, p. C8.

Mathias, H., and Patterson, R. (1985). *Electronic cinematography*. Belmont, CA:

Wadsworth.

Matsushita, M. (1988). Coordinating international trade with competition policies. In E. U. Petersmann and M. Hilf (Eds.), *The new GATT round of multilateral trade negotiations* (pp. 395-432). Deventer, Netherlands: Kluwer.

Matsushita Elec. Industrial Co. v. Zenith Radio, 475 U.S. 574 (1986).

McAvoy, K. (1993, December 13). Reed Hundt stand-up chairman. *Broadcasting & Cable*, p. 9.

McAvoy, K., and West, D. (1995, March 20). Newt Gingrich: The great liberator for cybercom. *Broadcasting & Cable*, p. 6.

McClellan, C. (1996, September 30). Network heads want digital standard. *Broadcasting & Cable*, pp. 30, 31.

McConnell, C. (1994, October 31). Don't sacrifice HDTV for standard-definition. *Broadcasting & Cable*, p. 53.

McConnell, C. (1995a, January 16). More, not less, time needed for HDTV switch. *Broadcasting & Cable*, p. 103.

McConnell, C. (1995b, April 10). Turning data streams into revenue streams. *Broadcasting & Cable*, p. 32.

McConnell, C. (1995c, December 4). FCC gets advanced-TV recommendation. *Broadcasting & Cable*, p. 26.

McConnell, C. (1996a, April 22). Broadcasters ready for digital switch. *Broadcasting & Cable*, pp. 10, 14.

McConnell, C. (1996b, May 13). FCC plans mandatory ATV standard. *Broadcasting & Cable*, p. 16.

McConnell, C. (1996c, July 15). Cable joins high-tech coalition against digital TV standard. *Broadcasting & Cable*, p. 9.

McConnell, C. (1996d, July 29). UHF spectrum: Telcom's hot new property. *Broadcasting & Cable*, pp. 20, 22.

McConnell, C. (1996e, October 21). Broadcasters arm for ATV fight. *Broadcasting & Cable*, pp. 6-7, 12.

McConnell, C. (1996f, December 2). FCC wants to move quickly on DTV. *Broadcasting & Cable*, p. 6.

McConnell, C. (1996g, December 2). Broadcasters seek to keep chs. 60-69. *Broadcasting & Cable*, p. 20.

McConnell, C. (1996h, December 2). Have Grand Alliance standard, will travel. *Broadcasting & Cable*, p. 7.

McConnell, C. (1997, March 24). Broadcasters sweeten DTV pot. *Broadcasting & Cable*, pp. 6, 10.

McConnell, C., and West, D. (1995, December 4). Dick Wiley: Delivering on digital. *Broadcasting & Cable*, pp. 32-40.

McConville, J. (1996, February 5). EchoStar jumps into the fray. *Broadcasting & Cable*, p. 54.

McCormick, N. (1997, June 30). UK awards digital frequencies. *Broadcasting & Cable*, p. 52.

McKnight, L., and Neil, S. (1987). The HDTV war: The politics of HDTV standardization. *Third International Colloquium on Advanced Television Systems: HDTV '87* (pp. 5.6.1.-5.6.17). Montreal, Canada: Canadian Broadcasting Corporation.

McKnight, L., Neuman, W. R., Reynolds, M., O'Donnell, S., and Schneider, S. (1988). *The shape of things to come: A study of subjective responses to aspect ratio and screen size.* Cambridge, MA: Massachusetts Institute of Technology, The Media Laboratory.

McMann, R. H., Jr. (1982, April). *High definition television—An overview.* Paper presented at the annual convention of the National Association of Broadcasters, Las Vegas, NV.

Mentley, D. E., and Castellano, J. A. (1990). Forecast of the HDTV market. *HDTV World Review, 1*(4), 10-15.

Millstein, J. E. (1983). Decline in an expanding industry: Japanese competition in color television. In J. Zysman and L. Tyson (Eds.), *American industry in international competition* (pp. 106-141). Ithaca, NY: Cornell University Press.

Ministry of Posts and Telecommunications. (1989). *Advanced television (ATV): The promise and the challenge.* Tokyo: Author.

Ministry of Posts and Telecommunications. (1992). *Major policies of the Japanese broadcasting administration.* Tokyo: Author.

Ministry of Posts and Telecommunications. (1996). *Standardization of telecommunications in Japan.* Tokyo: Author.

MITI and trade issues—Looking closer. (1989, February-March). *HDTV Newsletter*, pp. 11-15.

Mitsusada, H. (1995, July 3). TV makers aim to fatten profits with flat-panels. *The Nikkei Weekly*, p. 8.

Moore, B. C. (1973). The FCC: Competition and communication. In M. J. Green (Ed.), *The monopoly makers* (pp. 35-73). New York: Grossman.

Morgan, G. (1992, September). A compressed market? *Advanced Television Markets*, pp. 10-11.

Morgan, G. (1993). Picking up the bits. *ATM Montreux*, p. 36.

Morgan, G. (1995, June). EU plans its digital rules. *Advanced Television Markets*, pp. 1-2.

Morita, A., Reingold, E. M., and Shimomura, M. (1986). *Made in Japan.* New York: E. P. Dutton.

Mosco, V. (1979). *Broadcasting in the United States.* Norwood, NJ: Ablex.

MPT speeds up digital terrestrial to year 2000. (1997, April). *Advanced Television Markets*, p. 1.

Muramatsu, M. (1991). The "enhancement" of the Ministry of Posts and Telecommunications to meet the challenge of telecommunications innovation. In S. Wilks and M. Wright (Eds.), *The promotion and regulation of industry in Japan* (pp. 286-308). New York: St. Martin's Press.

Muscarà, P., and Causin, V. (1995, November). Telepiù Mólto Più. *Cable and Satellite Europe*, pp. 40-41.

Nakae, K. (1994). NHK's Hi-Vision operation in the Lillehammer Olympic games. *Perspectives on wide screen and HDTV production* (pp. 29-34). Washington, DC: National Association of Broadcasters.

Nakamura, Y. (1988, April 22). *HDTV: Past, present and future.* Transcript from the Advanced Television Seminar. Seattle, WA: KCTS-TV.

Nakatani, H. (1988, April 22). *NHK at work for the successful development of DBS and Hi-Vision.* Transcript from the Advanced Television Seminar. Seattle, WA: KCTS-TV.

Nash, N. (1996, January 15). Holding Compuserve responsible. *The New York Times*, p. D4.

National Association of Broadcasters. (1989). *NAB guide to advanced television systems.* Washington, DC: Author.

National Broadcasting Company. (1964). *The national color television audience.* New York: Author.

National Telecommunications and Information Administration. (1989). *Advanced television, related technologies, and the national interest.* Washington, DC: U.S. Department of Commerce.

NBC unveils new HDTV standard. (1988, October 17). *Broadcasting,* p. 31.

Negrine. R. (1988). Satellite broadcasting: An overview of the major issues. In R. Negrine (Ed.), *Satellite broadcasting: The politics and implications of the new media* (pp. 1-21). London: Routledge.

Negroponte, N. (1995). *Being digital.* New York: Alfred A. Knopf.

Nethold/Canal+ deal keeps Hughes out. (1996, October). *Cable and Satellite Europe,* pp. 6-7.

Neuman, W. R. (1988, April). *The mass audience looks at HDTV: An early experiment.* Paper presented at the annual convention of the National Association of Broadcasters, Las Vegas, NV.

Neuman, W. R. (1990). *Beyond HDTV: Exploring subjective responses to very high definition television.* Cambridge, MA: Massachusetts Institute of Technology, The Media Laboratory.

Neuman, W. R., Crigler, A., Schneider, S. M., O'Donnell, S., and Reynolds, M. (1987). *A study of television sound.* Cambridge, MA: Massachusetts Institute of Technology, The Media Laboratory.

Neuman, W. R., and O'Donnell, S. (1992). *Where do we stand?* Cambridge, MA: Massachusetts Institute of Technology, The Media Laboratory.

New HDTV estimates: $12 million or less. (1990, October 29). *Broadcasting,* p. 33.

NHK Science and Technical Research Laboratories. (1993). *High definition television: Hi-Vision technology.* New York: Van Nostrand Reinhold.

Niblock, M. (1991). *The future for HDTV in Europe.* Manchester, England: European Institute for the Media.

Nickelson, R. L. (1990). HDTV standards—Understanding the issues. *Telecommunication Journal, 57,* 302-312.

Ninomiya, Y. (1995, April). The Japanese scene. *IEEE Spectrum,* pp. 54-57.

Ninomiya, Y., Ohtsuka, Y., Izumi, Y., Gohshi, S., and Iwadate, Y. (1987). An HDTV broadcasting system utilizing a bandwidth compression technique—MUSE. *IEEE Transactions on Broadcasting,* BC-33, 130-160.

Nippon Hoso Kyokai. (1987, November 18). Comments of NHK—the Japan Broadcasting Corporation—submitted to the Federal Communications Commission, In the Matter of Advanced Television Systems and Their Impact on the Existing Television Broadcast Service, MM Docket No. 87-268.

Nippon Hoso Kyokai. (1988, January 19). Reply comments of NHK—the Japan Broadcasting Corporation, before the Federal Communications Commission, MM Docket No. 87-268.

Nippon Hoso Kyokai. (1989). *Outline of NHK.* Tokyo: Author.

Nippon Hoso Kyokai. (1995). *NHK factsheet '95.* Tokyo: Author.

Nippon Hoso Kyokai. (1996). *NHK factsheet '96.* Tokyo: Author.

Noble, G. W. (1992, September). *The politics of HDTV in Japan.* Paper presented at the annual meeting of the American Political Science Association, Chicago, IL.

Nugent, N. (1994). *The government and politics of the European Union* (3rd ed.). London:

Macmillan.

Ogan, C. (1992). Communications policy options in an era of rapid technological change. *Telecommunications Policy, 16,* 565-567.

Ohanian, T. A., and Phillips, M. E. (1996). *Digital filmmaking.* Boston: Focal Press.

O.J. Eur. Comm. No. L 311/28, 6.11.1986, *Council Directive of 3 November 1986 on the adoption of common technical specifications of the MAC/packet family of standards for direct satellite television broadcasting.*

O.J. Eur. Comm. No. L 142/1, 25.5.1989, *Council Decision of 27 April 1989 on high-definition television.*

O.J. Eur. Comm. No. L 363/30, 13.12.1989, *Council Decision of 7 December 1989 on the common action to be taken by the Member States with respect to the adoption of a single worldwide high-definition television production standard by the Plenary Assembly of the International Radio Consultative Committee (CCIR) in 1990.*

O.J. Eur. Comm. No. C 40/101, 17.2.1992, *Opinion of the Economic and Social Committee on the Directive proposal for common transmission standards of satellite television signals.*

O.J. Eur. Comm. No. L 137/17, 20.5.1992, *Council Directive of 11 May 1992 on the adoption of standards for satellite broadcasting of television signals.*

O.J. Eur. Comm. No. C 139/4, 2.6.1992, *Proposal for a Council Decision on an Action Plan for the introduction of advanced television services in Europe.*

O.J. Eur. Comm. No. L 196/48, 5.8.1993, *Council Decision of 22 July 1993 on an Action Plan for the introduction of advanced television services in Europe.*

O.J. Eur. Comm. No. C 209/1, 3.8.1993, *Council Resolution of 22 July 1993 on the development of technology and standards in the field of advanced television services.*

O.J. Eur. Comm. No. C 128/54, 9.5.1994, *Opinion of the European Parliament issued on 19 April 1994.*

O.J. Eur. Comm. No. C 181/3, 2.7.1994, *Council Resolution of 27 June 1994 on a framework for Community policy on digital video broadcasting.*

O.J. Eur. Comm. No. L 281/51, 23.11.1995, *Directive 95/47/EC of the European Parliament and the Council of 24 October 1995 on the use of standards for the transmission of television signals.*

Okai, H. (1987). Towards the realization of HDTV—Situation in Japan. *Third International Colloquium on Advanced Television Systems: HDTV '87* (pp. 5.4.1-5.4.9). Montreal, Canada: Canadian Broadcasting Corporation.

Okimoto, D. I. (1986). The Japanese challenge in high technology. In R. Landau and N. Rosenberg (Eds.), *The positive sum strategy: Harnessing technology for economic growth* (pp. 541-567). Washington, DC: National Academy Press.

Okimoto, D. I. (1989). *Between MITI and the market: Japanese industrial policy for high technology.* Stanford, CA: Stanford University Press.

Omura, T., and Sugimoto, M. (1987, March 29-30). *Plans for HDTV development in Japan.* Address to the NAB Engineering Conference and Management Conference.

On eve of CCIR Assembly. (1986, May 9). *Communications Daily,* p. 3.

Ono, Y. (1990, April 18). *Direction of HDTV in the U.S. as influenced by international development.* Handout from the Advanced Television Seminar. Seattle, WA: KCTS-TV.

Opinion Research Corporation. (1951). A survey of audience reaction to RCA color television. In J. T. Cahill, R. T. Werner, R. B. Houston, and E. R. Beyer, Jr., *Petition of Radio Corporation of America and National Broadcasting Company, Inc. for approval of color standards for the RCA color television system* (pp. 610-661). New York: Radio Corporation of America.

Opinion Research Corporation. (1953). A survey of audience reaction to RCA color tele-vision. In J. T. Cahill, R. T. Werner, R. B. Houston, and E. R. Beyer, Jr., *Petition of Radio Corporation of America and National Broadcasting Company, Inc. for approval of color stan-dards for the RCA color television system* (pp. 41-86). New York: Radio Corporation of America.

Organisation for Economic Co-operation and Development. (1972). *The industrial policy of Japan.* Paris: Author.

Oudin, M. (1995). Preface. *The wider view report '95* (pp. 5-6). Brussels: Vision 1250.

Outlook for broadcasting policy and digitalization of broadcasting in multimedia age. (1995, May 1). *MPT News,* pp. 1, 4.

Passell, P. (1989, August 11). The uneasy case for subsidy of high-technology efforts. *The New York Times,* pp. A1, D3.

Patterns of media growth. (1990, August). *Screen Digest,* pp. 177-184.

Peeters, A., and van Merwijk, C. (1995a). *Films in letterbox versus beeldvullende films: Een enquête.* Hilversum, Netherlands: Nederlandse Omroep Stichting, Afdeling Kijk- en Luisteronderzoek.

Peeters, A., and van Merwijk, C. (1995b). *Letterbox en ondertiteling in het actieve beeld: Experiment en groepdiscussie in verband met PAL Plus.* Hilversum, Netherlands: Nederlandse Omroep Stichting, Afdeling Kijk- en Luisteronderzoek.

Peterson, C. I. (1994, July). *From continuous wave to HDTV: Relations between state and pri-vate capital in the development of new communication media.* Paper presented at the con-ference of the International Association for Mass Communication Research, Seoul, South Korea.

Peterson, J. (1993). Towards a common European industrial policy? The case of high def-inition television. *Government and Opposition, 28,* 496-511.

Philips opts to leave standards to market. (1993, June 25). *Broadcast,* p. 6.

Philips profits slashed. (1992, August 14). *Broadcast,* p. 14.

Pitts, K. A. (1992). How acceptable is letterbox for viewing widescreen pictures? *IEEE Transactions on Consumer Electronics, 38,* xliii-li.

Pitts, K. A., and Hurst, N. (1989). How much do people prefer widescreen (16 x 9) to standard NTSC (4 x 3)? *IEEE Transactions on Consumer Electronics, 35,* 160-169.

Pollack, A. (1989, September 30). The setback for advanced TV. *The New York Times,* pp. 31, 35.

Pollack, A. (1994, September 15). Japanese taking to wide-screen TV. *The New York Times,* p. D1.

Pollack, A. (1997, March 11). Japan says it will move up introduction of digital television by a few years. *The New York Times,* p. C6.

Pool, I. de Sola. (1983). *Technologies of freedom.* Cambridge, MA: Harvard University Press.

Port, O., and Burrows, P. (1994, January 24). R&D, with a reality check. *Business Week,* pp. 62-64.

Postman, N. (1992). *Technopoly: The surrender of culture to technology.* New York: Knopf.

Powell, B., and Itoi, K. (1994, March 7). I didn't really say that, did I? *Newsweek,* p. 47.

Powers, K. H. (1994). Framing the camera image for aspect-ratio conversions. *Perspectives on wide screen and HDTV production* (pp. 16-19). Washington, DC: National Association of Broadcasters.

Poynton, C. A. (1990). The current state of HDTV. *HDTV World Review, 1*(2), 44-51.

Prentiss, S. (1990). *HDTV: High definition television.* Blue Ridge Summit, PA: TAB Books.

Prestowitz, C. V., Jr. (1989). *Trading places.* New York: Basic Books.

Radiocommunication Assembly. (1995, June 29). *Draft revision of recommendation ITU-R BT.709: Parameter values for the HDTV standards for production and international programme exchange* (Doc. 11/1006-E).

Rawsthorn, A. (1995, January 23). Media futures: Sony launches PALplus. *Financial Times,* p. 13.

Reeves, B., Detenber, B., and Steuer, J. (1993, May). *New televisions: The effects of big pictures and big sound on viewer responses to the screen.* Paper presented at the annual conference of the International Communication Association, Washington, DC.

Reeves, B., Lang, A., Kim, E. Y., and Tatar, D. (1997, May). "The effects of screen size and message content on attention and arousal." Paper presented at the annual conference of the International Communication Association, Montreal, Canada.

Reeves, B., Lombard, M., and Melwani, G. (1992, May). *Faces on the screen: Pictures or natural experience?* Paper presented at the annual conference of the International Communication Association, Miami, FL.

Refined HDTV cost estimates less daunting. (1990, April 9). *Broadcasting,* pp. 40-41.

Regelman, K. (1994, September 19). HDTV boosters not amused by new American system. *Variety,* p. 70.

Renaud, J. L. (1992, February). Pressure for W-HDTV at WARC. *Advanced Television Markets,* p. 1.

Renaud, J. L. (1995a). The rise of the wide screen. *Advanced Television Markets* (ATM DIGIMEDIA/NAB '95), pp. 14-18.

Renaud, J. L. (1995b, February). Widescreen builds its market. *Television Business International,* pp. 68-69.

Renaud, J. L., and Morgan, G. (1990, May). Nothing but Mac trouble. *Cable and Satellite Europe,* pp. 58-62.

Renaud, J. L., and Schilling, M. (1993). Japan pioneers a path to HDTV. *Advanced Television Markets* (special issue for the 1993 Montreux International TV Exhibition), pp. 27-28.

Research firm sees bright future for HDTV. (1982, February 1). *Broadcasting,* p. 88.

Richards, E., and Burgess, J. (1989, November 16). Congressional group fears cuts in high-tech research. *The Washington Post,* pp. E1, E7.

Robert R. Nathan Associates. (1988). *High definition TV's potential economic impact on television manufacturing in the United States.* Washington, DC: Electronic Industries Association.

Roberts, J. L. (1996, August 26). The disc wars. *Newsweek,* pp. 42-43.

Roche, M. T. (1987). Télévision haute définition, un enjeu planétaire. *Médiaspouvoirs,* No. 7, 139-143.

Rogers, E. M. (1995). *Diffusion of innovations* (4th ed.). New York: The Free Press.

Roizen, J. (1986, September). Dubrovnik impasse puts high-definition TV on hold. *IEEE Spectrum,* pp. 32-37.

Rosenbloom, R. S., and Cusumano, M. A. (1987). Technological pioneering and competitive advantage: The birth of the VCR industry. *California Management Review, 29*(4), 51-76.

Rosenthal, E. D. (1995, July). HDTV: A thriving niche. *Videography,* pp. 50-54.

Rowen, H. (1992, October 11). Clinton's approach to industrial policy. *The Washington Post*, pp. H1, H4.

Russomanno, J. A., Trager, R., and Everett, S. (1993, May). *The narrow view of wide screen: Public acceptance of tomorrow's television*. Paper presented at the annual conference of the International Communication Association, Washington, DC.

S. 952, 101st Cong., 1st Sess. (1989).

S. 1001, 101st Cong., 1st Sess. (1989).

Samuel, P. (1990). High-definition television: A major stake for Europe. In J. F. Rice (Ed.), *HDTV: The politics, policies, and economics of tomorrow's television* (pp. 17-26). New York: Union Square Press.

Sanger, D. E. (1989, March 21). Japanese test illustrates big lead in TV of future. *The New York Times*, pp. A1, D10.

Savage, J. G. (1989). *The politics of international telecommunications regulation*. Boulder, CO: Westview.

Schiller, H. I. (1992). *Mass communications and American empire* (2nd ed.). Boulder, CO: Westview.

Schilling, M. (1992a, April). Japan's HDTV channel struggles to fill schedule. *Television Business International*, p. 81.

Schilling, M. (1992b, November). The changing face of Japanese television. *Television Business International*, p. 41.

Schmidtleitner, L. (1994, April). The winter games: Sport as a team operation. *Diffusion*, p. 35.

Schoof, H., and Brown, A. W. (1995). Information highways and media policies in the European Union. *Telecommunications Policy, 19*, 325-338.

Schreiber, W. F. (1991). *Fundamentals of electronic imaging systems: Some aspects of image processing* (2nd ed.). New York: Springer-Verlag.

Schubin, M. (1990, December). The "yellow peril." *Videography*, pp. 25-30.

Sculley, S. (1993, May 31). The Grand Alliance becomes a reality. *Broadcasting*, p. 59.

Seel, P. B. (1992, August). *The high-stakes game of high-definition television: United States policy concerning the development and standardization of high-definition television technology*. Paper presented at the annual convention of the Association for Education in Journalism and Mass Communication, Montreal, Canada.

Seel, P. B. (1993). *Television wars: Local effects of competition between multinational telecommunication corporations*. Bloomington, IN: Indiana University, Indiana Center on Global Change and World Peace.

Seel, P. B. (1995a). High-definition television and United States telecommunication policy (Doctoral dissertation, Indiana University, 1995). *Dissertation Abstracts International, 56*, 2468A.

Seel, P. B. (1995b, April). *Media convergence and the formation of the HDTV Grand Alliance*. Paper presented at the annual convention of the Broadcast Education Association, Las Vegas, NV.

Senitt, A. (Ed.). (1988). *World radio TV handbook*. London: Billboard Limited.

Severin, W. J., and Tankard, J. W. (1992). *Communication theories: Origins, methods, and uses in the mass media* (3rd ed.). New York: Longman.

Sharp, M., and Pavitt, K. (1993). Technology policy in the 1990s: Old trends and new realities. *Journal of Common Market Studies, 31*, 129-151.

Sherlock, M. J. (1990). HDTV and the financial effects on broadcasters. In J. F. Rice (Ed.),

HDTV: The politics, policies, and economics of tomorrow's television (pp. 71-79). New York: Union Square Press.

Shimizu, T. (1989). *High definition television: Comparison of research and development strategies.* Unpublished master's thesis, Massachusetts Institute of Technology, Cambridge.

16/9 broadcasting: Present and future services. (1996, May). *Wider View* (special issue Cannes Festival 1996), p. 2.

Slaa, P. (1991). HDTV as a spearhead of European industrial policy. *Telematics and Informatics, 8,* 143-154.

Slack, J. D. (1984). *Communication technologies and society: Conceptions of causality and the politics of technological intervention.* Norwood, NJ: Ablex.

Society of Motion Picture and Television Engineers. (1989, January 12). *Statement of the Society of Motion Picture and Television Engineers to the ANSI Board of Standards Review: Appeal of SMPTE 240M—signal parameters—1125/60 high-definition production system.* White Plains, NY: Author.

SOFRES. (1989). L'attrait des nouvelles normes de télévision auprès du public. In R. Forni and M. Pelchat, *La télévision à haute définition: Tome II. Contributions des experts et résultats d'un sondage d'opinion sur les nouvelles normes de télévision* (Rapport de l'office parlementaire d'évaluation des choix scientifiques et technologiques) (pp. 383-394). Paris: Economica.

Sohn, G. B., and Schwartzman, A. J. (1995). *Pretty pictures or pretty profits.* Washington, DC: Media Access Project.

Solomon, R. (1990). HDTV: Digital technology's moving target? *Intermedia, 18*(2), 58-61.

Spiero, J. P. (1994, April). The winter games: Sport as a team operation. *Diffusion,* pp. 34-35.

Staelin, D. H., Bozdogan, K., Cusumano, M. A., Ferguson, C. H., Filippone, S. F., Reintjes, J. F., Rosenbloom, R. S., Solow, R. M., and Ward, J. E. (1988). *The decline of U.S. consumer electronics manufacturing: History, hypotheses, and remedies.* Cambridge, MA: Massachusetts Institute of Technology, Consumer Electronics Sector Working Group.

Stelmach, L. B. (1993). Viewers' reactions to pan and scan and letter-box images. *Proceedings: Vol II.* Ottawa, Canada: International Workshop on HDTV '93.

Stern, C. (1995a, April 10). Fox sees digital as more than HDTV. *Broadcasting & Cable,* p. 30.

Stern, C. (1995b, December 11). White House wants TV auction in 2002. *Broadcasting & Cable,* p. 10.

Stern, C. (1996a, February 5). Spectrum auction still looms. *Broadcasting & Cable,* p. 12.

Stern, C. (1996b, February 5). What is spectrum worth? *Broadcasting & Cable,* p. 16.

Stern, C. (1996c, March 18). Broadcasters see support waning for auction. *Broadcasting & Cable,* p. 12.

Stern, C. (1996d, March 25). White House says auctions will raise $32 billion. *Broadcasting & Cable,* p. 19.

Stern, C. (1996e, April 1). No doubt about digital. *Broadcasting & Cable,* p. 5.

Stern, C. (1996f, April 22). Dole: Broadcasters "bullying Congress." *Broadcasting & Cable,* p. 6.

Stow, R. L. (1993). Market penetration of HDTV. In S. M. Weiss and R. L. Stow, *NAB 1993 guide to HDTV implementation costs* (appendix II). Washington, DC: National

Association of Broadcasters.

Streeter, R. G. (1987). An update on the television and film aspects of HDTV. *SMPTE Journal*, 96, 1108-1111.

Stüdeman, F. (1996, July 18-24). Tiny duchy with a grand media design. *The European*, p. 19.

Sudalnik, J. E., and Kuhl, V. A. (1994). *High-definition television: An annotated multidisciplinary bibliography, 1981-1992*. Westport, CT: Greenwood.

Sugimoto, M. (1987). The NHK strategy for HDTV services. *Third International Colloquium on Advanced Television Systems: HDTV '87* (pp. 5.1.1-5.1.17). Montreal, Canada: Canadian Broadcasting Corporation.

Sugimoto, M. (1988, April 22). *The potential impact of HDTV on society*. Transcript from the Advanced Television Seminar. Seattle, WA: KCTS-TV.

Sukow, R. (1993a, February 8). More testing likely in HDTV's future. *Broadcasting*, p. 36.

Sukow, R. (1993b, February 15). Now there are four. *Broadcasting*, p. 6.

Survey of Current Business. (1990, March), pp. S-16, S-17.

Takahashi, M. (1985, October 21). Survey of 1000 households. *Nikkei High Tech Report*, pp. 5-7.

Tannas, L. E., Jr. (1989, October). HDTV displays in Japan: Projection-CRT systems on top. *IEEE Spectrum*, pp. 31-33.

Tanton, N. E., and Stone, M. A. (1988). HDTV displays: Subjective effects of scanning standards and domestic picture sizes. *High definition television* (pp. 96-103). Selected papers presented by EUREKA 95 participants at the International Broadcasting Convention, Brighton, England.

Tatsuno, S. M. (1990). *Created in Japan*. New York: Harper and Row.

Taylor, J. P. (1983, November 21). Latest HDTV idea: CBS's 2-channel transmission system simultaneously serving 3 kinds of receivers. *Television/Radio Age*, pp. 107-111, 256, 260.

Telecommunications Act of 1996, Pub. L. No. 104-104, § 336, 110 Stat. 56, 107-108(1997).

Television Bureau of Advertising. (1987). *Trends in television*. New York: Author.

Television Digest with Consumer Electronics (White Paper). (1995, October 9), p. 4.

Terry, H. A., and Krasnow, E. G. (1992). In defense of the "Broadcast Policy-Making System" model. *Journal of Broadcasting & Electronic Media*, 36, 479-480.

Testing, testing ... HDTV. (1989, October 2). *Broadcasting*, pp. 37-38.

The Atlanta Olympic Games in 16/9. (1996, June). *Wider View*, pp. 3-6.

The competitive status of the U.S. electronics industry. (1984). Washington, DC: National Academy Press.

The costs of coloring a television station. (1967, January 2). *Broadcasting*, pp. 36-38.

The European. (1992, December 23-27), p. 37.

The Flat Panel Display Task Force. (1994). *Building U.S. capabilities in flat panel displays*. Washington, DC: U.S. Department of Defense.

The government's guiding hand: An interview with Ex-DARPA Director Craig Fields. (1991, February-March). *Technology Review*, pp. 35-40.

The market for 16/9 receivers in Europe. (1995, April). *The Wider View*, p. 8.

The mood of Montreux: Pressing for a world HDTV standard. (1983, June 6). *Broadcasting*, pp. 37-39.

The President's Commission on Industrial Competitiveness. (1985). *Global competition: Vol. 2. The new reality*. Washington, DC: Author.

The Task Force on Women, Minorities, and the Handicapped in Science and Technology. (1988). *Changing America: The new face of science and engineering.* Washington, DC: Author.

Thorpe, L. J. (1990a, April 18). Transcript from the Advanced Television Seminar. Seattle, WA: KCTS-TV.

Thorpe, L. J. (1990b). Moving forward: Challenges and opportunities. In J. F. Rice (Ed.), *HDTV: The politics, policies, and economics of tomorrow's television* (pp. 35-46). New York: Union Square Press.

Thurber, D. (1990, November 9). *High-definition TV equipment goes on sale in Japan.* Associated Press newswire.

Trade Expansion Act of 1962, Pub. L. No. 86-794, §102(2), 76 Stat. 872 (1962).

Tsebelis, G. (1990). *Nested games: Rational choice in comparative politics.* Berkeley, CA: University of California Press.

Tunstall, J., and Palmer, M. (1990). *Liberating communications.* Oxford: Blackwell.

Twelve that would be HDTV. (1988, October 17). *Broadcasting,* pp. 38-39.

Tyson, L. D. (1993). *Who's bashing whom: Trade conflict in high-technology industries.* Washington, DC: Institute for International Economics.

Udagawa, H. (1988, November). NHK: A tale of dreams. *Tokyo Business Today,* pp. 46-48.

U.S. Congress, Office of Technology Assessment. (1983). *International competitiveness in electronics.* Washington, DC: U.S. Government Printing Office.

U.S. Congress, Office of Technology Assessment. (1990). *The big picture: HDTV and high-resolution systems.* Washington, DC: U.S. Government Printing Office.

U.S. Congress, Office of Technology Assessment. (1992). *U.S.-Mexico trade: Pulling together or pulling apart?* Washington, DC: U.S. Government Printing Office.

U.S. Department of Commerce. (1990). *U.S. industrial outlook.* Washington, DC: U.S. Government Printing Office.

U.S. Department of State. (1985). *A worldwide high definition television (HDTV) studio/production standard: Technical and economic considerations.* Washington, DC: Author.

U.S. faces uphill fight on HDTV. (1986, May 12). *Broadcasting,* p. 46.

U.S. gets its way on HDTV at CCIR. (1989, May 29). *Broadcasting,* p. 55.

U.S. House of Representatives. (1989a). *High definition television: Hearings before the Subcomm. on Telecommunications and Finance of the House Comm. on Energy and Commerce,* 101st Cong., 1st Sess. 14 (statement of Robert A. Mosbacher), 152 (statement of Craig I. Fields).

U.S. House of Representatives. (1989b). *High definition television: Hearing before the Comm. on Science, Space, and Technology,* 101st Cong., 1st Sess. 90 (statement of Craig I. Fields).

U.S. House of Representatives. (1989c). *Public policy implications of advanced television systems, Staff of the Subcomm. on Telecommunications and Finance of the House Comm. on Energy and Commerce,* 101st Cong., 1st Sess. (Comm. Print 101-E).

U.S. House of Representatives. (1990). H. Rept. 481, pt. 1, 101st Cong., 2d Sess.

U.S. House of Representatives. (1992). *Report prepared by the Subcomm. on Technology and Competitiveness transmitted to the Comm. on Science, Space, and Technology,* 102d Cong., 2d Sess.

U.S. House of Representatives. (1993). *High-definition television, 1993: Hearings before the Subcomm. on Telecommunications and Finance,* 103rd Cong., 1st Sess.

U.S. industry adopts NHK parameters for HDTV. (1985, March 25). *Broadcasting,* pp. 68,

70.

U.S. Senate. (1958). *Allocation of TV channels: Report of the Ad Hoc Advisory Comm. on Allocations*, 85th Cong., 2d Sess. 221-226.

U.S. Senate. (1989a). *Commercialization of new technologies: Hearing before the Senate Comm. on Commerce, Science, and Transportation*, 101st Cong., 1st Sess. 24 (statement of Robert A. Mosbacher).

U.S. Senate. (1989b). *High definition television: Hearing before the Subcomm. on Science, Technology, and Space of the Comm. on Commerce, Science, and Transportation*, 101st Cong., 1st Sess. 4 (statement of Craig I. Fields).

U.S. Senate. (1989c). S. Rept. 159, 101st Cong., 1st Sess.

Utsumi, Y., Isobe, N., and Hasegawa, T. (1995). New phase of HDTV in Japan—HDTV production compatible with current broadcasts. *Symposium record of the 19th international television symposium and technical exhibition* (pp. 351-368). Montreux, Switzerland: The Symposium.

Vision 1250—Focal point for 1250 HD production. (1995, April). *Wider View*, p. 12.

Wallace, D. (1995, October 13). Pass the popcorn. *The Denver Post*, weekend section, p. 20.

Wassiczek, N, Waters, G. T., and Wood, D. (1990). European perspectives in the development of HDTV standards. *Telecommunication Journal, 57*, 313-321.

Weiss, S. M., and Stow, R. L. (1993). *NAB 1993 guide to HDTV implementation costs*. Washington, DC: National Association of Broadcasters.

Wenders, W. (1994, April). The aesthetics of HDTV. *Diffusion*, pp. 44, 45.

West, D. (1995a, April 10). Being the best means HDTV to Bob Wright. *Broadcasting & Cable*, p. 28.

West, D. (1995b, June 19). HDTV gauntlet thrown down in Montreux. *Broadcasting & Cable*, p. 37.

West, D., and McConnell, C. (1996, April 15). Joseph A. Flaherty: The brave new, brand-new world of television. *Broadcasting & Cable*, pp. 32-38.

Westerink, J. H. D. M., and Roufs, J. A. J. (1989). Subjective image quality as a function of viewing distance, resolution, and picture size. *SMPTE Journal, 98*, 113-119.

Whitney, C. G. (1996, December 4). Embarrassed, France halts Thomson sale. *The New York Times*, pp. C1, C9.

Willard, G. E., and Cooper, A. C. (1985). Survivors of industry shake-outs: The case of the U.S. color television set industry. *Strategic Management Journal, 6*, 299-318.

Winston, B. (1989). HDTV in Hollywood: Lights, camera, inaction. *Gannett Center Journal, 3*(3), 123-137.

Wober, J. M. (1985). *The changing screen—May bigger be better?* London: Independent Broadcasting Authority, Research Department.

Wober, J. M. (1989). *Tomorrow's television—Today's opinions*. London: Independent Broadcasting Authority, Research Department.

Wood, D., and Tejerina, J. (1990). The current EBU analysis of the CIF and CDR HDTV production formats. *Fourth International Colloquium on Advanced Television Systems: HDTV '90* (pp. 4A.2.2-4A.2.15). Montreal, Canada: Canadian Broadcasting Corporation.

Word from ATSC: HDTV's future to be written in 1,125 lines. (1987, June 1). *Broadcasting*, p. 57.

Working Party 5. (1988). *Economic factors and market penetration: The Working Party 5 report to the FCC Planning Subcommittee on Advanced Television Service*. Washington, DC:

Author.

Yamamura, K., and Vandenberg, J. (1986). Japan's rapid-growth policy on trial: The television case. In G. R. Saxonhouse and K. Yamamura (Eds.), *Law and trade issues of the Japanese economy* (pp. 238-283). Seattle: University of Washington Press.

Zampetti, A. B. (1994). *Globalisation of industrial activities: A case study of the consumer electronics sector.* Paris: Organisation for Economic Co-operation and Development.

Zenith Radio Corp. v. Matsushita Elec. Indus. Co., 513 F. Supp. 1100 (1981).

Zenith's first digital sets won't do cable HDTV. (1997, May 19). *Broadcasting & Cable,* p. 15.

Zou, W. Y., and Kutzner, J. A. (1997). Practical implementation of digital television: Update 1996. *SMPTE Journal, 106,* 233-242.

■ Index